International Virtual Conference on Emerging Research Trends in Structural Engineering 2020

AA007060

IOP Conference Series: Materials Science and Engineering
Volume 989

Chenai, India
16 - 17 July 2020

Editors:

V. Vasugi
M. Helen Santhi

ISBN: 978-1-7138-3301-7
ISSN: 1757-8981

Printed from e-media with permission by:

Curran Associates, Inc.
57 Morehouse Lane
Red Hook, NY 12571

Some format issues inherent in the e-media version may also appear in this print version.

Institute of Physics
Dirac House, Temple Back
Bristol BS1 6BE UK

Phone: 44 1 17 929 7481
Fax: 44 1 17 920 0979

techtracking@iop.org

Additional copies of this publication are available from:

Curran Associates, Inc.
57 Morehouse Lane
Red Hook, NY 12571 USA
Phone: 845-758-0400
Fax: 845-758-2633
Email: curran@proceedings.com
Web: www.proceedings.com

TABLE OF CONTENTS

IOP Conf. Series: Materials Science and Engineering **989** (2020) 011001 doi:10.1088/1757-899X/989/1/011001

Preface

Civil engineering field is one of the most ancient industries in today's world and also it is the ever growing sector. The construction industry counters with new challenges in this fast growing world and tries to cope with a unique and most deserving solution to the problem. The structural engineering stream is ever growing with the advanced and technology coming and implementation of the same. The civil engineering is mainly concerned with the man's needs and protected shelter to work and with various changing conditions structural engineering is also being advanced to cope with the fast growing world.

The International Conference on Emerging Research Trends in Structural Engineering(ERTSE-2020), held at Vellore Institute of Technology, Chennai, India on July 16 & 17, 2020, was jointly organized by a group of pioneers of academic research on the various aspects of Structural Engineering supported by an eminent group of international scientific committee members active in this field. The industry partner for the conference was NORMET India Pvt.Ltd, Chennai. The conference was conducted in the online mode due to the complete lockdown that was followed in most of the countries because of COVID 19 and no clue about end prediction of it. With the development of advanced technology, the virtual conference is increasingly used to bring people in various locations together. The conference was hosted entirely over the Internet using Zoom platform designed specifically for the virtual experience.

There were 12 keynote lectures for duration of 45 minutes each by eminent academicians and researchers in the thrust areas of structural engineering. Out of them 8 were by foreign speakers and the remaining, by Indian speakers. Totally 59 papers were presented in the conference and the authors were given 10 minutes for presentation and 3-5 minutes for discussion. Totally 312 people from India and abroad (USA, UK, Finland, Sri Lanka, Malaysia, Albania, Kuwait, Ethiopia and Oman) participated in the conference which includes final year students from B.Tech Civil Engineering and M.Tech Structural Engineering, Research Scholars, Faculty members and Industry practitioners. All the sessions were well received by all the participants and all the sessions were very interactive. The conference provided a hand on experience and knowledge about the emerging research trends that is happening in the structural engineering and its application.

Dr.V.Vasugi & Dr.M.Helen Santhi

Organizers

IOP Conf. Series: Materials Science and Engineering **989** (2020) 011001 doi:10.1088/1757-899X/989/1/011001

EDITORS

Dr. V. Vasugi,
Associate Professor,
School of Civil Engineering,
Vellore Institute of Technology Chennai, Chennai-600127,
Tamil Nadu, India
Email:vasugi.v@vit.ac.in

Dr. M.Helen Santhi,
Professor,
School of Civil Engineering,
Vellore Institute of Technology Chennai, Chennai-600127,
Tamil Nadu, India
Email:helensanthi.m@vit.ac.in

COMMITTEE

Chairman: Dr.K.Muthumani, Professor & Dean, School of Civil Engineering, Vellore Institute of Technology Chennai

Organizers: Dr. V.Vasugi and Dr.M.Helen Santhi, School of Civil Engineering, Vellore Institute of Technology Chennai

Technical Committee

Dr. Chandu Bolisetti, Idaho National Laboratory, Idaho, US
Dr.T.Jothi Saravanan, Indian Institute of Technology Bhubaneswar, Odisha, India
Dr.Baskar Gopalan, Shinas College of Technology Oman
Dr.Ranjith Dissanayake, University of Peradeniya, Srilanka
Dr. C. Jayasree, Kuwait Institute for Scientific Research, KUWAIT
Dr. Hüseyin Bilgin, EPOKA University, Albania
Dr.Priyadharshini Perumal, University of Oulu, Finland
Er. Rajprabhu Thangappa, North Carolina State University, USA
Dr. Binaya Patnaik, Gambella University,Ethiopia
Er. KRS.Narayan, REROUTE Reliance Industries Limited, India
Er. Sourish Nanda, Normet India Pvt. Ltd
Dr.S.Karthiyaini, Vellore Institute of Technology Chennai
Dr.M.Shanmuga Sundaram, Vellore Institute of Technology Chennai
Dr.M.Ramesh Kannan, Vellore Institute of Technology Chennai

IOP Conf. Series: Materials Science and Engineering **989** (2020) 011002 doi:10.1088/1757-899X/989/1/011002

Peer review declaration (ERTSE-2020)

All papers published in this volume of *IOP Conference Series: Materials Science and Engineering* have been peer reviewed through processes administered by the Editors. Reviews were conducted by expert referees to the professional and scientific standards expected of a proceedings journal published by IOP Publishing.

- Type of peer review: Single-blind
- Conference submission management system: www.tiny.cc/oea2qz /

  E-mail submission (ertse2020@gmail.com)
- Number of submissions received: 35
- Number of submissions sent for review: 35
- Number of submissions accepted:32
- Acceptance Rate (Number of Submissions Accepted / Number of Submissions Received X 100): 91.4 %
- Average number of reviews per paper: 3
- Total number of reviewers involved: 6
- Any additional info on review process: The reviewers reviewed the papers by considering the relevance to the conference themes, paper format requirements, plagiarism and language. The decisions made were (i) Recommended for publication, (ii) Recommended with revision, and (iii) Not recommended. The papers which required revision were sent back to the authors to incorporate the corrections and changes as suggested by the reviewers. The revised papers were again reviewed and recommended for publication.
- Contact person for queries: vasugi.v@vit.ac.in & helensanthi.m@vit.ac.in

Status of building information modelling (BIM) in construction practice-UK context

Tanvir Qureshi[1,2] **and Muhannad Al Hasani**[3]

[1]Department of Geography and Environmental Management, The University of the West of England, Bristol, UK
[2]Department of Civil & Mineral Engineering, University of Toronto, Toronto, Canada
[3]Al-Jazira Building & Construction, Salwa Rd, Wholesales intersection, Doha, Qatar

[1,2]Email:tanvir.qureshi@uwe.ac.uk

Abstract. This paper presents the level of Building Information Modelling (BIM) practices in the current UK construction industries. A survey was undertaken which contains a series of a self-explanatory questionnaire. The survey analysis provides an understanding of the level of BIM in practice and how BIM can be incorporated in the future construction industries from theory to practice. Findings suggest that most big companies are in the process of adopting BIM at some level of practice, while small companies are dropped behind. BIM is also very much depending on the clients' demand and the project complexity. Overall, the infrastructure construction industry is moving forward in the direction of complete BIM in practice.

1. Introduction

Technology, in the 21st century, has much progressed, and there are many commercial software tools available for calculation, drawing and designing the infrastructure and construction projects. Building information modelling (BIM) is a combination of tools in a platform which integrates all parameters of a infrastructure construction project (project management, architectural plans, structural designs, electrical designs, plumbing plans, sustainable energy factors, environmental factors, supervision, and modification) in a centralised system to build a model in various level of detail information. The BIM system is designed to integrate and manage data required for a project development without any error and fragmentation trouble, where participants of the project development have necessary access [1,2]. The BIM stores all information, developed by all individual stakeholder involved in a project, requires for actuating any project from conceptual level to completion. Efficiency in structural design, project management, the materials quantity surveying, detection of conflicts, necessary amendments, and upgradation according to the requirements of the stakeholders during the life cycle can be significantly improved with the adaptation of BIM in a project.

BIM is capable of multi-dimensional representations and analyses. This ranges from the 3D modelling that includes all building component information to scheduling (4D), estimation (5D), sustainability (6D), and facility management (7D) [3]. The multi-dimensional integration capabilities allow the combination of all technical and quantitative service to develop and maintain a project in a single platform.

The UK Government Construction Strategy (GCS) announced BIM requirements in the Cabinet Office in May 2011. Milestones were set to achieve Level 2 BIM maturity in all construction levels to deliver significant benefits. Level 2 BIM includes the methodology of working in a 3D model work

IOP Conf. Series: Materials Science and Engineering **989** (2020) 012001 doi:10.1088/1757-899X/989/1/012001

environment with description. Data required for Level 2 can be designed using isolated BIM tools [4]. Enterprise resource planning (ERP) software packages can provide commercial data for BIM. With updated data, level 4D scheduling and 5D cost estimating is also possible to integrate into the model using separate BIM tools. COBie UK 2012 developed a piece of digital information for a building in a spreadsheet data format. A constructive data exchange platform at Level 2 BIM can be linked with COBie UK 2012 data and managed by any company running the construction project efficiently. Also, British standards institution (BSI) road map written by the B/555 team, mentioning standards and documents need for Level 2 BIM maturity requirements.

This study carried out among construction managers, engineers, architects, and BIM specialists in the UK. The objective is to develop a clear understanding concerning BIM dynamic competency materials and tools, market maturity, practical market performance benchmarking, knowledge modelling, and BIM levels achieved across the construction industry. The data and information used in this paper is a part of a final year dissertation project on BIM by M. Al-Hasani in the Anglia Ruskin University in the UK [5].

2. Methodology

A survey was conducted to assess BIM stages in the UK construction industries. To achieve this, a survey questionnaire was prepared, and information was collected from different levels of target professionals and companies. The poll was mostly self-explanatory and electronically sent to various participants where possible. Add to that, and data was collected via a set of interviews with the contractors responsible for managing current construction projects, and with subject matter experts.

Market surveys (questionnaires) along with practical interviews, provide specific understanding about current BIM acceptation levels in the construction world. These guided us to outline practical advancements and principal obstructions for BIM level improvement at the 5D level (i.e., quantity survey integration). Finally, based on these findings, a general recommendation and future direction are outlined for efficient BIM adaptation in the construction industries.

3. Result and discussion

The survey data on BIM is statistically analysed to interpret the results. There were 26 participants representing different construction organisations leading practical project developments. There were 10 self-explanatory questions with multiple choice and 1 open-ended questions.

Question 1 inquired the information regarding the type of organisation that the participant was employed during the survey. The statistical data in the pie chart is presented in figure 1. The most significant target groups were contractors and those that were involved in practical construction projects (57%), second were those project's managers (16%), followed by (15%) quantity surveyors to understand BIM development towards 5D cost estimation. Next, there were civil/structural engineering (7%), architects (3%), and building service engineers (2%). The contractors are the key in implementing BIM in the project development phase. Hence the overall survey data is expected to be a strong balance between practical field and technical engineering practice in the current state.

Question 2 deals with the process of how different organisations use BIM and exchange models between different operators. The survey data is plotted in figure 2. It is reassuring to note that the majority use established common or shared database platforms (57% to 58%). Further investigation of the survey data asserts that people are also using other forms of data-sharing platforms and are relatively divided. There was only 8% found not to be using any modelling or common data-sharing platforms or clouds.

Question 3 collected information related to the principal reason for BIM adoption by organisations. The statistical data in the bar diagram is presented in figure 3. One of the significant findings of this was that a majority of participants (66%) agreed that they adopt BIM when there is an established model. The influence of clients in this matter was also evident as 34% reported that they prefer to adopt BIM when clients request them to do so. Although there were positive values for adopting BIM in most cases, around 20% of contractors said they have not used BIM to date. There was also a small percentage (4%) of contractors or participants we met who claimed that BIM was their standard approach to all projects. This is a positive sign for BIM, considering companies have quickly responded before BIM is required for any project.

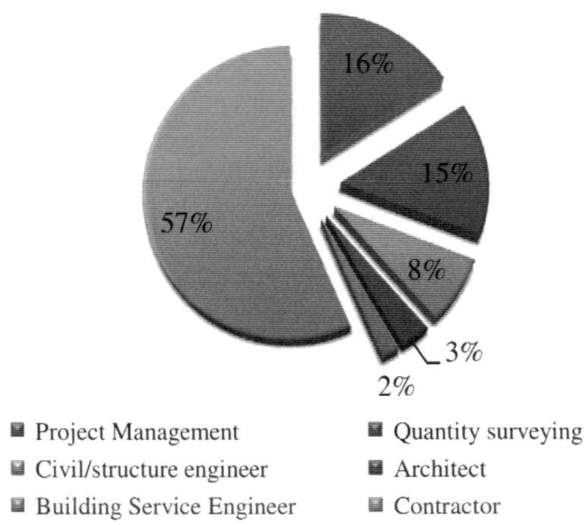

Figure 1. Q1: Principle work by the company.

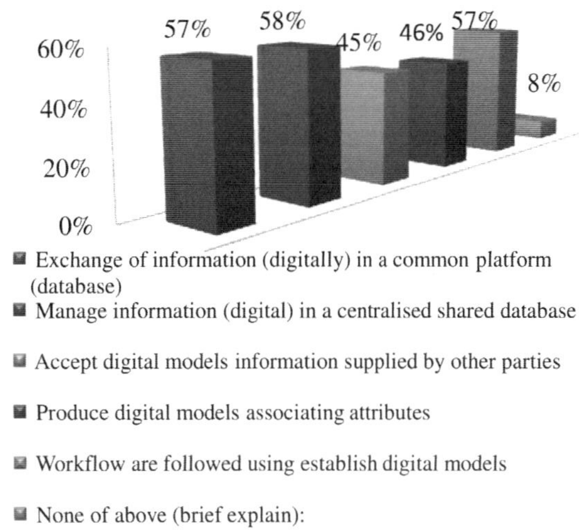

Figure 2. Q2: How the company uses BIM?

The availability of BIM expertise within the organisations was surveyed in question 4. The distribution of response is presented in figure 4 as a pie chart. Current data suggested that most of the organisations, 38% of companies use external BIM specialties. The second largest number of participants (34%) reported using a combination of in-house and external services, around 20% were found to have been using completely inside expertise. Another 8% of companies were found to still be in the trial period for BIM implementation and establishing BIM adopting processes. Add to that most advancement in BIM is noted in large size companies dealing with complex and large infrastructure construction projects.

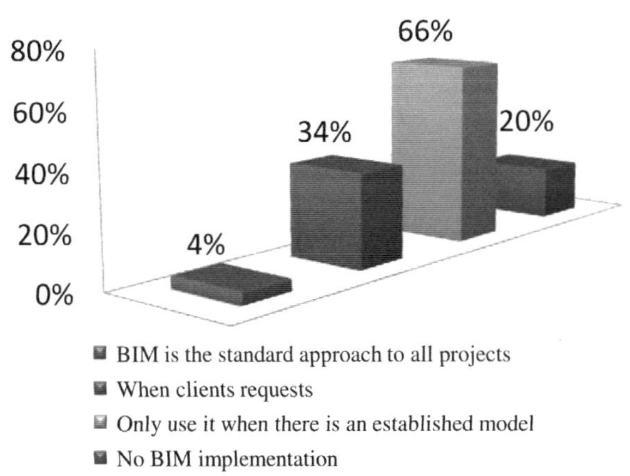

Figure 3. Q 3: BIM implementation approach in organisations.

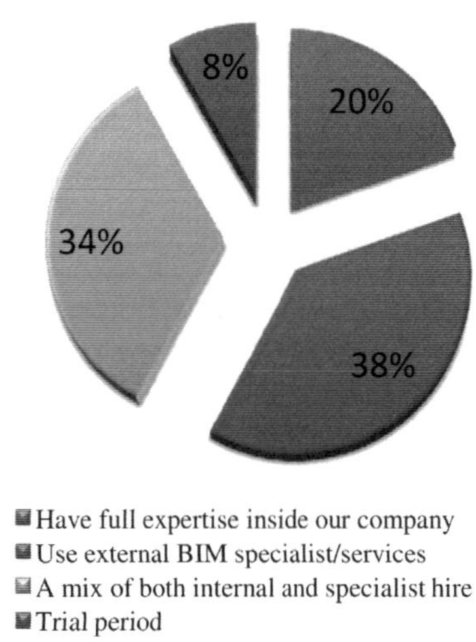

Figure 4. Q4: BIM service available in organisation.

Question 5 investigates the BIM system maturity level in delivering a practical project in the field. The survey data is presented in figure 5. It could be noted while a significant proportion (44%) already reported delivering at level 2 and level 1 BIM, only 1% were found to have adopted complete BIM. In practical construction practices, there were still 19% that were found to be using level 0 BIM. This survey information indicated that there is an apparent inclination of adopting BIM in the construction industries, which is developing over time.

Question 6 collected the information regarding BIM dimension productivities every organisation has accrued within its project implementation stage. A major proportion of participants (66%) reported using 3D digital models linked with attributes followed by 44% who were using 3D models at least in their project delivering workflow (figure 6). It can also be noted that a large proportion of respondents adopted 4D, 5D, and even 6D models along with their established 3D model practice, based on the complexity and demand of the project. Modern software-based model utilisation in the workflow has become the norm in current construction industries worldwide.

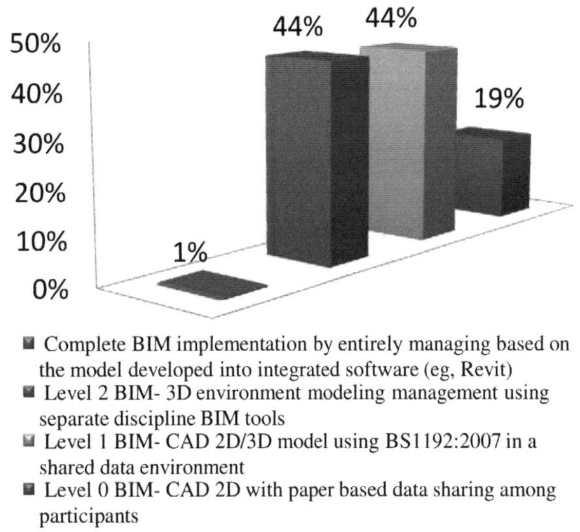

Figure 5. Q5: Level of BIM practice in the organisation.

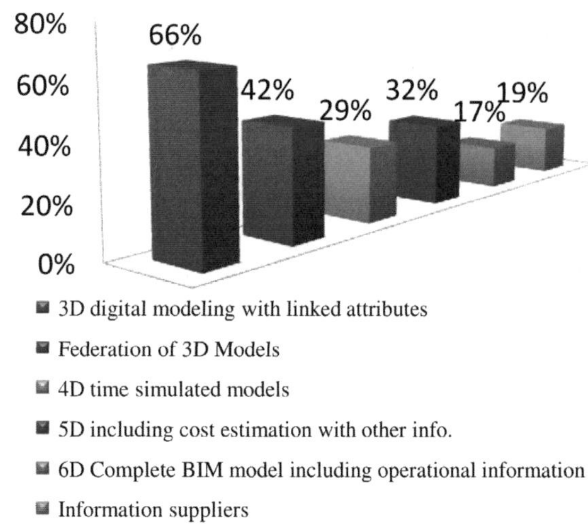

Figure 6. Q6: BIM dimension maturity in practice.

Question 7 continues with the theme to discover organisation-wide policies on education and training of their technical experts for acquiring BIM adaptation skills. The survey data is presented in figure 7. While many participants did not respond to this question, it was clear that these organisations/contractors do not have a structured approach to education and training facilities for BIM. Out of the remaining respondents, one-third (30%) acknowledged that they had a wide policy in their organisation along with a structured approach to BIM education and training programs. Similar proportions have reported that they were developing education and training schemes. Around 25% have reported having an approach to train and educate staff at the project level and 18% had absolutely no approach at all. However, 14% confirm that they have facilities for special staff that are particularly interested in BIM model development work. With these negative findings, it could be suggested that most of the 23 respondents appear to use mixed methods (internal and external) for education and training.

Question 8 was for the target audience, those who were not using BIM to date. A remarkable 80% intend to adopt BIM in the workflow chain in a year or two (figure 8). Surprisingly 20% of participants indicated that they have no plan for adopting BIM. This is a positive indication considering that even if they do not have any plans for BIM at this stage, they have an idea that BIM implementation will be required in the near future.

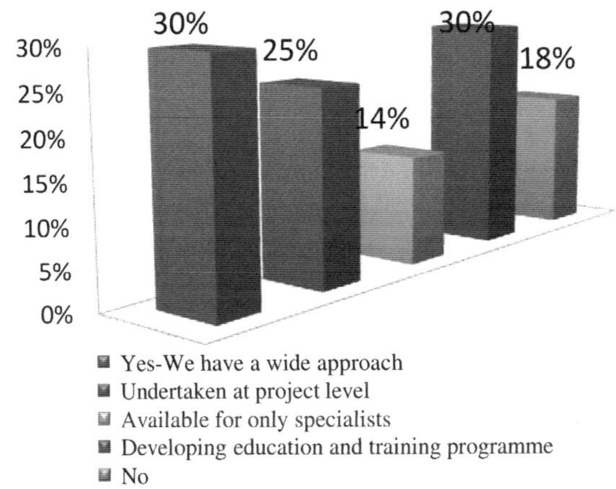

Figure 7. Q7: Does your organisation have any structured education or training for BIM?

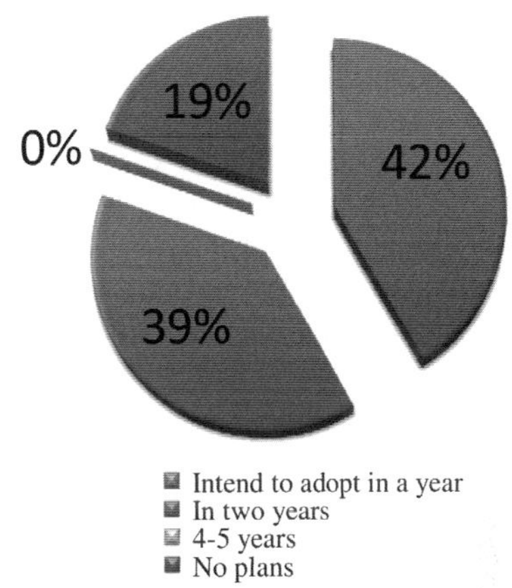

Figure 8. Q8: Future adoption of BIM.

Question 9 inquiry about COBie: UK2012 utilisation by organisations for information exchange (figure 9). The answer to this question in the practical construction field was found to be mostly negative (78%). However, others (22%) were found to be using COBie: UK2012 for information exchange.

Question 10 was asked to discover the method that organisations use for dealing with a standard data source environment. The survey data is presented in figure 10. A large group (39%) was found working in a standard data environment online. However, it was found that almost 36% have been found to be not using any particular method. However, a majority (45%) use PAS1192: 2013 followed by 36% that use BS1192: 2007. Furthermore, 17% were found to be using a personal data sharing environment. At this stage, it could be concluded that the role of BIM in the delivery stage must be improved in the future.

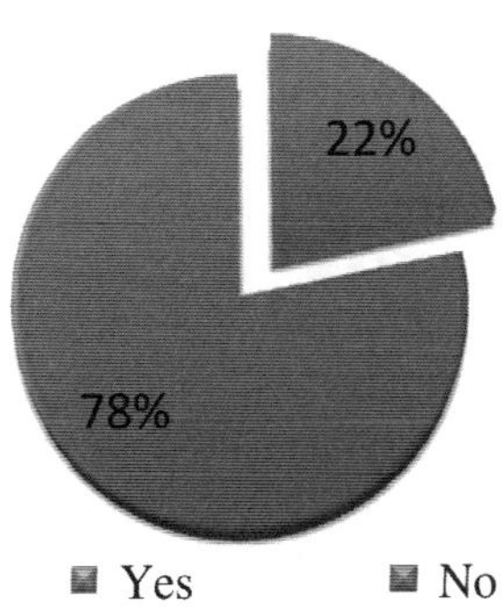

Figure 9. Q9: COBie: UK2012 use for information exchange.

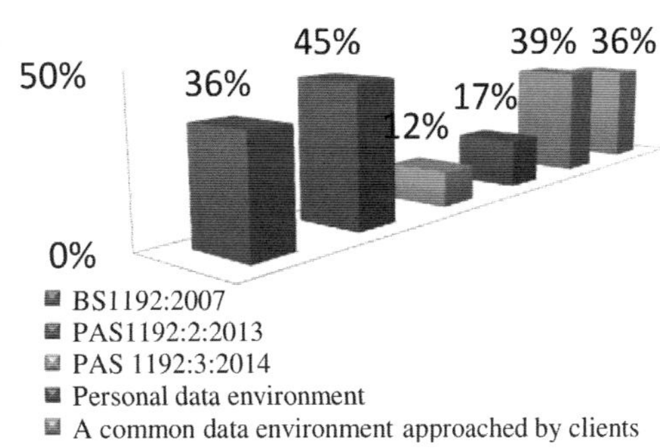

Figure 10. Q10: BIM methods use by organisation for managing.

Question 11 was an open question without any pool. The question obtained information on the BIM upskilling scheme available for employees in organisations. Not every respondent answered the open questions. Yet, out of those that answered, the majority indicated that a section of the company is dealing with this matter. Established company employees have reported that all staff will be undertaking a degree of BIM training for minimum awareness only, and other more detailed training, including BIM APs and BIM CPs. There were further indications of training BIM specialists at the project level for upskilling BIM technicians in the organisation.

The field survey indicates that BIM development is still in the developmental stage and organisational approaches towards it were found to depend on five key parameters, as seen in figure 11. These parameters dictate the implementation of BIM in infrastructure construction projects. However, the survey was conducted in 2015 and there is a dire need for similar survey at present time for future studies. The demand of BIM practice in construction is increasing. The future development of smart and connected infrastructure and utilisation of advanced sensory materials as well as sensors systems for structural health monitoring application is expected to require a complete BIM integration in a project at an early stage.

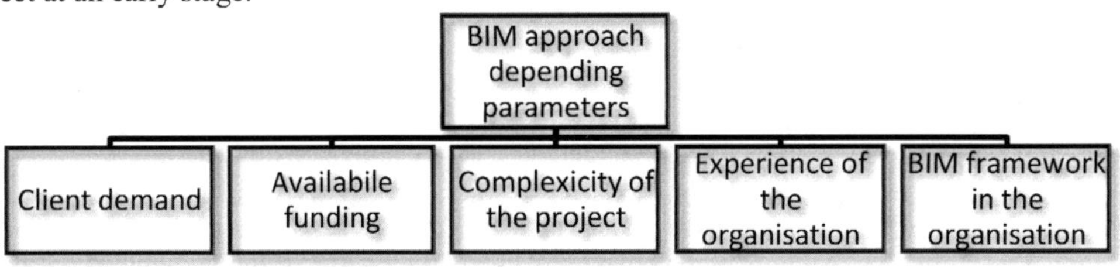

Figure 11. Key parameters in approaching BIM.

4. Conclusions

The BIM system incorporation and development in the construction industries of the UK is at its starting phase. Although government strategies are approaching to achieve level 2 BIM into practice, current construction practice still does not reflect the scenario. There are five key parameters noted form the survey data analysis that influences the BIM adaptation: client demand, funding, the complexity of the project, experience of the organisation, and BIM framework. Large modern construction companies have adopted BIM much faster, while many of them are considering BIM based on client demand and fund availability. Small companies are not comfortable adopting BIM at a good pace, as it takes a longer time to change their workflow style.

Acknowledgment

The authors are grateful to Prof. Adrian Bell of Anglia Ruskin University (UK) for his contribution and guidance in the project.

5. References

[1] Cyon Research 2003 The building information model A look at Graphisoft's virtual building concept
[2] Khemlani L 2004 The IFC building model: A look under the hood AECbytes feature 1-10
[3] Bakchan A, Faust K and Leite F 2019 Seven-dimensional automated construction waste quantification and management framework: Integration with project and site planning *Resources Conservation and Recycling* **146** 462-474
[4] Out-Law 2011 Building information modelling legal news and guidance from Pinsent Mansons
[5] Al-Hasani M 2015 How building information modelling influences the future of quantity surveying BSc (Hons) in Quantity Surveying dissertation Anglia Ruskin University UK

Parametric study of position of an outrigger-belt system through pushover analysis

Aakash Gupta[1,3], Rajat Mohan[2] and Debabrata Podder[1]

[1] Department of Civil Engineering, National Institute of Technology Meghalaya, India
[2] Department of Civil Engineering, ADGITM, Guru Gobind Singh Indraprastha University, New Delhi, India

[3]Email: ceaakash@gmail.com

Abstract. This study revolves around the high-rise RCC buildings with lateral force resisting system (LFRS). While constructing a multi-storey building, it is necessary to survey various salient features such as lateral load resistance capacity and lateral stiffness. The LFRS is used to introduce the lateral force resistances in the building and the LFRS used for this study are core and outrigger-belt system. Since the core is applied to the whole building at once, thus application related to it are easier, but when another LFRS (outrigger belt system in this study) which is dedicated to a certain floor, is added in the building, then its position can affect the performance of the building. There has been a lot of theories regarding the optimum positions which show how the position of these storey dependent LFRS can affect the overall performance of the buildings. When a multi-storey building is prone to high intensity earthquake load or heavy wind load, the outrigger-belt and core system would resist the rotation and prevent excessive sway caused by lateral loads. Thus, if proper analysis is performed on the building for optimum position of these systems, then the overall performance can be increased significantly. For this purpose, the buildings of 20-, 25-, and 30-storeys are analysed by Pushover Analysis with two load patterns using Midas GEN. The study is also performed in buildings with different stiffness conditions to understand the overall behaviour of the building for optimum positions. A total of 285 building models are tested using two different load patterns. These buildings differ from each other in terms of position of outrigger, stiffness and number of storeys.

1. Introduction

The demand for construction of high-rise buildings has drastically increased in the 21st century, due to the surge of population and to prevent scarcity of land area. These high-rise buildings also attracted the attention for new businesses and sometimes act as a centre of attraction. To satisfy the social, economic and environmental needs, advanced analysis in the field of building technology, building material and computational structural analysis of components of buildings have been carried out by architects and engineers to create a highly efficient and safe high-rise building. In various studies, it is found that when the height of a building structure increases, an adverse effect is observed when lateral loads drastically increases, and it causes lateral displacement or sway effect on the tall building. The high-rise buildings are considered highly affected by lateral loads and need to be designed as per required of a particular seismic zone. Buildings susceptible to heavy wind loads are subjected to an oscillation effect, which may cause discomfort to the occupants. Whereas, when an earthquake occurs, the ground motion would lead to the vibration of superstructure, causing an inertial force to act on the building. This may cause severe damage to structural and non-structural damage to the super-structure and could be highly dangerous for the occupants. Thus, to overcome these challenges, different lateral force resisting system (LFRS) are used in buildings. Outrigger is a special kind of a horizontal LFRS, which is placed between two structural systems, i.e. exterior structural framework and core system, to prevent swaying and overturning moments. The benefit of using outrigger system in a tall building at various levels is that it helps in increasing the stiffness to a building, thus increasing the overall strength of the structure, and resists the building deformation occurring due to lateral loads. Outrigger system is used as a load-resisting system in slender buildings, where they reduce storey drift and improve the comfort level of occupants during different conditions of lateral forces. From different studies conducted, it was found that the high-rise buildings containing an outrigger system would experience a major reduction in the overturning moment of about 40%, and a major reduction in the sway effect. Belt system is being provided in a building to improve the lateral system efficiency. When outriggers are used in frame systems, the belts system helps in transferring the gravity load to mega frame systems, thus helping in reducing the shear-lag effect and evenly distributing the loads across multiple columns of a building, which also distributes the load uniformly on the foundation as well. The outrigger arrangement can be introduced in form of heavily reinforced walls, deep girders, bracings etc. [1]. The outrigger-belt system used in this study was of heavily reinforced wall of specific arrangement as shown in Fig. 1, which is a simplified diagram to depict the outrigger-belt arrangement. Similar kind of arrangement of reinforcement can be seen in one of the cases discussed by Choi et al. in ref [8].

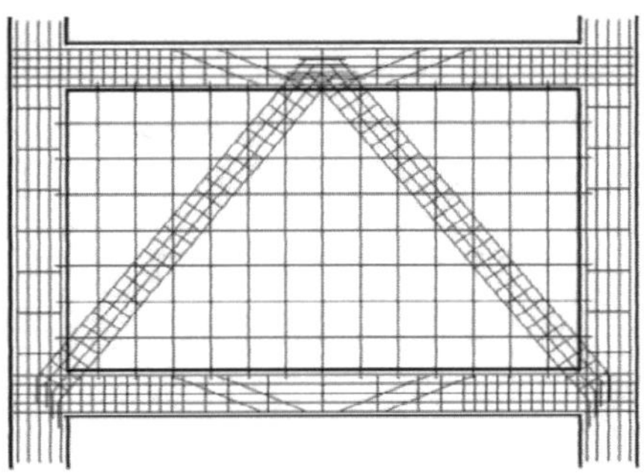

Figure 1. Outrigger-belt arrangement used in the study.

2. Literature review

The main focus of this study is to determine the optimum position of the outrigger belt arrangement in a high-rise RCC building and to study the effect of change in position if the stiffness of the building is varied along height, or as whole. Various researchers have performed several structural analyses and modeling on related topics. A framed tube Structural system is being modeled using computational method and an effect of outrigger-belt system on a three-dimensional building structure is being modeled using a rotational spring under lateral load conditions [1]. While conducting various studies, it is found that optimum location of the outrigger system is dependent upon a type of loading on a building, moment of inertia and axial stiffness acting on the exterior walls [1], [6], [8]. As a result, when there is an increase in axial stiffness, the optimum location of outrigger system will move in the downward direction. A research was conducted in which an ideal location of outrigger-belt system was determined using various arrangements of single and double outrigger levels on various sizes, height and shape of a building [2], [5]. Reference [2], conducted an analysis on buildings with various stories such as 28, 42 and 57-storey buildings using the finite element method, where it was observed that placing the double outrigger system at middle height, i.e. 42-strorey, would help in maximizing the reduction of lateral deflection in the building. When a multi-storey building is subjected to seismic loading like earthquake loads, the behavior of outrigger location is studied and the efficacy of placing an outrigger in a tall building is examined [3], [4]. From their investigation on a 20-storey and 30-storey building, both finite element method and component-mode synthesis method were used for building analysis and found that the component-mode synthesis method is effective in minimizing the analysis time and addition of outrigger-system helps in improving the stiffness of the building in [3]. In reference [4], an analysis on G+3 building was performed to evaluate its performance under the seismic loads acting upon the building. After conducting a non-linear pushover analysis on G+3 building, it was observed that curve, formed between base shear and displacement, was linear initially and then started to diverge from linearity as the columns and beams of the building was subjected to inelastic forces of action in [4]. This indicates that static loading is the key factor in governing the pushover computational analysis and failure of building structure during earthquake loads may occur as a consequence of using bad quality of construction materials. Reference [5] conducted structural analysis of a outrigger braced frame system, to obtain the optimal location of outrigger-belt and core system of a building subjected to lateral loads. After conducting various studies, it was observed that the outrigger system placed in tall building helps in improving the stiffness of structure and makes the structure more resilient against severe lateral loads, by reducing 33% and 66% drift and top displacement using one outrigger-system and two outrigger-system respective in [5]. Reference [6], conducted several studies regarding the analysis of 40-storey tall building for ideal position of outrigger system, using Sap2000 software. The author conducted a 40-storey building analysis, performance analysis have been done by placing a central shear wall and found that stability and stiffness of a structure increases by placing an outrigger system when earthquake of magnitude 7.8 was acting on building in [6]. After performing analysis on three-dimensional building using ETABS, the conclusion drawn was that after placing an outrigger and decreasing the dimension of columns at certain levels, the building stiffness improves and the structure becomes more comfortable for occupants [1-3]. Various analyses are being conducted to prepare a comparison between the optimal location of outrigger and time-history analysis, so as to assess the behaviors of structural analysis on building [7]. After several observations, the results of analysis on 20 and 25 storey buildings was taken out and the optimum location came out to be 10 and 14 simultaneously, showing that outrigger has influence on behavior of building susceptible to lateral loads in [7]. Reference [8], performed analysis of a high-rise building and studied the effects of building behavior by placing an outrigger system, including effect of using structural systems like differential column shortening. After performing several analyses, it was observed that by reducing the core overturning, it helps in reducing flexural and shear forces across the foundation, and tall buildings with mega-columns gives an aesthetic and flexible system to the occupants [1], [2], [8]. It should be noted that the conventional outrigger connects the core of the building to its exterior column, whereas the virtual outrigger or belt system connects the exterior columns with each other, forming a belt type system [8]. Thus, through all the literature discussed, it is evident that outrigger systems can improve the performance of the building and the optimum position of these LFRS can improve the performance and to make it simpler.

3. Methodology and nomenclature

The building of 20-, 25-, and 30-storeys were modelled in Midas GEN software and they were analysed through Pushover Analysis. The Pushover Analysis is a static non-linear analysis procedure, which is well known to give the seismic vulnerability of a structure. This analysis is known to give good results, although it is also known as an analysis which is highly dependent on the type of loading selected. The building models are tested in 2 phases, which are explained properly in the later portion of this section. In brief, the models in Phase 1 either have a constant stiffness or the stiffness varies along the height. The variation of stiffness along the height is achieved by varying the cross-sectional area of beams and columns after every floor, which can be observed in Fig. 2, as in Midas GEN, different cross-sections, are represented by different colours, thus showing different models from Phase 1 with constant stiffness and varying stiffness.

As for nomenclature of building models, three types of models have been considered, which is SC, 3C and 5C, where SC is a building with the constant cross-section area of columns and beams, 3C is a building having cross-sectional area of columns and beam that changes after every 3 storeys and 5C is a building having cross-section of beams and columns that change after 5 storey levels. So, according to this nomenclature, a SC25 model means a building contains 25 storeys with same cross-sectional areas, while a 3C30 model means that a building has 30-storeys and its cross-sectional area would decrease after every three storeys. In Phase 1, core and outrigger-belt (COB) thickness is 200mm for all models, thus it is not mentioned in the name. Although in Phase 2, the COB thickness varies to vary overall stiffness of the buildings, thus if A is used as a suffix, it means wall thickness is 200mm and if B is mentioned then it stands for 300mm thickness of the COB. So, B-3C30-24 building model means 30-storey building which has a COB thickness of 300mm and the cross-sectional area of beam and column reducing at every three levels, with outrigger at 24th storey.

The terms LC-1 and LC-2 is used in the study are related to the load cases used for the pushover analysis, which corresponds to load case 1 and load case 2, respectively. The load case 1 is a static load case derived from the equivalent static method derived using the IS1893 [9], it is the most commonly used load case for pushover analysis. Whereas the load case 2 corresponds to the N2 method, which is known as one of the most accurate methods of analysis [10]. All 285 models are pushed using these two load cases for the pushover analysis and their corresponding results noted accordingly.

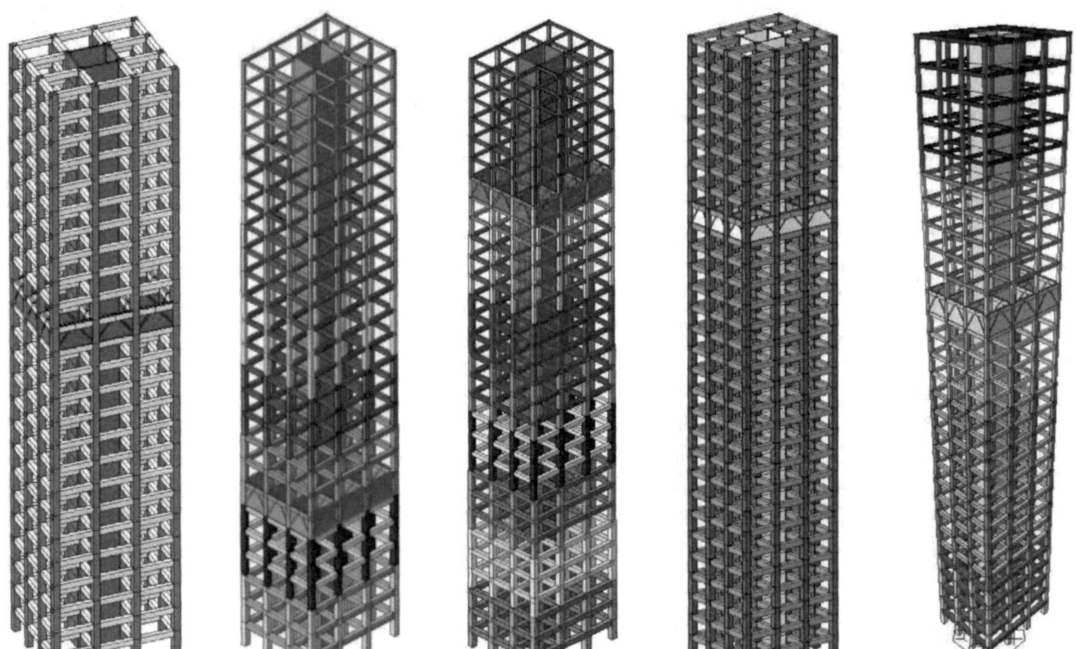

Figure 2. Isometric view of: (a) SC25-14, (b) 3C25-07, (c) A-3C30-24 or 3C30-24, (d) A-SC30-22 or SC30-22, (e) 5C30-20.

In Phase 1 of this study, there are overall 225 models, which can be divided into 3 categories; SC, 3C, and 5C. Differences in these are mentioned in previous part of nomenclature. These categories are later divided into subcategories based on number of storeys, and the buildings are of 20-, 25-, and 30-storey and such category has as many models as its number of storey which differs from each other on the basis of the storey at which the outrigger-belt arrangement is place. The storey to storey height is

3.2m that gives the height of 20-storey building as 64m and 30-storey buildings as 96m. Thus, they satisfy all the criteria set by IS16700 [11]. These buildings are then analysed through Pushover analysis using the two load cases mentioned above. The results corresponding to each model with varying positions of outrigger-belt system is recorded. The parameters selected are base shear (BS) and roof displacement (RD). These are converted into unitless and relative numbers and the plotted in form of graph.

In Phase 2, the change in optimum positions is observed in the models when the COB thickness is introduced. So as to study the effect on the optimal location of outrigger-belt system caused by the change in overall stiffness of buildings. This phase conducts change in the thickness of COB in models, which results in change in overall stiffness. 30 storey models are analysed in this phase, which contain an analysis of SC and 3C models for determining the optimum position and the effects due to change in stiffness corresponding due to COB thickness. This phase helps in giving precision and accurate data, and also provides a series of graphs which help in determining the overall view of change in displacement occurring after using outrigger-belt. So, SC30 and 3C30 models are selected from phase one and are renamed as A-SC30 models and A-3C30 models, respectively. Suffix A is used as the core and outrigger-belt arrangement used in them are of 200mm thickness. Along with them, 60 more models are formed by using a thickness of 300 mm for COB, which are added with a suffix of B. So, it can be said that the thickness of COB is increased by 50% to study the change in optimum position due to change in overall stiffness. The results obtained in this phase are also converted in unitless and relative form, in a similar manner as Phase 1. It should be noted that all models that are analysed here have the same plan and same plan dimensions. It should also be notes that effect of "strong columns and weak beams" are also taken in consideration for modelling.

4. Results

The results obtained from the Phase 1 and Phase 2 is converted to graphs. The obtained results are presented in unitless and comparative manner. The individual dots in a graph represent an individual building model corresponding to relative storey at which outrigger-belt is present along with the base shear (BS)/ roof displacement (RD) for that model. Graphs have been plotted by using the data collected from the analysis of all 285 building models. It should be noted that the Pushover Analysis load cases, LC-1 and LC-2 are applied in X-direction of loading which is also a stiffer direction in this study. The values on Y-axis of all the graphs represent the storey level in a relative manner which corresponds to the building having outrigger-belt arrangement at that level. Whereas the X-axis of the graph corresponds to RD or BS depending on the graph is being observed. In Phase 1 of the study, a total of 225 models were analysed through Pushover Analysis. The three category models that they have are of SC, 3C, and 5C, every category has 75 models each, out of which 20, 25, and 30 models are of 20-, 25-, and 30-storeys, that differ from each other on the basis of storey level of outrigger-belt.

Table 1. Optimum storey positions of outrigger-belt system for models of phase 1.

Building models	LC-1		LC-2	
	BS	RD	BS	RD
SC20	9 to 12	5 to 6	9 to 12	6 to 7
SC25	9 to 13	3 to 7	10 to 14	5 to 8
SC30	11 to 16	3 to 7	11 to 16	6 to 8
3C20	10 to 15	7 to 15	10 to 15	9 to 15
3C25	13 to 19	18 to 23	13 to 18	14 to 21
3C30	18 to 22	22 to 27	16 to 23	19 to 26
5C20	10 to 14	6 to 14	9 to 14	7 to 14
5C25	12 to 17	4 to 8	12 to 16	5 to 15
5C30	18 to 23	22 to 27	17 to 23	19 to 26

The Phase 2, in which the change in optimum position is being observed with change in stiffness of the building, although, in this phase the change in stiffness of the buildings are introduced by changing the thickness of COB. The models in series SC30 and 3C30 from Phase 1 are use in this phase with the names of A-SC30 and A-3C30 respectively. The 60 new models in this phase correspond to series B-SC30 and B-3C30. All new models are also analysed by two load cases and they corresponding BS and RD are noted and graphs are plotted in Figures 12-13. Thus, for this stage, the figure under observation is figures 5, 8, 12, 13. The optimum positions are inferred and written in table 1 and table 2 for Phase 1 and Phase 2, respectively.

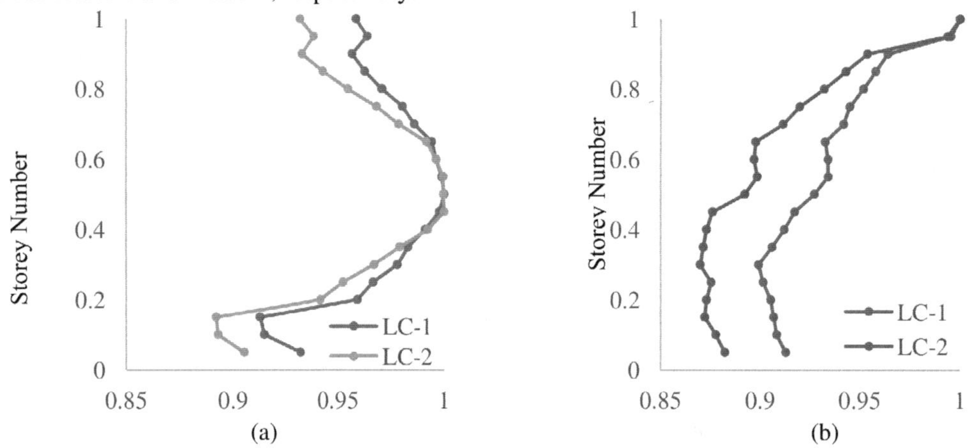

Figure 3. Graph of buildings in SC20 showing variation in (a) BS, (b) RD.

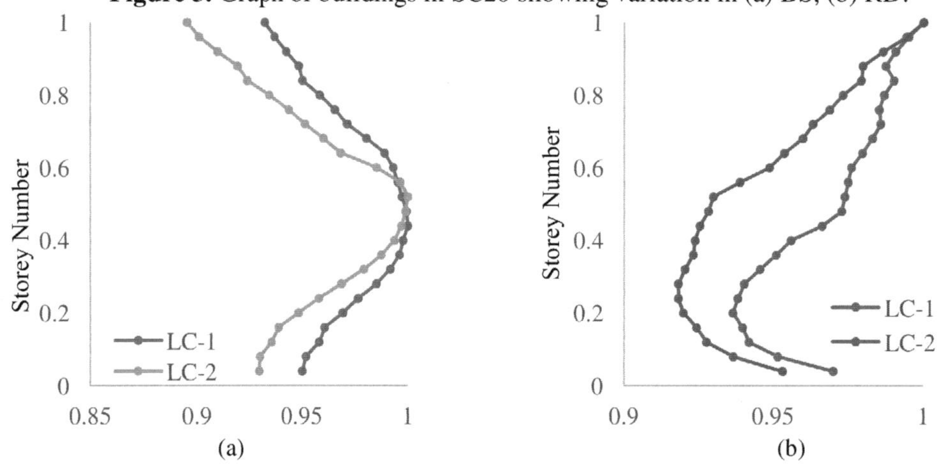

Figure 4. Graph of buildings in SC25 showing variation in (a) BS, (b) RD.

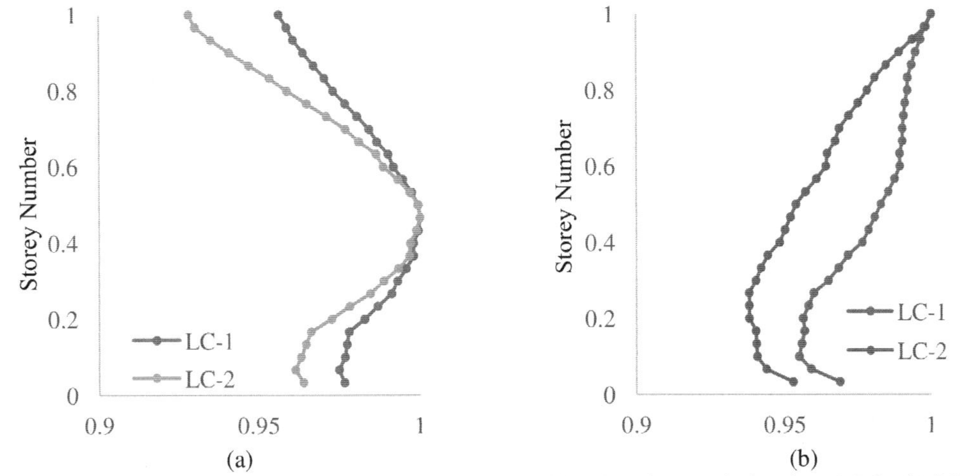

Figure 5. Graph of buildings in SC30 or A-SC30 showing variation in (a) BS, (b) RD.

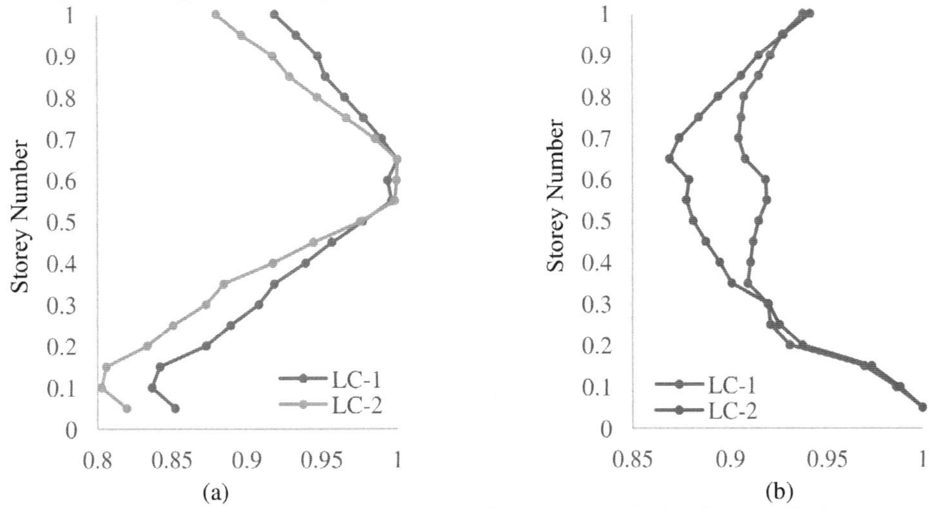

Figure 6. Graph of buildings in 3C20 showing variation in (a) BS, (b) RD.

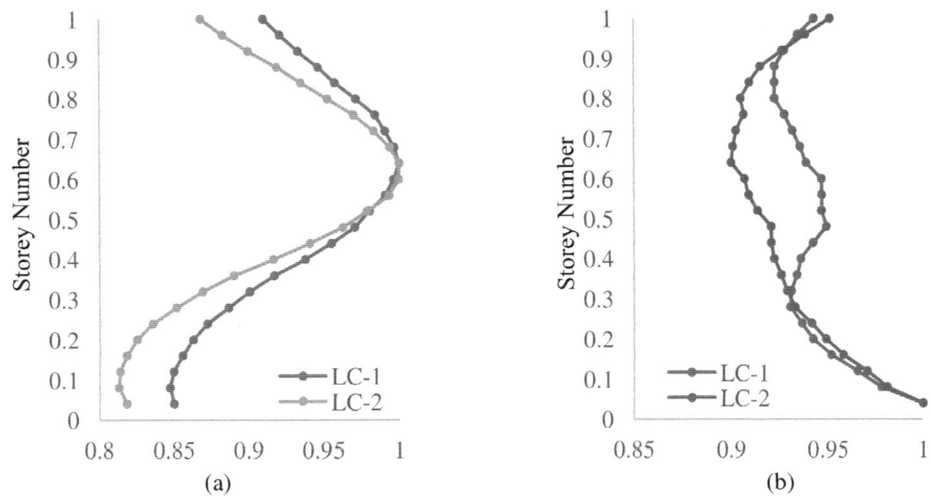

Figure 7. Graph of buildings in 3C25 showing variation in (a) BS, (b) RD.

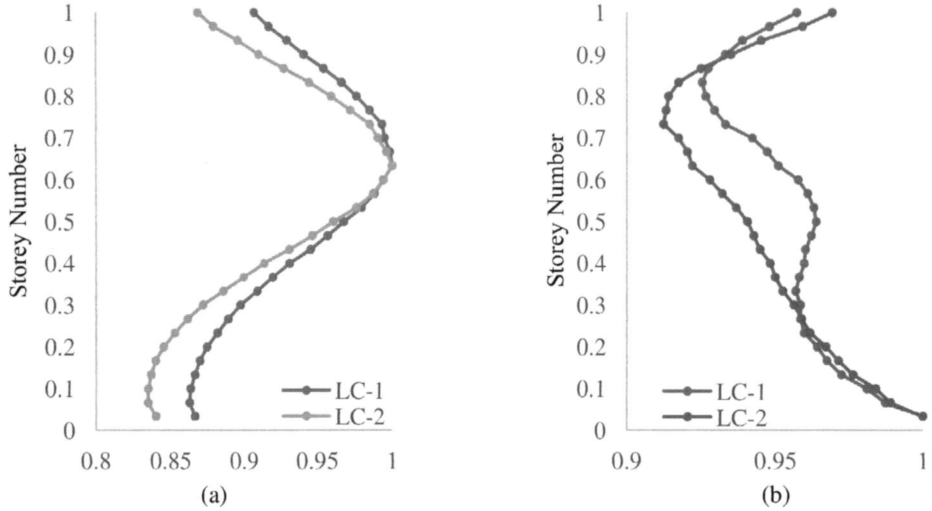

Figure 8. Graph of buildings in 3C30 or A-3C30 showing variation in (a) BS, (b) RD.

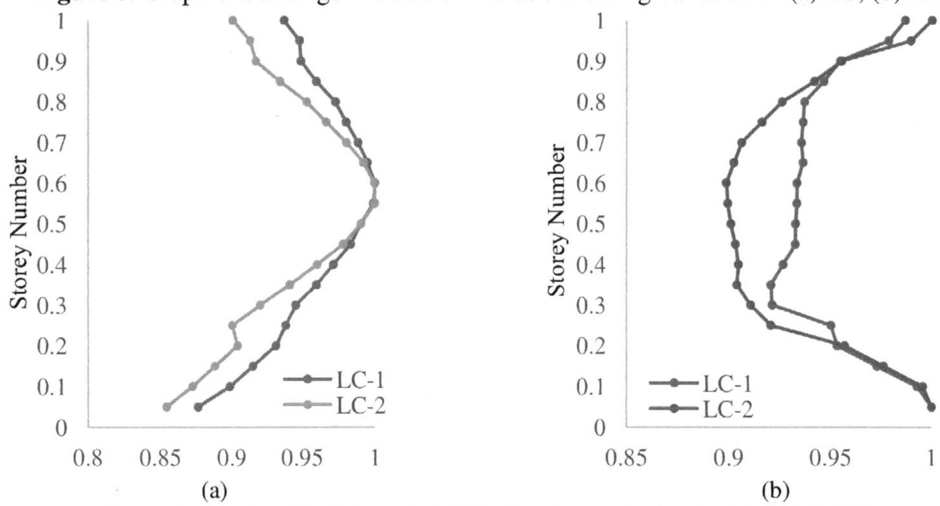

Figure 9. Graph of buildings in 5C20 showing variation in (a) BS, (b) RD.

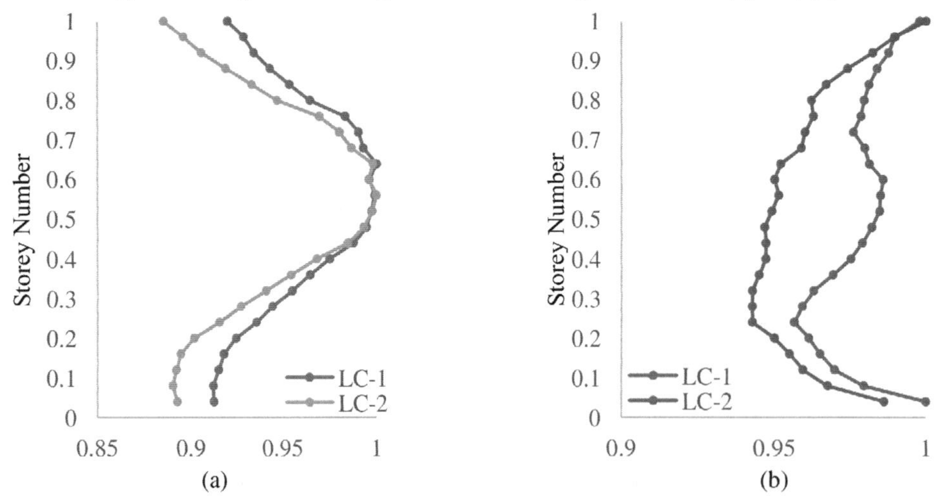

Figure 10. Graph of buildings in 5C25 showing variation in (a) BS, (b) RD.

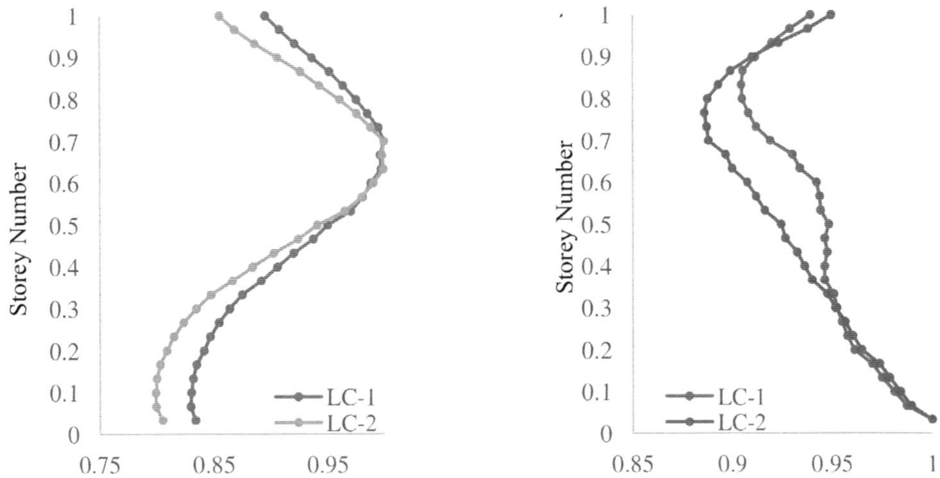

(a) (b)

Figure 11. Graph of buildings in 5C30 showing variation in (a) BS, (b) RD.

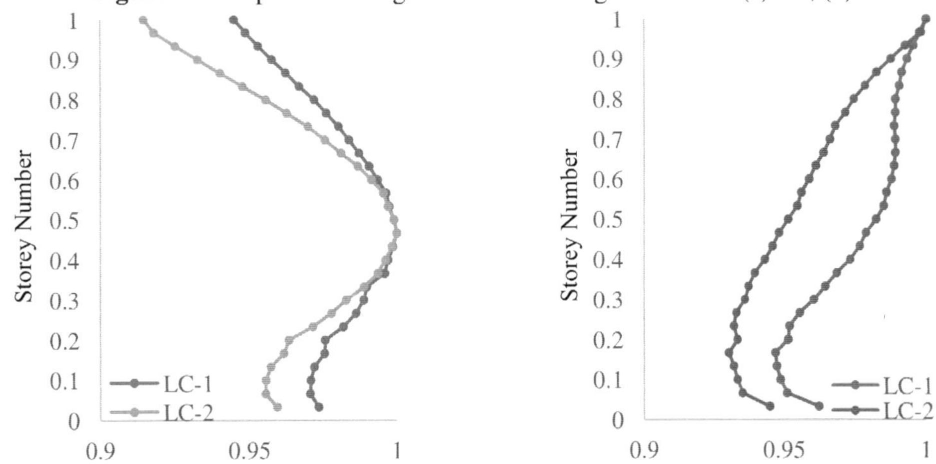

(a) (b)

Figure 12. Graph of buildings in B-SC30 showing variation in (a) BS, (b) RD.

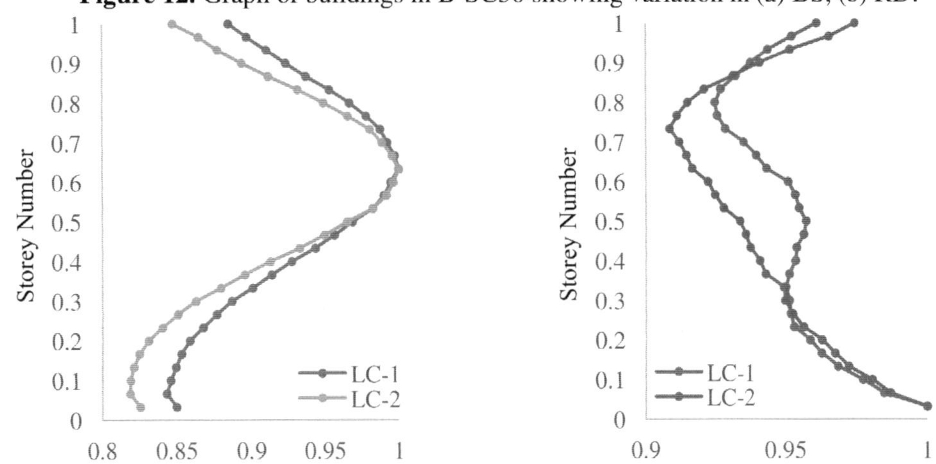

(a) (b)

Figure 13. Graph of buildings in B-3C30 showing variation in (a) BS, (b) RD.

Table 2. Optimum storey positions of outrigger-belt system for models of phase 2.

Building models	LC-1		LC-2	
	BS	RD	BS	RD
A-SC30	11 to 16	3 to 7	11 to 16	6 to 8
A-3C30	18 to 22	22 to 27	16 to 23	19 to 26
B-SC30	12 to 17	3 to 7	11 to 17	3 to 8
B-3C30	17 to 21	21 to 26	16 to 22	18 o 25

5. Conclusions

The study is carried out for optimum position of outrigger-belt system in the building in which the stiffness of the building is varied with height and as a whole. The following conclusions can be drawn from the study:

- The graphs generated show that with change in position of outrigger and belt system, the performance of the building can be affected and optimum positions can improve the performance by 5-10%.
- If the buildings have a constant stiffness throughout the height (ignoring the additional stiffness introduced due to outrigger-belt system), then the optimum positions based are usually at 0.40-0.50H for base shear and 0.15-0.30H for roof displacement.
- If the stiffness of the building is decreasing with the increase in height, then the optimum position based on base shear shifts a little upwards to a region of 0.55-0.70H of the building, whereas if the roof displacement is taken as a parameter, then the 0.65-0.80H becomes the optimum positions. Although the same buildings, lower region of 0.20-0.35 also gave good (and sometimes better) results.
- It can be inferred that if the overall stiffness of the building is increased at once (for example in Phase 2), then it will surely increase the performance and decrease the roof displacement of the building, but it may not have any major impact on the optimum position of the outrigger-belt system.

Here H stands for the total height of the building and data is mentioned from the bottom of the building. Since fewer models were used in assessing the effect of outrigger-belt system in these high-rise buildings, thus, author would like to suggest more work in this area using a bigger data set.

6. References

[1] Kamgar R and Rahgozar R 2017 Determination of optimum location for flexible outrigger systems in tall buildings with constant cross-section consisting of frame tube, shear core, belt truss and outrigger using energy method *International Journal of Steel Structures* **17** 1-8

[2] Fawzia S and Fatima T 2016 Optimum position of steel outrigger system for high rise composite buildings subjected to wind loads *Advanced Steel Construction* 12(**2**) 134-153

[3] Tavakoli R, Kamgar R and Rahgozar R 2017 Seismic performance of outrigger-braced system based on finite element and component mode synthesis methods *Iranian Journal of Science and Technology Transactions of Civil Engineering* **1** 1-9

[4] Poluraju P and Rao P V S N 2011 Pushover analysis of reinforced concrete frame structure using SAP 2000 *International Journal of Earth Sciences and Engineering* 4(**6**) 684-690

[5] Salman K, Kim D, Maher A and Latif A 2020 Optimum control on Structural Integrity using outrigger braced frame system under lateral loads *Journal of Structural Integrity and Maintenance* 5(**1**) 40-50

[6] Jagadheeswari A S and Christy C F 2016 Optimum position of outrigger belt truss in tall buildings subjected to earthquake and wind loads *International Journal of Earth Sciences and Engineering* 9(**3**) 373-377

[7] Haghollahi A, Ferdous M B and Kasiri M 2012 Optimization of outrigger locations in steel tall buildings subjected to earthquake loads *15th World Conference on Earthquake Engineering* LISBOA

IOP Conf. Series: Materials Science and Engineering **989** (2020) 012002 doi:10.1088/1757-899X/989/1/012002

[8] Choi H S and Joseph L 2012 Outrigger system design considerations *International Journal of High Rise Buildings* 1(**3**) 237-246
[9] IS 1893 Part 1 2016 *Criteria for earthquake resistant design structures. Part 1: General provisions and buildings (Sixth revision)* Bureau of Indian Standards New Delhi India
[10] Pinho R, Marques M, Monteiro R, Casarotti C and Delgadob R 2013 Evaluation of nonlinear static procedures in the assessment of building frames *Earthquake Spectra* **29** 1459–1476
[11] IS 16700 2017 *Criteria for Structural Safety of Tall Concrete Buildings* Bureau of Indian Standards New Delhi India

IOP Conf. Series: Materials Science and Engineering **989** (2020) 012003 doi:10.1088/1757-899X/989/1/012003

Structural performance of nano-silica based blended CFST stub circular column

Himanshu Bisht[1], Aakash Gupta[1,2], Dodda Srinivas[1] and Dibyendu Adak[1]

[1]Department of Civil Engineering, National Institute of Technology Meghalaya, Shillong, Meghalaya, India.

[2] Email: ceaakash@gmail.com

Abstract. The objective of this study is to find the optimum percentage replacement of cement with the nano-silica in a concrete mix to improve the structural performance Concrete Filled Steel Tube (CFST) stub circular column under the axial compressive load. The CFST stub columns with circular cross-sectional area of conventional concrete mix (without nS) and nano-silica blended concrete mix (with nS-CFST) had been experimentally and numerically studied. The amorphous silica is a major component of the cementitious materials, and it reacts with the calcium hydroxide, which is comprised of the calcium silicate hydration. The addition of this nano-silica affects the different properties of the concrete and increases its performance. On the other hand, CFST is known for the composite behavior generated between the hollow tube of steel and a core constructed of concrete, which improves the collected performance of the member. In present study, the concrete mix of M25 had been modified with 4%, 6%, 8% and 10% of nano silica. The structural behavior of stub columns (with / without nS) are assessed under axial compressive load. Also, the CFST stub columns were then modelled into ABAQUS for finite element analysis. The nano-silica blended concrete in CFST stub columns showed an improved the structural performance than the conventional cement concrete stub circular CFST column.

1. Introduction

The partial replacement of cement with micro and nano-materials in a concrete mix is a common practice. It improves various properties of the concrete mix, which includes tensile strength, compressive strength, early age strength, etc. One of the most commonly used materials for partial replacement is nano-silica (nS). If a nano-silica is introduced in the concrete mix, it tends to increase its compressive, flexure and tensile strength [2]. The cement mortar mix which has nano-silica have more homogeneity binder, fewer pores, and more adhesion at interfacial zone [10]. Due to its ultrafine nature, the role of nano-silica is not limited to only the pozzolanic or filler effect, but it also increases the speed of kinetics of hydration [7]. It helps in improving the strength of the concrete by giving a boost to the generation of the nano-crystals of Calcium-Silicate-Hydrate (C-S-H) gel, and these nano-crystals occupy the nanopores as well as the micropores or voids that were left vacant in the cement-based concrete [2][5][8][10]. Nano-particles like nS, act as nuclei in the cement paste, as it has high activity, thus it promotes and accelerates the cement hydration process [2-3]. The percentage replacement of cement with the nano-silica to get maximum performance is highly dependent on the source of nano-particles, dispersing technique, w/c ratio, size of nano-silica etc. [11]. The compressive strength of the cement paste containing 5% silica was increased by 64% for 1 day, 39% for 7 days and 35% for 28 days in [1], it increased by 17.5% in 28days in [3], it increased by 5% and 11% at 7 and 28 days at 0.55 water binder ratio in [6], and it increased by 45% in 7 days in [5]. The compressive strength of the concrete mix was reported to increase by 42% in 28 days when 6% of Nano-Silica was added in [4]. When w/b ratio of 0.55 was used then the increment in compressive strength of 77% and 87% was observed at 10% Nano-silica in [6]. In reference [9], when 2.5%, 5% and 7.5% cement content was replaced with nano-silica, then it gave maximum compressive strength at 5% replacement, which was corresponding to an increase 15%. Reference [10], tested the samples at 1%, 3%, 5%, 7% and 10% replacement of cement with 19nm nano-silica content, which in gave maximum compressive strength at 7% on 3,7 and 28 days with a maximum improvement of 55.7% at days. When there is an excess of nano-silica in the concrete mix, then the excess portion has no activation, and it takes the place of the cement in the form of an inert particle, which tends to decrease in the compressive strength [9-10].

Concrete-Filled-Steel-Tube or CFST comprises of a hollow steel section of required shape filled with plain concrete or concrete along with reinforcements. The steel tube in CFST acts as a permanent formwork that annihilates the use of temporary wooden formwork, which reduced the construction cost. This placement of the steel in the perimeter is considered as the most efficient use of the steel since the contribution in the moment of inertia and flexure capacity is maximum [12][16]. CFST is often used in high-rise buildings and sometimes low-rise industrial buildings that are exposed to robust loading or where efficient structural systems are required. Its composite action benefits both concrete and steel, as in this arrangement, the strength and ductility of the concrete core are improved because of the continuous confinement provided by the steel tube, whereas the local buckling of the steel tube is impeded by the concrete by avoiding tube to buckle inwards [16-17]. A lot of modifications to the CFST has been proposed in different studies to improve strength or get smaller cross-sections. Attaching circular stirrups along the surface of the column was a better alternative to increasing the thickness [13]. Sometimes another steel member is also inserted into the steel tube along with the concrete, and other cases are also used in studies. CFST columns with circular and square cross-sections are the most commonly used, followed by the hexagon, ellipse and other shapes. In short columns, confinement effects are observed to be more when compared with the slender columns under the axial loading [14].

In this study is focused on the structural properties of the conventional and nano silica modified blended CFST stub columns. To discuss it more straightforwardly, this study is conducted in three different phases. The first phase is about testing the standard M25 concrete mix along with the concrete in which there is a replacement of different percentages of cement with the colloidal nano-silica (nS). For this purpose, cubes and cylindrical specimens are cast and tested for its axial compressive strength. In Phase 2, the standard concrete mix and nano-silica blended concrete mixes are then used as core materials in the steel tubes with a circular cross-section to form nano-silica blended CFST or as termed in this study "nS-CFST". In Phase 3, numerical models are developed for the nS-CFST specimens from Phase 2.

2. Experimental details

A concrete of M25 grade has been taken as a reference concrete mix for this study. The basic ingredients of the present study are Portland Pozzolana Cement of 43-grade, mild steel tube of a circular cross-sectional area, and nano-silica of particle size of 8-15nm and pH of 9-10. The amount of cement replaced with nS is 4%, 6%, 8%, and 10% by its weight. All the related values of the design mix containing the amount of cement, fine aggregate, coarse aggregate, and water are mentioned in table 1. This concrete mix and nano-silica blended concrete mix is used to cast specimens in the form of cubes of sides 150mm and cylinders of diameter 150mm and height 300mm, which are tested for axial compressive strength. This testing of cubes and cylinders are covered in the Phase 1.

Table 1. Design mix values for the different percentages of nano-silica.

Parameter with its value for 1m^3 volume						
Cement (kg)	nS (kg	% of nS	Water (litre)	Fine Aggregate (kg)	Coarse Aggregate (kg)	w/b Ratio
395	0	0	219	675	1100	0.48
379.2	15.8	4	219	675	1100	0.48
371.3	23.7	6	219	675	1100	0.48
363.4	31.6	8	219	675	1100	0.48
355.5	39.5	10	219	675	1100	0.48

In Phase 2, the nano-Silica blended CFST or nS-CFST are tested along with the standard CFST specimens. They have the core of the mixes represented in table 1 from the Phase 1. The tube sections selected for analysis is the circular cross-sectional shapes. These specimens are also tested for axial compressive load, and their behaviors are noted down. After phase 2, the numerical models of nS-CFST specimens are developed to study the behavior under compressive. The finite element numerical techniques are used for the analysis of behavior of the modelled stub columns under axial compressive load. The finite element software used for the modelling of specimens is ABAQUS, in which steel tube is modeled using S4R elements as they are known to give accurate results for steel members, whereas the concrete core is modeled utilizing the three-dimensional eight-node solid (C3D8R) element. The bottom end of the sample is kept fixed and the compressive load is applied from the top cross-sectional area. The load is applied with the kinematic coupling using the modified RIKS method. This method is available within the software library and it is considered very accurate. The non-linear geometry parameter is also included to incorporate large displacements in the analysis. The load in the modified RIKS method is proportionally applied in several load increments. In this, the equilibrium iteration is performed in every load increment and the equipoise path is traced in the space of load-displacement.

3. Results

The cubes and cylinders of nano-silica blended specimens and control specimens from Phase 1 are tested under the axial compressive load using the universal testing machine, in which the displacement is incremented with time, and corresponding loads are noted. The specimens are tested at 14 and 28 days under axial compressive load and the output of the testing is mentioned in table 2. Based on the output of the compressive test, it can be inferred that the increase of strength through colloidal nano-silica was very significant. The observations were almost the same in both cubical and cylindrical specimens. It gave the maximum improvement of 44.7% in compressive strength when 6% of cement was replaced by nS. The compressive strength decreased slightly at 8% replacement, and a significant drop can be observed in the case of a 10% replacement. This highest increase in compressive stress of 36% was observed at 6% of nS.

The nS-CFST specimens and standard CFST specimen of the circular cross-sectional area that are tested in Phase 2 reflected almost the same behavior. The optimum percentage of nano-silica in nS-CFST specimens are observed when 6% of the cement was replaced by the nano-silica by mass. The stress developed at 8% was almost similar to the stress at 6%. Although, when the replacement of cement with nS is increased to 10%, there is a very extreme drop that can be observed in the nS-CFST specimens. The values of maximum stresses generated by the circular nS-CFST specimens as well as the standard specimen can be observed in in Fig. 1 - 5 and also in table 3. The table clearly shows a huge drop in stress values at 10% replacement of nano-Silica. The stress-strain behavior of the specimens throughout the loading duration can be observed in the graphs represented in Fig. 1 - 9.

Table 2. Compressive strength of nano-silica blended concrete.

Days	% nS	Stress (MPa)	
		Cube	Cylinder
14	0%	23.07	18.53
	4%	27.32	22.97
	6%	32.23	27.68
	8%	31.83	26.14
	10%	28.17	23.90
28	0%	26.52	21.3
	4%	30.25	25.71
	6%	36.19	30.76
	8%	34.82	29.17
	10%	31.24	26.55

Results from Phase 1 and Phase 2 show a collective outcome that once the amount of colloidal nano-silica crosses a certain percentage, there will be a huge drop in the stress values. This satisfies for both the phases. Reference [9,10] discusses the same behavior, as when there is an excess of nS in the concrete mix, the nS particles left after the reaction acts as the inert particles which reduce the performance of the specimen. Through the stress-strain graphs of nS-CFST presented in fig. 1 - 4, it is inferred that even after the nS-CFST specimens have reached its maximum value or the ultimate strength, there is a very large plastic region, which is one of the normal behaviors observed in the CFST members, as it is known to delay the failure of the members.

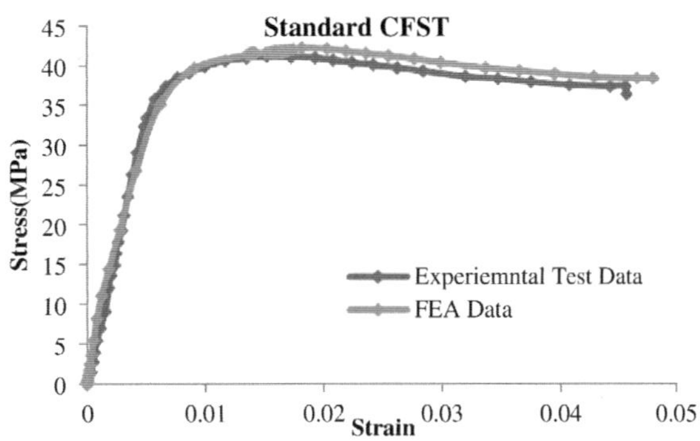

Figure1. Stress-strain curves of standard CFST.

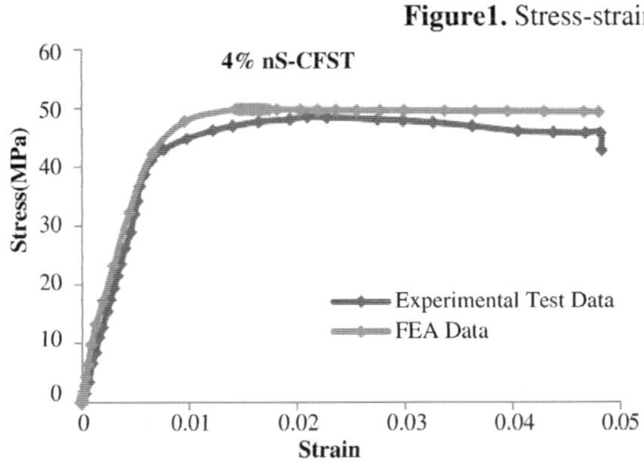

Figure 2. Stress-strain curve of nS-CFST with 4% replacement.

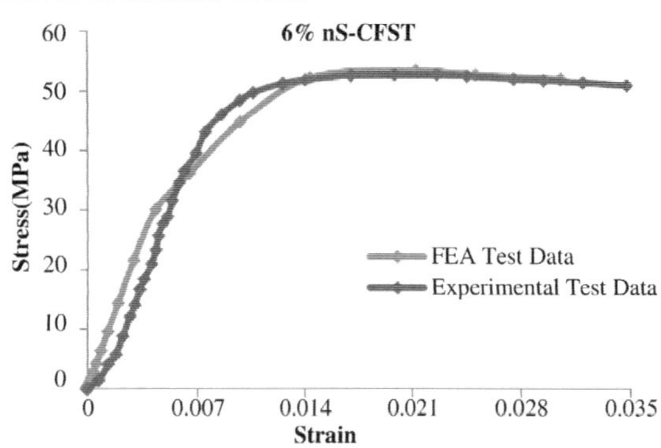

Figure 3. Stress-strain curves of nS-CFST with 6% replacement.

IOP Conf. Series: Materials Science and Engineering **989** (2020) 012003 doi:10.1088/1757-899X/989/1/012003

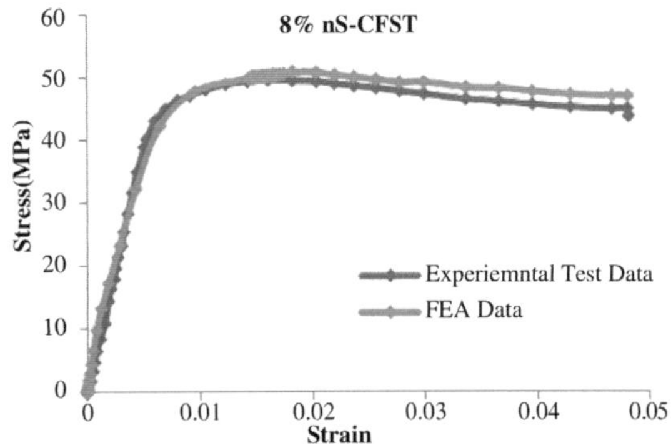

Figure 4. Stress-strain curves of nS-CFST with 8% replacement.

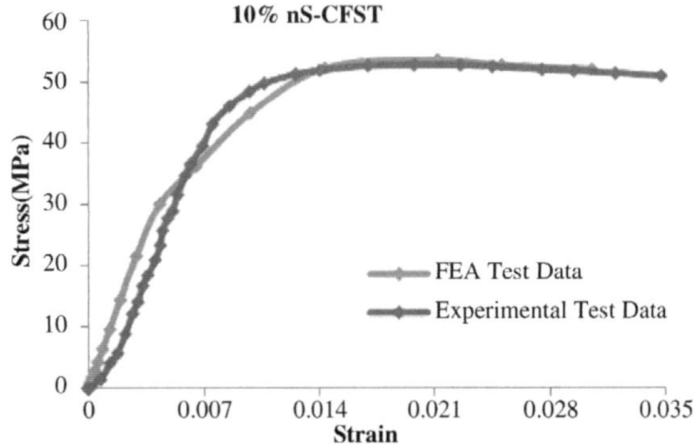

Figure 5. Stress-strain curve of nS-CFST with 10%replacement.

(a) (b)

Figure 6. The behavior of circular nS-CFST under the axial compressive load for (a) 4% nS Test Specimen, (b) 4% nS FEA Model.

(a) (b)

Figure 7. The behavior of circular nS-CFST under the axial compressive load for (a) 6% nS Test Specimen, (b) 6% nS FEA Model.

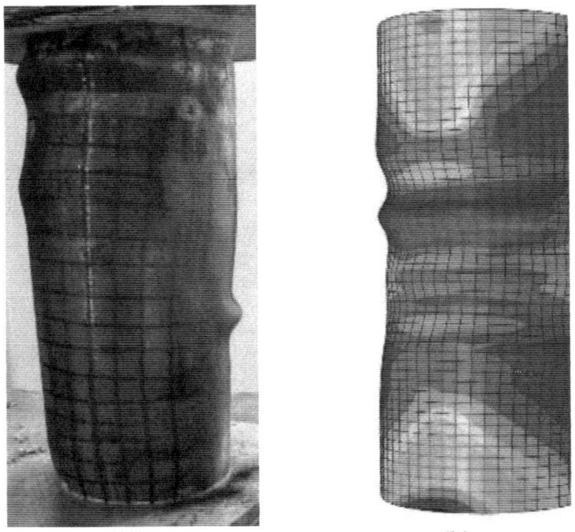

(a) (b)

Figure 8. The behavior of circular nS-CFST under the axial compressive load for (a) 8% nS Test Specimen, (b) 8% nS FEA Model.

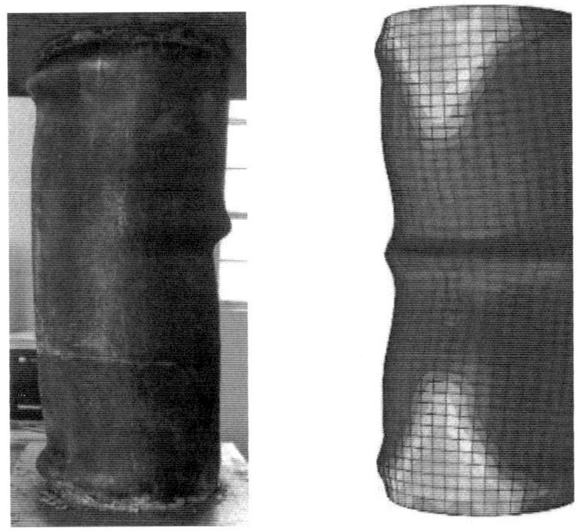

(a) (b)

Figure 9. The behavior of circular nS-CFST under the axial compressive load for (a) 10% nS Test Specimen, (b) 10% nS FEA Model.

Reference [12][18] explains how the cross-sectional areas have an effect on the stresses of the nS-CFST specimens as well. For this, the term "confinement factor" is used, and the formula used for the calculation of this factor is; $\xi = \frac{f_s A_s}{f_c A_c}$ In this f_s stands for the strength of steel, whereas f_c stands for strength of concrete, and A_c and A_S stands for the cross-sectional area of the concrete and steel respectively. As the name stands, it expresses the confinement effect on the concrete core generated by the steel tube. Higher the effect, better the performance of the member. So, the confinement factor for the experimental test specimens is also calculated and represented in a graph given in Fig. 10. The higher values of confining factors for specimens nS-CFST from the graph explains the better confinement effect. This can be used to compare nS-CFST of different cross-sectional areas. It can also be observed from Fig. 1-5 and table 3, that the increase in compressive strength of specimen was around 28% at 6%, which gave the best performance.

Table 3. Compressive Stress Developed In nS-CFST Stub Columns.

% nS	Maximum Stress (MPa) in nS-CFST
0	40.86
4	48.52
6	52.67
8	50.28
10	42.21

In Phase 3, the numerical models are successfully developed for the experimental nS-CFST specimens and standard CFST. Similar behavior can be observed in all the numerical models. The results of the stress-strain graphs obtained from the numerical models and experimental specimens are represented together in fig. 1-5. The output values show almost the same behavior as that of the specimens. The deformed profile of the numerical models is almost same to the experimental specimen, which can be observed in Fig.6-9, which validates the model.

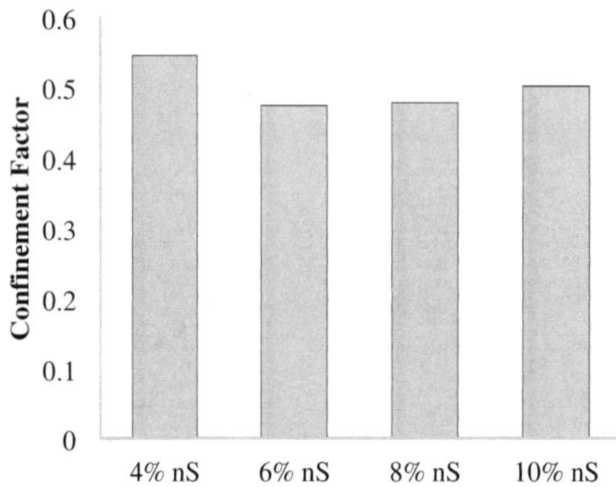

Figure 10. Confinement factors for nS-CFST specimens.

4. Conclusions

The study observes the effects of partial replacement of cement with the nano-silica (nS) in M25 mix and nS-CFST stub columns. The numerical models were also developed in the study. The following conclusions can be drawn from the study:

- The blended concrete mix 6% of nano-silica showed better strength than all other nano-silica based blended concrete mix and control cement mix.
- A similar trend could be seen in nS-CFST stub columns, the optimum amount of nano-silica in nano-silica blended concrete and CFST was observed at 6%, although the 8% also gave quite good results.
- A significant reduction in the compressive strength in the case of both nS blended concrete and nS-CFST specimen with more than 6% of nano-silica in the mix, which shows the optimum dosage of nS plays an important role and the drastic decrease in strength after that dosage.
- A numerical model for the nS-CFST was successfully developed, which shows approximately the same results given by the experimental specimens.

The conclusions describe the improvement of the performance of the blended concrete mix and nS-CFST corresponding to the percentage replacement of concrete with nano-silica. It could also be seen from the stress-strain curves that the nS-CFST models acted in a very ductile manner, which shows its ability to delay the ultimate failure.he first paragraph after a heading is not indented (Bodytext style).

5. References

[1] Singh L.P, Bhattacharyya S K and Ahalawat S 2012 Preparation of size controlled silica nano particles and its functional role in cementitious system *Journal of Advanced Concrete Technology* **10** 345-352

[2] Nazerigivi, Nejati H R, Ghazvinian A and Najigivi A 2018 Effects of SiO_2 nano-particles dispersion on concrete fracture toughness *Construction and Building Materials* **171** 671-679

[3] Li H, Xiao H, Yuan J and Ou J 2014 Microstructure of cement mortar with nano-particles *Composites Part B* **35(2)** 185-9

[4] Jo B, Kim C, Tae G and Park J 2007 Characteristics of cement mortar with nano-SiO_2 particles Constr Build Mater **21(6)** 1351-5

[5] Hou P, Kawashima S, Kong D, Corr D J, Qian J and Shah S P 2013 Modification effects of colloidal nanoSiO2 on cement hydration and its gel property Composites: Part B **45** 440-448

[6] Tobón J I , Paýa J, and Restrepo O J 2015 Study of durability of Portland cement mortars blended with silica nano-particles *Construction and Building Materials* **80** 92-97

[7] Said A.M., Zeiden M.S, Bassuoni M.T, and Tian Y. 2012 Properties of concrete incorporating nano-silica *Construction and Building Materials* **36** 838-844

[8] Najigivi, Khaloo A, Zad A I and Rashid S A 2013 Investigating the effects of using different types of SiO_2 nano-particles on the mechanical properties of binary blended concrete *Composites Part B: Engineering* **54** 52–58

[9] Arif M, Al-Hagri M G, Shariq M, Rahman I, Hassan A and Baqi A 2019 Mechanical properties and microstructure of micro- and nano-additives-based modified concrete composites: A sustainable solution *Journal of The Institution of Engineers (India): Series A*

[10] El-Baky S A, Yehia S and Brahim S K 2013 Influence of nano-silica addition on properties of fresh and hardened cement mortar in *5th International Conference on Nanocon, Brno, Czech Republic*: 16–8

[11] Varghese J, Gopinath A, Bahurudeen A and Senthilkumar R 2018 Influence of nano-silica on Characteristics of Cement Mortar and Concrete *Sustainable Construction and Building Materials* 839– 851

[12] He L, Zhao Y and Lin S 2018 Experimental study on axially compressed circular CFST columns with improved confinement effect *Journal of Constructional Steel Research* **140** 74-81

[13] Wu B, Zhang Q and Chen G M 2018 Compressive behavior of thin-walled circular steel tubular columns filled with steel stirrup-reinforced compound concrete *Engineering Structures* **170** 178-195

[14] Huang F, Yu X and Chen B 2012 The structural performance of axially loaded CFST columns under various loading conditions *Steel and Composite Structures* **13 (5)** 451-471

[15] Sakino K, Nakahara H, Morino S and Nishiyama I 2004 Behavior of centrally loaded concrete-filled steel-tube short columns *J. Struct. Eng.* **130(2)** 180-188

[16] Gupta P K, Sarda S M and Kumar M S 2007 Experimental and computational study of concrete filled steel tubular columns under axial loads *Journal of Constructional Steel Research* **63** 182–193

[17] Han L, Li W and Bjorhovde R 2014 Developments and advanced applications of concrete-filled steel tubular (CFST) structures: Members Journal of Constructional Steel Research 100 211–228

[18] Tao Z, Wang Z and Yu Q 2013 Finite element modeling of concrete-filled steel stub columns under axial compression J. Constr. Steel Res. 89 121–131

IOP Conf. Series: Materials Science and Engineering **989** (2020) 012004 doi:10.1088/1757-899X/989/1/012004

A study on the setting and flow behaviour of alkali activated slag/fly ash composites in ambient condition

Namitha S [1,2], Muhammed Nabil K[1], Mohammed Rafeeque N V [1], Reshma Sundhar [1], Thushara Raju[1] and Ramaswamy K P[1]

[1,2] Department of Civil Engineering, TKM College of Engineering, Kollam, Kerala, India.

[2]Email: namithasudan@gmail.com

Abstract. Alkali-activated binders (AAB) are considered as a promising sustainable future construction material, and promoted as an environmentally beneficial option when compared to Ordinary Portland Cement (OPC). Portland cement-based concrete is the most used material in the construction industry. In the production of OPC, harmful gasses like CO_2, NO_2, SO_2 and specks of dust are discharged into the atmosphere due to calcination of limestone and combustion of fossil fuel. Hence in the present scenario, it has become an emphatic necessity to replace Ordinary Portland Cement based concrete with AAB based system (geopolymer) which is more sustainable. The characteristics of geopolymer composites are predominantly influenced by some of the variables such as temperature and time of curing, alkaline solution to solid binder ratio, water content, type of alkaline activators used, molarity of alkaline solution, chemical composition of aluminosilicate precursors etc. Although geopolymer systems are good in terms of durability and strength, the main issue faced for its less popularity in the field application is its low workability, poor slump retention and too fast setting.

The aim of this experimental study is to investigate the compatibility between modern generation superplasticizers (SP) and various alkali activated binder systems prepared using industrial wastes and the evaluation on the effect of alkali binder ratio (a/b) on the fresh properties of alkali activated binder (AAB) systems. Considering the practical implications, it is the need of the hour to develop a geopolymer mix setting in ambient condition. This study focuses on setting and flow behaviour of the alkaline activated composites of slag and fly ash in ambient condition only. The slag/fly ash ratios of 100/0, 75/25, 50/50 and 25/75by mass and alkali-to-binder ratios of 0.4 to 1 with an increment of 0.1 were used for the study. The type of superplasticizer used for the study is Sulphonated Naphthalene Formaldehyde (SNF) with dosages of 0%, 1%, 2% and 3%. The setting time and the flow of the paste were determined using vicat apparatus and mini slump cone respectively. The penetration resistance of the setting paste was also measured using vicat apparatus. The test results revealed that even though the setting time increase marginally with SP dosage, the use of SNF was not much effective in prolonging the setting time of alkali activated pastes. There was tremendous increase in the initial and final setting time for pastes with an increase in a/b ratio.

1. Introduction

Cement manufacturing industry is the second largest producer of green house gases. Cement based concrete is most widely used in construction practices and is a significant contributor to the environmental issues. There is a strong demand for green concrete with better strength properties and durability without a compromise on the rheological characteristics. The alkali activated binder systems which are good in terms of durability and strength, however lacks in better rheological properties [6].

The reduction in a/b ratio helps to increase compressive strength, impermeability and durability simultaneously but causes reduction in workability [5,9]. The use of superplasticizers enhances the workability and rheological properties even at low a/b ratio [13]. The superplasticizers such as Sulphonated Naphthalene Formaldehyde (SNF), Lignosulphonates (WRRe), Sulphonated Melamine Formaldehyde (SMF), and Polycarboxylic Ether (PCE) are the most commonly researched admixtures to be used in Alkali Activated Materials (AAM). But increase in the dosage of superplasticizer is uneconomical, and the mixes with higher a/b ratios but with reasonable strength need to be developed [10].

Geopolymerisation is a chemical reaction between various alumino-silicate oxides with silicates under highly alkaline conditions forming polymeric Si–O–Al–O bonds and it indicates that any Si–Al material could become a source of geopolymerisation [3]. Micro-structural studies done by Nath and Sarker pointed that aluminosilicate (N–A–S–H) gel was formed in geopolymerisation reaction in the absence of calcium oxides but if the calcium oxide is incorporated in the form of GGBFS or OPC, into fly ash based geopolymer then calcium silicate hydrate (C–S–H) gel is formed along with the geopolymer gel [8]. The formation of C–S–H gel reduces the cost of heat curing in construction sites and can be used to construct self cured geopolymer composites [3]. Previous studies on the effect of alkali content, silicon to aluminium ratio, water content and the amount of superplasticizer on the workability of fly ash based geopolymers show that increasing the alkali content and silica content result in reduced workability [2]. The early strength development of the mortar at ambient condition is enhanced by incorporating GGBFS as precursor [12] whereas the addition of sulphonated naphthalene formaldehyde (SNF) admixtures increases the workability and shrinkage but reduces the compressive strength [1, 11]. So it is necessary to determine an optimum dosage of SP for each of the alkali to binder ratio used. The effect of superplasticizers on the fly ash based geopolymer directly depends on the type of alkali activator and the superplasticizer [15]. Besides superplasticizers, concentration of alkaline solution also has a significant role in the regulation of setting time, workability, and strength properties of the mixes [7]. Increase in the concentration of alkali reduces the setting time and workability of the mixes [8]. Therefore, studies are to be performed to identify an optimum alkali to binder ratio and corresponding SP dosage.

Only limited studies have been carried out in the area of geopolymer – chemical admixture interactions and about the fresh properties of geopolymer concrete (GPC) in ambient curing conditions [14]. Behaviour of AAB systems using different chemical admixtures need to be studied, and the compatibility of admixture are to be explored further. More scientific studies are needed to find a solution for the aforementioned issues. Thus, this study is aimed at studying the fresh properties such as setting and flow behaviour of alkali activated fly ash/ slag pastes developed in ambient air conditions.

2. Materials used

Ultrafine ground granulated blast-furnace slag (GGBFS) obtained from Astrra Chemicals, Chennai and fly ash obtained from a Ready Mix Concrete plant were used as precursor in the preparation of alkali activated geopolymer mixes. The specific gravity of slag and fly ash were 2.90 and 2.21 respectively. Activator solution is prepared by mixing sodium silicate solution and sodium hydroxide (NaOH) solution in the ratio 2.5 by mass. Sodium silicate solution of density 1.59 gm/cm^3 with SiO_2 to Na_2O ratio by mass of 2.4 and 40% solid content was used. NaOH pellets with 97-98% purity were mixed with distilled water to make a solution of 10M concentration. The superplasticizers investigated in the study were Ceraplast 300 which is a Sulphonated Naphthalene Formaldehyde (SNF) based admixture conforming to ASTM C 494 - 98 Type F and IS 9103-1999. The superplasticizer dosages used in the study are expressed as percentage by weight of the total amount of precursor binders used in the mix. The properties of superplasticizers are provided in table 1.

IOP Conf. Series: Materials Science and Engineering **989** (2020) 012004 doi:10.1088/1757-899X/989/1/012004

Table 1. Properties of SNF.

Appearance	Liquid
Colour	Brown
Solid content %	32.8
Specific gravity	1.18
pH	≥7
Chloride content	Nil

3. Methodology

In this study, the tests were conducted in geopolymer paste with alkali-to-binder ratios of 0.4 to 1.0 at increments of 0.10. The type of superplasticizer used for the study is Sulphonated Naphthalene Formaldehyde (SNF) with dosages of 0%, 1%, 2% and 3%. The alkali activator systems selected were slag/ fly ash in the ratios of 100/0, 75/25, 50/50, 25/75, and 0/100 by mass. The workability and setting time was tested using mini slump cone and Vicat apparatus respectively. The precursors were mixed with alkaline solution using Hobart mixer. Then the superplasticizers were added (to the part of alkaline solution) with alkaline solution followed by thorough mixing for three minutes. After proper mixing, both the tests were conducted on the paste simultaneously. The workability of fresh geopolymer paste was studied using mini slump cone test apparatus shown in Figure 1. The mini slump cone was 3D printed using nylon.

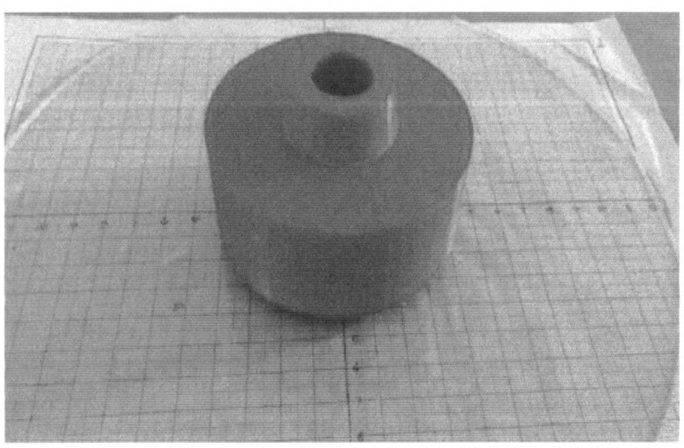

Figure 1. Mini slump cone.

Figure 2. Mini slump cone setup with poured paste.

The mini slump cone test apparatus was placed above the centre of a glass plate. The freshly prepared paste was poured into the mini-slump test setup which is a downscaled Abram's truncated conical mould having 19 mm top diameter, 38 mm bottom diameter and a height of 57 mm. The extra paste was removed and top surface levelled as shown in Figure 2. After lifting the mould vertically, diameters of paste spread in two perpendicular direction were measured and then the average value was taken as the final diameter of the spread [16]. The % spread was calculated from the measured value of average final diameter based on equation (1).

$$\% \, spread = \frac{Final\ Diameter\ - Initial\ Diameter}{Initial\ diameter} \times 100 \qquad (1)$$

As per IS 4031 Part 5 (2004) [4], Vicat apparatus was used to measure initial and final setting time of the paste. The time of penetration of Vicat needle is taken as basis for the conductance of these tests in the freshly prepared paste. The time taken by initial setting needle to penetrate a distance of

5±0.5mm from the bottom of the mould was taken as the initial setting time and the time when the annular ring of the final setting needle fails to make the impression on the paste was recorded as the final setting time of the paste [4].

4. Results and discussions

4.1. Mini slump cone test

The % spread of paste obtained for different mixes with varying a/b ratios and variation of flow with SP dosages of different mixes are shown in the Figure 3 to Figure 6.

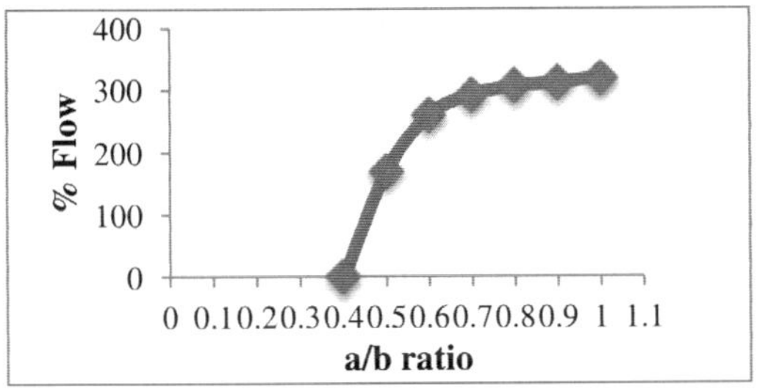

Figure 3. Variation of flow with a/b ratio of mix with slag/ fly ash 100/0.

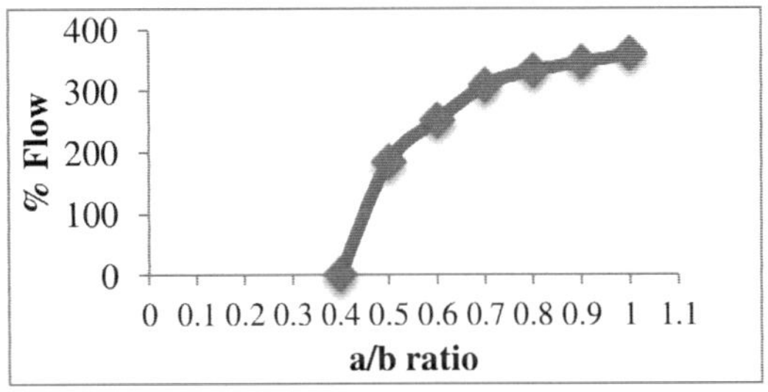

Figure 4. Variation of flow with a/b ratio of mix with slag/fly ash 75/25.

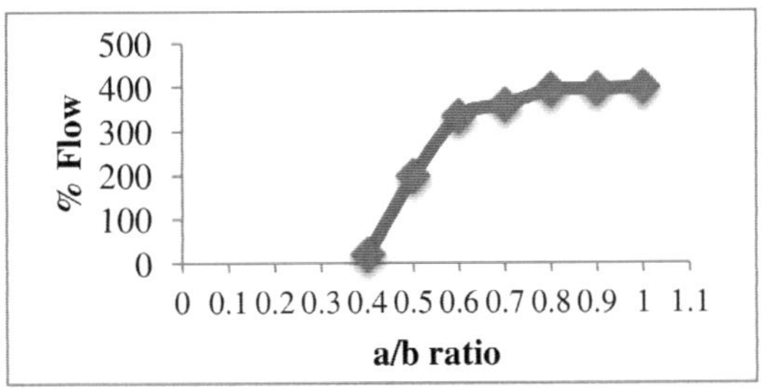

Figure 5. Variation of flow with a/b ratio of mix with slag/ fly ash 50/50.

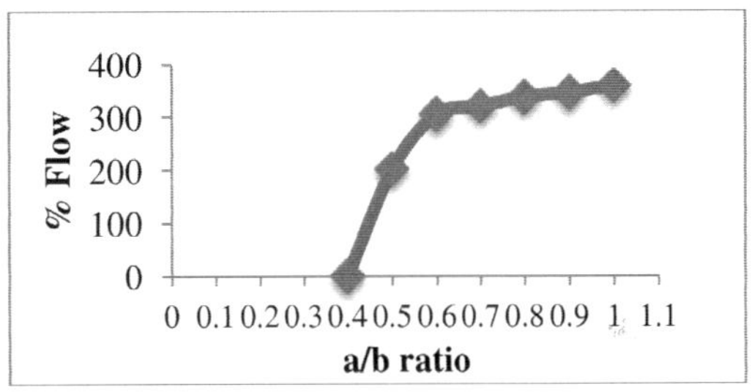

Figure 6. Variation of flow with a/b ratio of mix with slag/ fly ash 25/75.

The results (Figure 3 to 6) show that the spread of the paste that increase dramatically in the lower a/b ratio becomes almost constant at higher a/b ratios. Even though the maximum flow values differ, all the mixes were showing similar trend in the flow with an increase in alkali content. Also it was found that flow increases with increase in the fly ash content. The variation in flow with a/b ratio for paste with slag/ fly ash 50/50 and without SP is shown in the Table 2. It can be seen that flow increases with an increase in a/b ratio. Table 3 shows the variation in flow of paste with slag/ fly ash 100/0 at a constant a/b = 0.40 with an increase in superplasticizer dosage (from 0 to 3%). It is seen that flow increases with SP dosage at a constant a/b ratio. But, high dosage of SP was needed to make the paste flowable.

The variation of flow with SP dosage for each mix is indicated graphically in Figures 7 to 10. The flow was found to be increasing with SP dosages (minimal increase). However, the effectiveness of SP in increasing the flow was found to be lesser when compared to a/b ratio. The mixes with slag/ fly ash 100/0, and 25/75 with alkali to binder ratio of 0.40 and SP 2% satisfies the range of workability 110±5% as per IS 4031(Part 7) – 1988.

Table 2. Variation of flow with a/b ratio for paste with slag/fly ash 50/50.

Mix	Mini Slump Flow
Mix with Slag/ fly ash 50/50, SP dosage = 0 % and a/b = 0.4	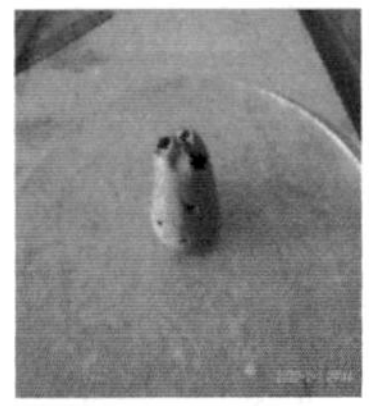
Mix with Slag/ fly ash 50/50, SP dosage = 0 % and a/b = 0.5	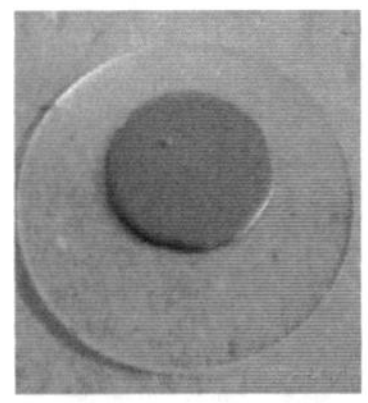
Mix with Slag/ fly ash 50/50, SP dosage = 0 % and a/b = 0.6	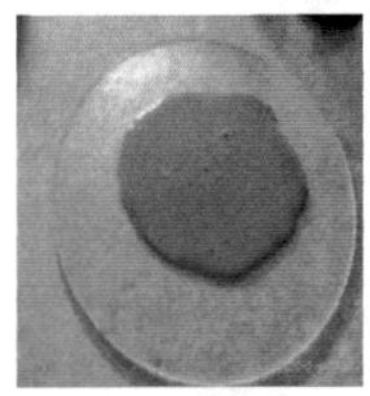
Mix with Slag/ fly ash 50/50, SP dosage = 0 % and a/b = 0.7	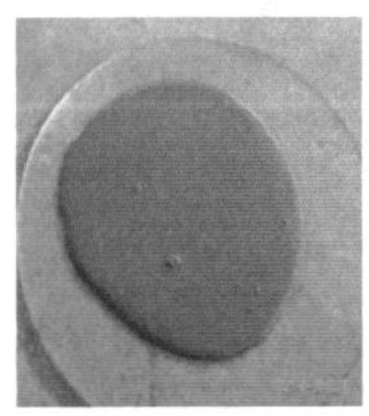
Mix with Slag/ fly ash 50/50, SP dosage = 0 % and a/b = 0.8	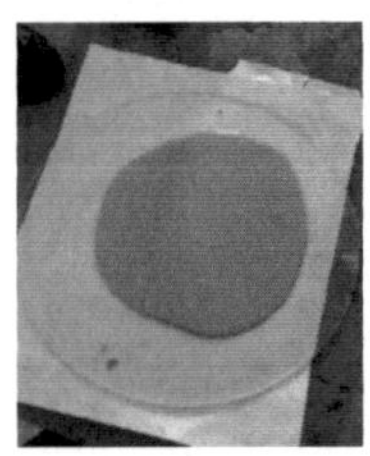
Mix with Slag/ fly ash 50/50, SP dosage = 0 % and a/b = 0.9	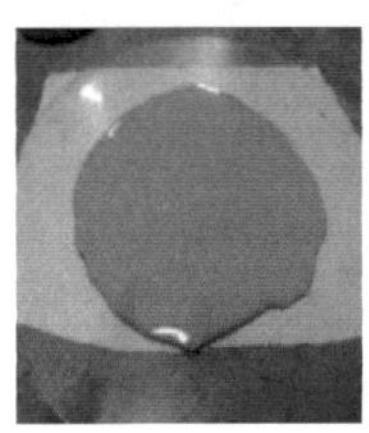

Table 3. Variation of flow with SP dosage for paste with slag/ fly ash 100/0 at a/b of 0.40.

Mix	Mini Slump Flow
Slag/ fly ash 100/0, a/b = 0.4 and SP dosage = 0 %	
Slag/ fly ash 100/0, a/b = 0.4 and SP dosage = 1 %	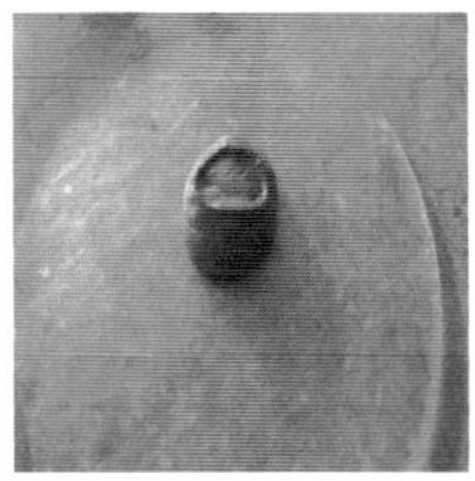
Slag/ fly ash 100/0, a/b = 0.4 and SP dosage = 2 %	
Slag/ fly ash 100/0, a/b = 0.4 and SP dosage = 3 %	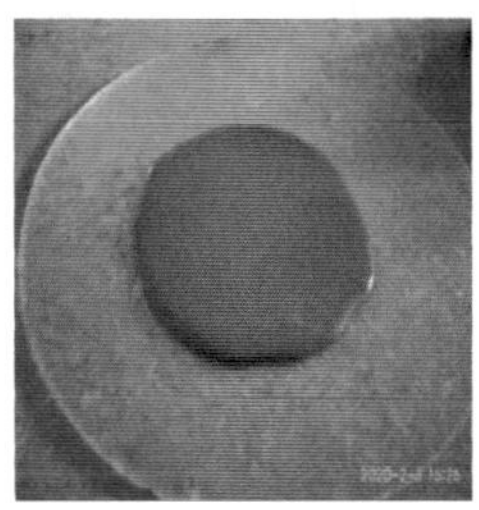

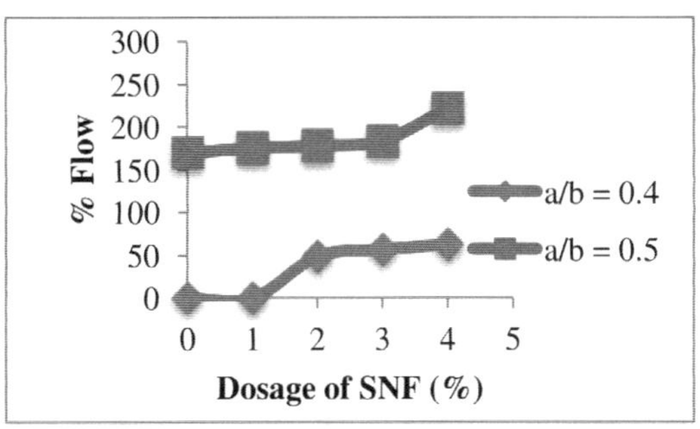

Figure 7. Variation of flow with SP dosage of mix with Slag/ Fly ash 100/0.

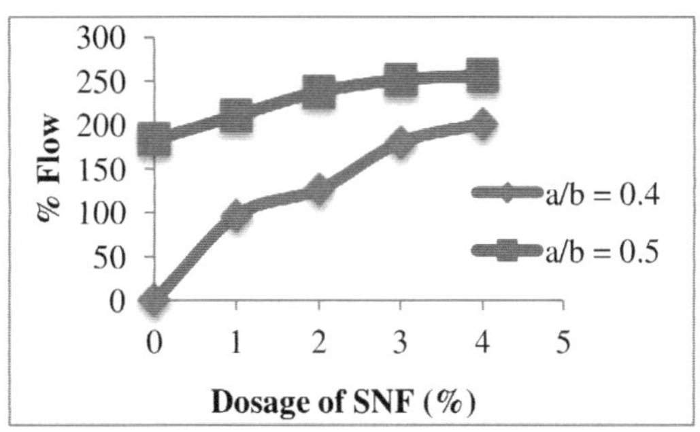

Figure 8. Variation of flow with SP dosage of mix with Slag/ Fly ash 75/25.

IOP Conf. Series: Materials Science and Engineering **989** (2020) 012004 doi:10.1088/1757-899X/989/1/012004

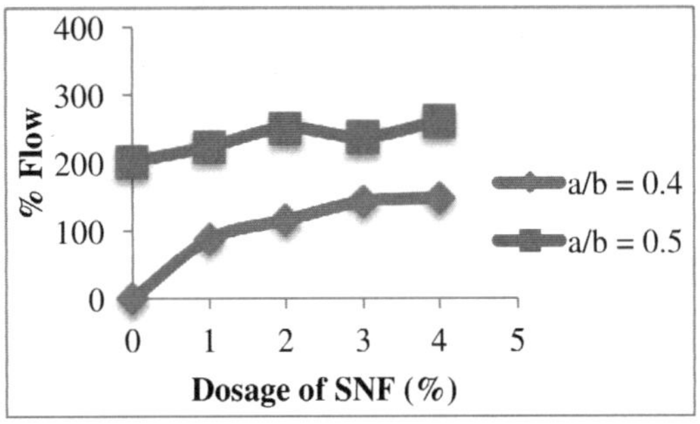

Figure 9. Variation of flow with SP dosage of mix with Slag/ Fly ash 50/50.

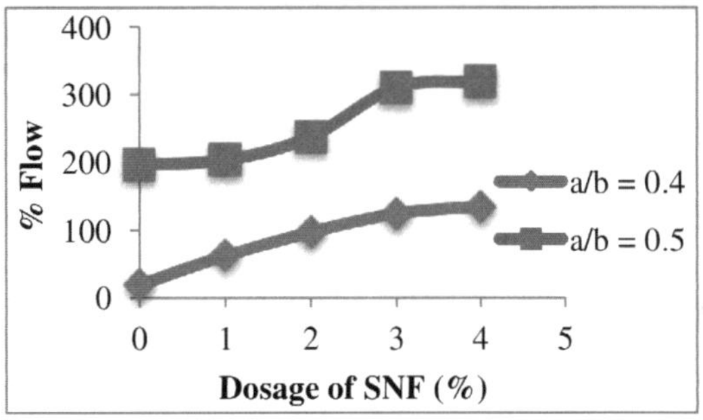

Figure 10. Variation of flow with SP dosage of mix with Slag/ Fly ash 25/25.

4.2 Initial and final setting time

The variation of initial and final setting time for different paste mixes with a/b ratio is shown in Figure 11 and Figure 12. The variation of setting times for different mixes with varying SP dosage is represented in Figures 13 to 20. From the study, it was found that both initial and final setting time increases tremendously with increase in the a/b ratio. Also, the use of SNF was found to be less effective in prolonging the setting time of alkali activated geopolymer paste as there was only a marginal rise in setting time of paste with increase in SP dosage. It is notable that the mix with 25% slag and 75% fly ash has not attained final setting even after 24 hours in ambient conditions (at a/b ratio of 1).

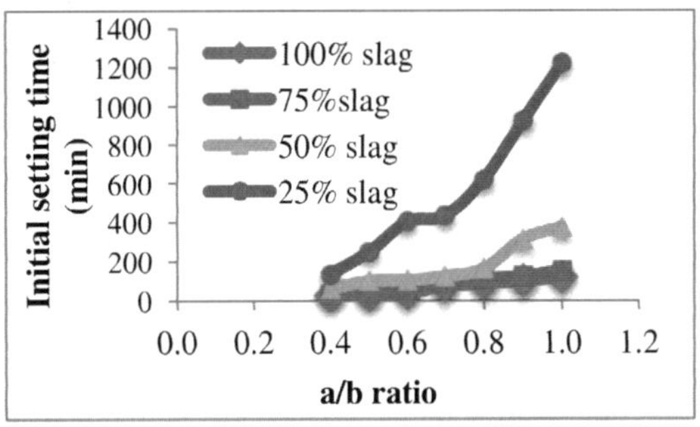

Figure 11. Variation of IST with a/b ratio.

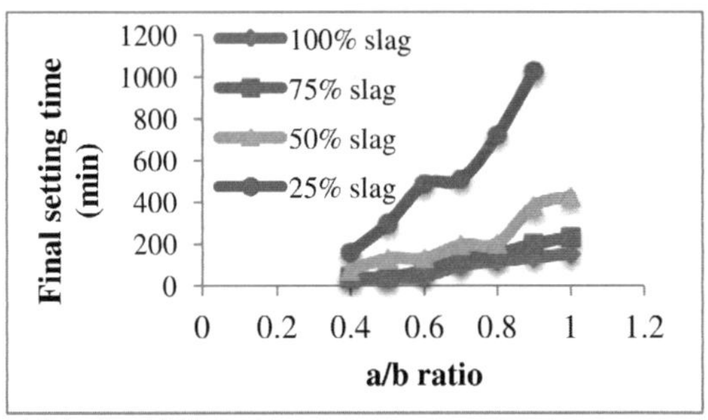

Figure 12. Variation of FST with a/b ratio.

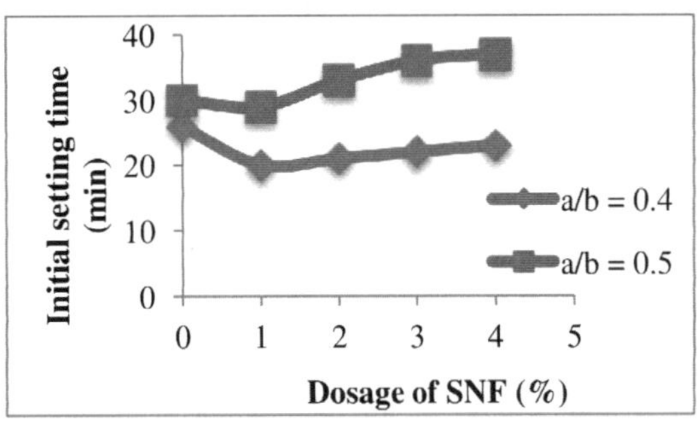

Figure 13. Variation of IST with SP dosage for paste with slag / flyash 100/0.

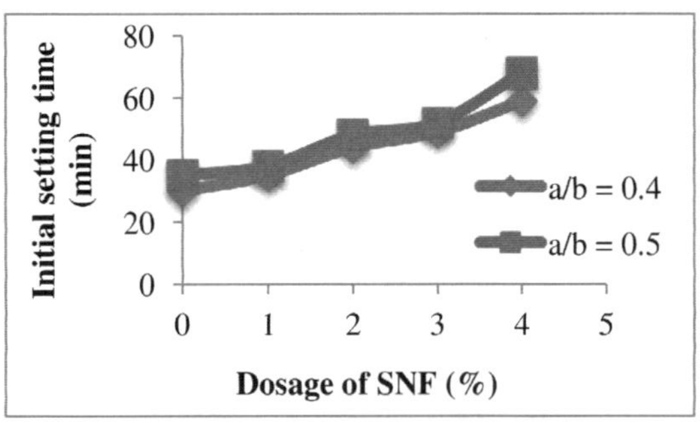

Figure 14. Variation of IST with SP dosage for paste with slag / flyash 75/25.

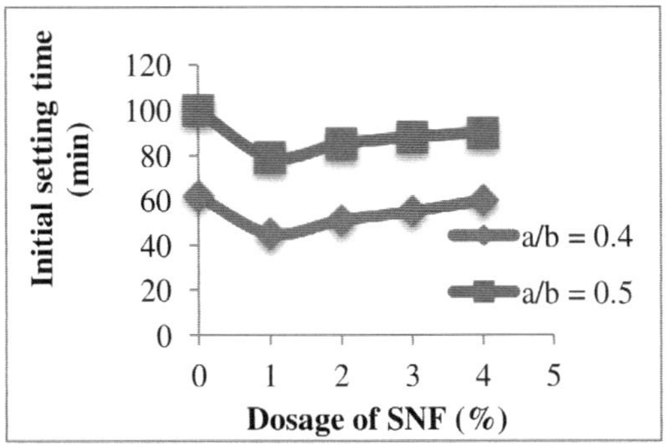

Figure 15. Variation of IST with SP dosage for paste with slag / flyash 50/50.

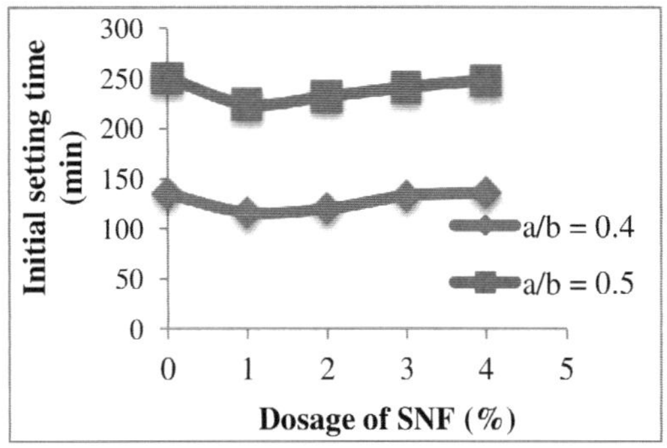

Figure 16. Variation of IST with SP dosage for paste with slag / flyash 25/75.

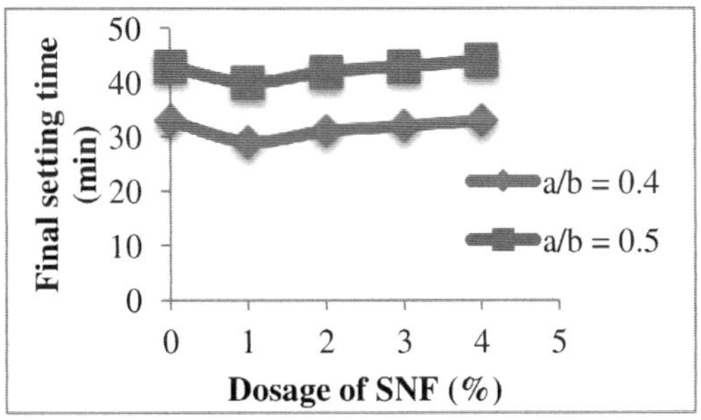

Figure 17. Variation of FST with SP dosage for paste with slag / flyash 100/0.

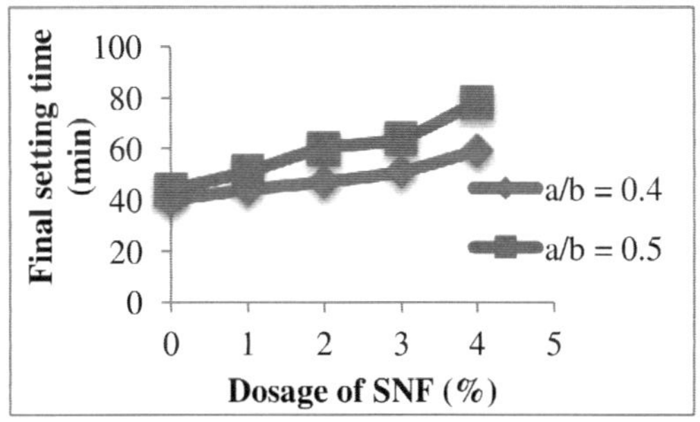

Figure 18. Variation of FST with SP dosage for paste with slag / flyash 75/25.

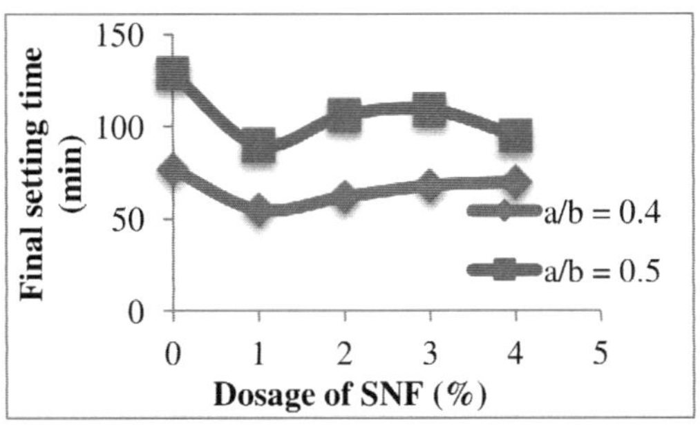

Figure 19. Variation of FST with SP dosage for paste with slag / flyash 50/50.

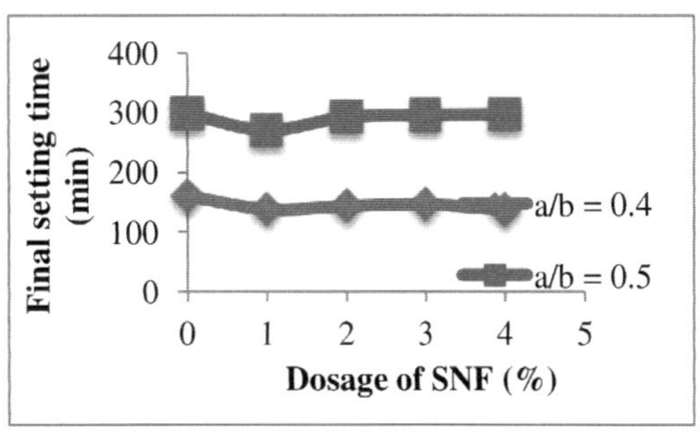

Figure 20. Variation of FST with SP dosage for paste with slag / flyash 25/75.

5. Conclusions

This study is conducted to examine the effect of dosage of superplasticizers (0 to 3%) on the rheological properties of alkali activated fly ash and slag composites with varying a/b ratios. Investigations to understand the effect of superplasticizers and alkali content on the behaviour of geopolymer paste were carried out using mini slump cone and vicat apparatus.

From the experimental program conducted on the geopolymer paste, it was found that as a/b ratio increases both final and initial setting time of the geopolymer paste was found to be increasing tremendously. The use of SNF is not found to be much effective in prolonging the setting time of alkali activated geopolymer paste. As SP dosage increases, increase in the setting time is marginal only and hence a superplasticizer compatible for the geopolymer composition need to be developed. It was observed that mix with a/b ratio 1.0 with 25% slag and 75% fly ash has not attained final setting even after 24 hours. Few geopolymer mixes had initial setting time more than 150 min and final setting time less than 350 min.

The mini slump cone test conducted on the paste shows that as a/b ratio increases, flow measured was increasing and the increase was found to be nominal at higher a/b ratios (> 0.7) and then it becomes constant. The a/b ratio above which the flow becomes a constant can be defined as saturation a/b ratio. There was a minimal increase in flow of the geopolymer paste with SNF dosage. Also, it was observed that as a/b ratio increases, stiffening of alkali activated paste get prolonged.

6. References

[1] Bakharev T, Sanjayan J G and Cheng Y B 2000 Effect of admixtures on properties of alkali-activated slag concrete *Cement and Concrete Research* **30(9)** 1367–1374

[2] Pacheco-Torgal F, DMoura, Yining Ding and Said Jalali 2011 Composition, strength and workability of alkali-activated metakaolin based mortars *Construction and Building Materials* **25** 3732–3745

[3] Ghosh K and Ghosh P 2012 Effect of Na_2O/Al_2O_3, SiO_2/Al_2O_3 and W/B Ratio on setting time and workability of flyash based geopolymer *International Journal of Engineering Research and Applications (IJERA)* **2** 2142-2147

[4] Indian Standard IS 4031 (Part 5) 1988 – *Methods of physical tests for hydraulic cement - determination of initial and final setting times* Bureau of Indian Standards New Delhi

[5] John L Provis 2017 Alkali-activated materials *Cement and Concrete Research* **114** 40– 48

[6] Laskar S M and Talukdar S 2017 Preparation and tests for workability, compressive and bond strength of ultra-fine slag based geopolymer as concrete repairing agent *Construction and Building Materials* **154** 176–190

[7] Mithanthaya I R, Marathe S, Rao N B S and Bhat V 2017 Influence of superplasticizer on the properties of geopolymer concrete using industrial wastes *Materials Today: Proceedings* **4** 9803-9806

[8] Nath P and Sarker P K 2014 Effect of GGBFS on setting, workability and early strength properties of fly ash geopolymer concrete cured in ambient condition *Construction and Building Materials* **66** 163-171

[9] Nematollahi B and Sanjayan J 2014 Effect of different superplasticizers and activator combinations on workability and strength of fly ash based geopolymer *Materials and Design* **57** 667-672

[10] Chindaprasirt P, Chareerat T and Sirivivatnanon V 2007 Workability and strength of coarse high calcium fly ash geopolymer *Cement & Concrete Composites* **29** 224–229

[11] Palacios M and Puertas F 2005 Effect of superplasticizer and shrinkage – reducing admixtures on alkali – activated slag pastes and mortars *Cement and Concrete Research* **35** 1358- 1367

[12] Palacios M, Phillip F G Banfill and Puertas F 2008 Rheology and setting of alkali activated slag *ACI Materials Journal* **105** 140-148

[13] Puertas F, Palomo A, Fernandez Jimenez A, Izquierdo J D and Granizo M L 2003 Effect of superplasticisers on the behavior and properties of alkaline cements *Advances in Cement Research* **15** 23-28

[14] Nailia R Rakhimova and Ravil Z Rakhimov 2019 Literature review of advances in materials used in development of alkali-activated mortars, concretes and composites *Journal of Materials in Civil Engineering* **31(11)** 1-13

[15] Xie J and Kayali O 2016 Effect of superplasticiser on workability enhancement of Class F and Class C fly ash-based geopolymers *Construction and Building Materials* **122** 36–42

[16] Zhijun Tan, Susan A Bernal and John L Provis 2017 Reproducible mini-slump test procedure for measuring the yield stress of cementitious pastes *Materials and Structures* **50(235)** 1-12

Rheological and strength behaviour of red mud - slag geopolymer composites

Sreelekshmi S[1,2], Remya S Chandran[1], Sera Rachel Varghese[1], Naeem[1], Thushara Raju[1] and Ramaswamy K P[1]

[1]Department of Civil Engineering, TKM College of Engineering, Kollam, Kerala, India.

[2]Email: sreelekshmi2802@gmail.com

Abstract. Alkali activated concrete (AAC) is accepted as impending alternatives to conventional Ordinary Portland cement concrete in order to limit CO_2 emissions as well as beneficiate a number of wastes into useful products. This study aims to explore the possibilities of using red mud, which is a waste product of aluminum manufacturing industry, as a geopolymer solid binder. The fresh properties such as setting time, workability and penetration resistance and the hardened properties like compressive strength and ultra-sonic pulse velocity (UPV) test of various red mud mixes along with fly ash (FA) and slag are found out. Also some of the specimens were immersed in water to check if there is any leaching. From the study it was identified that the mix with 25% red mud and 75% slag gives a better workability and strength characteristics.

1. Introduction

Ordinary Portland cement (OPC) is among one of the most consumed materials in the construction industry, which is highly energy consumed and their production causes intensive pollution to the atmosphere [1]. Therefore the replacement of OPC from concrete is likely indispensable. Alkali activated materials are identified as cost effective alternatives and they have a potential to be used as a major component of a sustainable future global construction material. They are not expected for the complete replacement of Portland cement across its applications, for reasons like those related to supply chain limitations, practical challenges in some modes of application, and the need for improved quality control of formulation and curing. Fly ash containing high amounts of unburnt carbon, slag and red mud with high alkalinity are waste materials with less reuse rate, which can be effectively utilized to make green concrete by using it in developing geopolymer systems.

Geopolymers are amorphous alumino-silicate synthesized by the polycondensation of geopolymeric precursor and alkali polysilicates. It involves a chemical reaction between Silicone (Si)-Aluminium (Al) minerals under alkaline conditions that results in a 3D polymeric chain and ring structure with Si-O-Al-O bonds. It uses industrial by-products and waste products like fly ash, ground granulated blast furnace slag (GGBFS), rice husk ash, metakaolin, red mud, etc. A combination of sodium hydroxide (NaOH) and sodium silicate (Na_2SiO_3) is the most common alkali activator used for the process. Geopolymer concrete (GPC) is one such eco-friendly and novel alternative to cement concrete with satisfactory strength and durability characteristics [2].

Based on alumina silicate sources, there are about 6 groups of GPC, viz, Fly Ash-based, Metakaolin-based, Slag-based, Rice Husk Ash-based, high calcium wood ash-based and a combination of any two of the aluminosilicates [3]. The advantages of using class F fly ash when compared to unmodified Portland cement are many, like an increase in the late compressive strength (after 28 days), an increase in the resistance to alkali silica reaction, sulphate attack resistance, workability, pore refinement and less heat generation during hydration and decreased water demand. But a well-defined range of values cannot be provided – to describe the environmental savings in a global scenario, which might be achieved only through the use of alkali-activated materials (full utilization of fly ash) instead of conventional cements and concretes. So it is necessary that the development and optimizations of both one part and two-part binder systems are done based on an ever broader range of raw materials and activators [4]. Also, there is no any specific binder combination which will have all the desired properties at once, therefore the ground breaking research is the only path forward to reach large scale deployment of this technology [5].

GGBFS is a by-product from the iron and steel industry, using blast furnace, obtained as a glassy, granular product which is then dried and ground into a fine powder. GGBFS-FA blends were recommended for AAC as a commercially promising and viable alternative to Portland Cement Concrete. The incorporation of GGBFS into AAC decreases the susceptibility to sulphate acid attack [6]. Slag is highly cementitious and contains calcium silicate hydrates (CSH) which enhances strength and improves the durability of concrete [7]. Depending on the amount of GGBFS in the cementitious material, the setting time of GGBFS cement becomes more than that of OPC. Use of GGBFS has many advantages like reduction in the risk of damage caused by alkali-silica reaction, higher resistance to chloride ingress, reduction in the risk of reinforcement corrosion and higher resistance to attacks by sulphate and other chemicals. Thus, the use of slag in conventional systems is proven but the use is often limited to 50% and these limitations can be overcome in geopolymer systems.

Another effective waste material that can be used to make the AAC is red mud. Disposal of red mud possess several environmental problems due to its high pH, underground water contamination due to alkali percolation, issues with safe storage, alkaline airborne dust emissions, requirement of vast land area for disposal, the presence of minor and trace amounts of heavy metals and radio nuclides and their subsequent seepage into ground water. It is reported that the addition of red mud to fly ash increases the alkalinity of the environment and results in the dissolution of silica and aluminium from the solid phases resulting in the promotion of the polycondensation process and thereby increasing the unconfined compressive strength [8]. The addition of red mud densifies the pore structure and contributes to the strength development of Controlled low strength material (CLSM), called as flowable fill, which is a kind of weak flowable mix used in the construction for non-structural elements like backfills and road bases. Red mud optimizes the microstructure by acting as microfiller and provides alkalis which in turn increases the pH value of pore solutions as well as accelerates the hydration of binary binders [9].

It is noted that the red mud - slag mixes require further researches and studies to know about their geopolymerisation properties and their subsequent use as a green building material. For this, the fresh properties, mechanical and durability characteristics of the developed alkali activated composites are to be investigated. This paper presents the fresh and strength properties of various mixes prepared from slag, fly ash and red mud.

2. Materials and methods

2.1. Materials

In this study, the materials used were Class F fly ash (FA), GGBFS, red mud, and alkaline solution. Low-calcium fly ash conforming to IS: 3812-2010 and GGBFS having a chemical composition of CaO (40%), SiO_2 (35%), Al_2O_3 (13%) and $Mg(OH)_2$ (8%) obtained from Astrra Chemicals, Chennai were used for preparing the geopolymer mix. For this study, slag was collected and sieved through IS sieve 90 micron, and the physical properties were obtained as, specific gravity of 2.9 and bulk density 1200 kg/m^3. Red mud, from HINDALCO, Belgaum, generated in the production of alumina from bauxite in the Bayer process with Fe_2O_3 (30-60%), Al_2O_3 (10-20%), SiO_2 (10-20%), Na_2O (2-10%), CaO (2-8%) and TiO_2 (0-25%) were used in this study. Mortar was prepared using uniformly graded manufactured sand passing IS 2 mm sieve. 8 Molar sodium hydroxide solution along with sodium silicate solution was used as alkali activators and alkaline solution to binder ratio adopted was 0.60. The water to solid ratio was 0.28, alkali silicate to alkali hydroxide ratio was 2.50 and activator modulus of 1.31.

2.2. Mix proportion

For preparing 8M NaOH solution, the solid NaOH pellets were mixed with distilled water and cooled at room temperature, and the Na_2SiO_3 solution was added to the mix, then properly stirred, and allowed to rest for a period of 24 hours before adding to the mixer. A Hobart mixer was used to mix the geopolymer solid binders along with the alkaline solutions for three minutes to get the proper mix. For this study, five types of mixes were used, which are shown in Table 1.

Table 1. Types of mixes.

Types of mix	Material proportion (%)
SGP	100% GGBFS
FSGP	50% FA + 50% GGBFS
R25S75GP	25% Red mud + 75%GGBFS
R50S50GP	50% Red mud + 50%GGBFS
R75S25GP	75% Red mud + 25%GGBFS

3. Experimental investigations

This study was carried out in two phases. The first phase of the study was conducted on paste to determine the rheology and fresh properties, which includes mini slump cone test for workability, setting time test to determine initial and final setting times and penetration resistance test. The work was continued on mortar in the second phase to evaluate the performance of mortar, which contains flow table test, compressive strength test (3, 7, 28 days), dynamic modulus of elasticity using ultrasonic pulse velocity (28 days) and pH test.

3.1. Tests on paste

The mini slump cone is just a mini version of the slump cone, used to determine the workability of cementitious paste. It is a truncated conical mould with dimensions as: 19 mm top diameter, 38 mm bottom diameter and a height of 57 mm, and is shown in Figure 1. The mould was placed in the centre of a circular piece of glass and filled with sample paste. The cone was then lifted, allowed one minute for spreading the paste, and then the average spread of paste was measured along any three diagonals and the average of the three values is taken as the final diameter of the paste. The workability in terms of % spread was then obtained by using the equation (1).

$$\%\ \text{spread} = \frac{(\text{Final diameter} - \text{Initial diameter})}{(\text{Initial diameter})} \times 100 \qquad (1)$$

Figure 1. Mini slump cone.

The initial setting time of paste is considered as the time period elapsed between the moment the water is added to the binders to the moment when the paste starts losing its plasticity. Vicat apparatus conforming to IS: 5513-1976 was filled to the brim of the mould with the prepared paste and the surface was leveled and the mould was gently shaked to expel the air, if any. The test block was immediately placed on the non-porous resting plate, under the initial setting needle. The rod with the needle was lowered and quickly released to penetrate into the paste in the mould. At first, the needle completely penetrated into the paste and then the procedure was repeated at regular intervals till the needle failed to penetrate into the mould for 5±0.5 mm. The time duration between the times of adding water to the binders to the time when the needle failed to penetrate the mould by 5±0.5 mm was recorded as the initial setting time of the paste and the time duration required for the annular ring to fail to make an impression on the wet paste was recorded as the final setting time.

3.2. Tests on mortar

The flow table test was conducted to get an idea about the workability of mortar. The flow table apparatus used in the study is shown in the Figure 2, and the top of the table and inside portion of mould were wetted with oil and cleaned before starting the test.

Figure 2. Flow table apparatus.

The mould was placed on the centre of table and filled it with mortar sample in three layers and compacted by tamping 25 times with a tamping rod of 16 mm diameter. After that, the top of the layer was levelled with a trowel and then the mould was removed from the mortar by a steady upward pull and the diagonals of the spread was measured after giving 25 blows to the apparatus in 15 seconds. The diameter of flow was measured along 3 directions and the average final diameter was obtained. Then the % flow of mix was obtained by using the equation (2).

$$\%flow = \frac{average\ diameter - original\ base\ diameter}{original\ base\ diameter} \times 100 \quad (2)$$

The mortar specimens prepared for testing compressive strength are shown in Figure 3. The cylindrical specimens of size 25mm diameter and 50mm height were used. The strength test was carried out on multi speed loading frame of 50kN capacity and the machine was modified by fixing a test rig. A loading rate of 0.5 kN/sec, starting load of 1 kN and stop load of 40% of starting load was used as the test settings. Before conducting the test, the exact dimensions and mass of the air cured specimens were measured, noted and given as input to the test. The load was applied gradually until the failure of the specimen and the load at the failure point were noted down. The load was released and the process was repeated for 3 specimens from each mix and the 3rd day, 7th day and 28th day mortar compressive strength were obtained.

Figure 3. a) Specimens in mould b) Demoulded specimens.

The strength and quality of mortar was also assessed by finding the velocity of an ultrasonic pulse passing through the specimen. Good quality specimens exhibit higher velocities which is also an indication of the quality of the material, while a specimen with cracks and voids might exhibit slower velocities. In this study, the transducer (of 150 kHz frequency) was held in contact with one surface of the specimen after 28th day, under test. After travelling a fixed path length L, the ultrasonic pulse was converted into an electrical signal by the second transducer which is kept in contact with the other surface of the specimen and an electronic timing circuit enables the transit time (T) of the pulse to be measured. The pulse velocity (V) was obtained by using the equation (3)

$$V = \frac{L}{T} \tag{3}$$

The dynamic modulus of elasticity of material (E_d) was found using the equation (4)

$$E_d = \frac{\rho(1+\mu)(1-2\mu)}{(1-\mu)} V^2 \tag{4}$$

Where ρ is the bulk density of the mortar in kg/m^3, μ is the Poisson's ratio (assumed value of 0.17 for mortar) and V is the ultrasonic pulse velocity in m/s. The bulk density was computed from the weighed mass and volume of each specimens. Also, the pH tests were conducted using a digital pH meter to know about the pH variations of the water in which the specimens are immersed to check for any leaching of alkali ions from the specimen.

4. Results and discussions

4.1. Flow behaviour of geopolymer paste

The flow of paste for various mixes measured, and the results are displayed in Table 2. The result suggests that as the content of red mud increases there is a massive decrease in the spread of paste. This might be due to the fact that the finer particles of red mud require more water molecules to wet and wrap the particles causing less free water molecules to exist in the paste, and the fluidity becomes less [10]. Slag paste and paste blended with slag and fly ash showed high % flow (296% and 300% respectively).

Setting time tests for paste mixes were carried out and the initial and final setting time of mixes were recorded. The setting time values of mixes obtained are shown in Table 2. From the study, it was found that both initial and final setting time increases tremendously with increase in the content of red mud. The higher Si content will catalyse the geopolymerization reaction and small amounts of calcium in red mud lowers the calcium hydration reaction, which will prolong the setting time [11]. The initial setting time of all the mixes were found to be greater than 30 minutes, which is optimum for the transportation and placing of mixes. The final setting time of the mix R25S75GP was found to be more than 1 hour and that of R50S50GP, R75S25GP and FSGP was found to be greater than 2 hours which are satisfactory for working with these mixes. It is also seen that the mix sets quickly when the slag content is high. Thus, incorporation of red mud is found to be beneficial in delaying both the initial and final setting times of paste.

The penetration resistance tests for the paste mixes were conducted along with the setting time tests and the penetration of the initial setting time needle at different times were noted and were plotted against time. It was found that the most resistive mix to penetration was SGP and the least resistive mix was found to be R75S25GP. Among the red mud mixes, R25S75GP was found to show satisfactory penetration resistance. The penetration resistance curves of various mixes are shown in Figure. 4. The stiffening process of various paste mixes can be understood from Figure. 4. It can be seen that the rate of stiffening for SGP and R25S75GP mixes are very fast whereas the stiffening rate is moderate for the mixes FSGP and R50S50GP.

Table 2. Setting times of geopolymer paste.

Mixes	Setting time		Mini slump test
	Initial setting time (min)	Final setting time (min)	Spread (%)
SGP	43	57	296.47
R25S75GP	57	70	260.27
R50S50GP	120	160	255.95
R75S25GP	205	285	170.27
FSGP	122	150	300.00

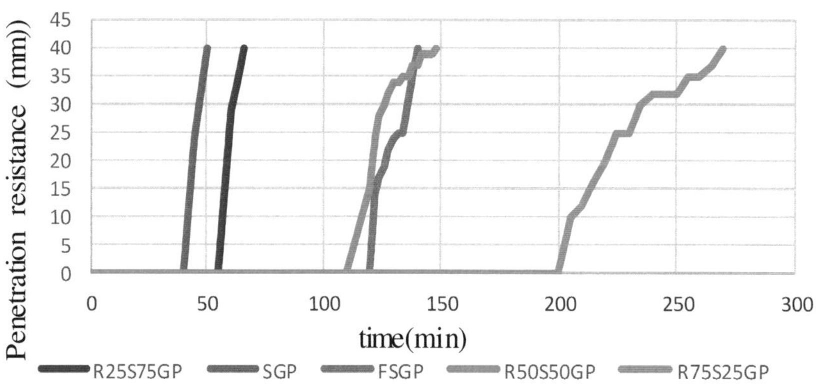

Figure 4. The penetration resistance curves of various mixes.

4.2. Behaviour of geopolymer mortar

The workability of various mortar mixes were found from the mortar flow table test. The flow of mortar observed for the various mixes obtained after conducting the test are shown in the Figure 5, and the flow results are shown in Table 3. Among all the mixes, it was found that the mixes R25S75GP, R50S50GP and the mix FSGP have good workability and satisfies 110 ± 5% as per IS 4031. From the test results and visual observations, the flow of the mix was found to be decreased with increase in the percentage of red mud. These observations were in alignment to the results from the mini slump cone test on paste. The decrease in flow of mortar with the inclusion of red mud may be due to its small particle size [10] and higher absorption of water.

Table 3. Flow of geopolymer mortar mixes.

Mixes	Spread or flow (%)
SGP	75
R25S75GP	108
R50S50GP	105
R75S25GP	50
FSGP	115

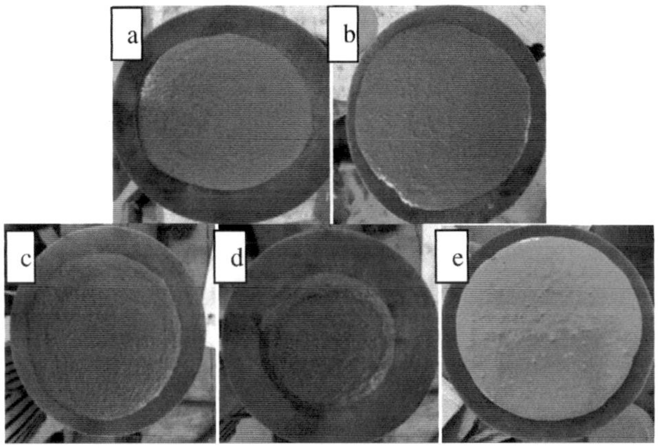

Figure 5. The flow pattern of various mortar mixes. (a) SGP, (b) R25S75GP, (c) R50S50GP, (d) R75S25GP, (e) FSGP.

Hardened properties like compressive strength and elastic modulus of both ambient cured specimens as well as water immersed specimens were found out. In the case of water immersed specimens, the pH of the water in which the specimens are submerged was monitored at intervals to check for any leaching from the specimens. Compressive strength tests were carried out on the specimens on 3rd day, 7th day and 28th day of curing and it was found that the strength increased with days of curing. It was also found that there is only a slight change in the strengths when the specimens were cured in water. The values of compressive strength obtained are given in Table 4. From the results, it can be seen that the mix with R25S75GP possess the greatest strength, in the range of 50MPa. A substantial decrement in the strength was observed when the red mud content goes beyond 25%. Shrinkage cracks occurred when the red mud content is excess in the AAC mix. The number of cracks increases and become larger when the red mud content exceeds 25%. As the red mud content increases, the amount of inert material which does not participate in polymerization also increases. They may hinder the polymerization reaction creating a weaker matrix. This might be the reason for the decreased strength of geopolymer when the red mud content exceeds 25% [12].

UPV tests were conducted on the specimens on the 28th day, and the dynamic elasticity modulus in GPa was found out using the equation (4). It was found that the elastic modulus of water immersed specimens were greater than that of air dried mixes. Although, normally, saturated specimens are expected to give a slightly higher UPV and E_d value, higher values obtained here (Table 5) may indicate that immersion curing may be more beneficial than air curing. Among the mixes, SGP was found to have the greatest elastic modulus. The mix of R25S75GP was also found to show a high value of elastic modulus, but is slightly lower than that of SGP. These two mixes showed higher compressive strength too.

IOP Conf. Series: Materials Science and Engineering **989** (2020) 012005 doi:10.1088/1757-899X/989/1/012005

Table 4. Compressive strength of geopolymer mortar specimens.

Mixes	Compressive strength (MPa)					
	Air dried mixes			Water immersed mixes		
	3rd day	7th day	28th day	3rd day	7th day	28th day
SGP	29.11	45.18	53.16	50.46	52.58	56.15
R25S75GP	39.52	42.66	48.34	58.83	59.11	59.75
R50S50GP	19.16	21.33	24.42	23.91	32.41	40.36
R75S25GP	12.85	14.02	17.13	13.63	15.46	17.08
FSGP	23.09	24.88	26.76	29.38	30.20	36.29

Table 5. Elastic modulus of mortar specimens.

Mixes	Elastic modulus, E_d (GPa)	
	Air curing	Immersion curing
SGP	18.90	40.50
R25S75GP	15.60	30.90
R50S50GP	1.90	5.30
R75S25GP	2.50	5.70
FSGP	11.60	18.70

The pH tests were conducted to determine if there were any leaching phenomena in the water immersed geopolymer mortar specimens. The pH of water was noted every week. Once the readings are taken, the solution was replenished with fresh water (volume equal to 4 times the volume of the specimens). Table 6 shows the pH variations of different geopolymer mixes. From the results obtained, it was found that leaching occurred (pH increased from 5.65 to 11-12) and alkaline substances might have leached out of the specimens into the water. The effect of this, on the microstructure of the specimens, demands another study.

Table 6. Results of pH test.

Mixes	Initial pH	3rd day pH	7th day pH	14th day pH	21st day pH
SGP	5.65	10.86	11.52	11.63	11.97
R25S75GP	5.65	11.54	11.74	12.07	12.31
R50S50GP	5.65	11.22	11.79	11.92	12.11
R75S25GP	5.65	11.74	11.85	11.96	12.04
FSGP	5.65	11.61	11.66	11.72	11.92

5. Conclusions

Geopolymers, which are rich in alkali silicates and alkali hydroxides, are being experimented worldwide to develop GPC as an alternative to OPC systems. Red mud is a waste product of aluminium manufacturing industry, which is rich in aluminosilicates, suitable for geopolymerisation. Red mud was chosen for this study as it possesses grave environmental problems as there are no e and environment friendly methods for its safe disposal. Based on the study conducted on slag and red mud mixes, following conclusions are drawn.

Based on the experiments conducted, it was found that, as the percentage of red mud increases in slag blended paste/mortar, the flowability of the mix and strength characteristics decreases, while the setting time of paste increases. So it is important that an optimum amount of red mud is to be added in the mix to balance out the workability criteria and strength criteria. Among the mixes, R25S75GP gives satisfactory workability and good compressive strength as well as better dynamic modulus of elasticity. When the red mud content become more than 25%, the number of cracks augmented and widened, which results in low strength mix and the dosage above 25% are not recommended for construction purposes. Hence this study suggests that the suitable percentage replacement of GGBFS in AAC with red mud is 25%. Further research is required to know about the influence of red mud and slag on the other mechanical properties like tensile strength, flexural strength, etc. so as to ensure the use of such industrial waste products to make green concrete, thereby finding a solution to the environmental problems related to the disposal of waste materials from the industries.

Acknowledgement
The authors thank the Centre for Engineering Research and Development (CERD) for the financial assistance given as part of Students Project in 2019-20.

6. References

[1] Melike E Bildirici 2019 Cement production, environmental pollution, and economic growth: evidence from China and USA *Clean Technologies and Environmental Policy* **21** 783–793

[2] Davidovits J 2005 Geopolymer, green chemistry and sustainable development solutions *Proceedings of the World congress geopolymer* Institute Geopolymere Saint Quentin France

[3] Ma C K, Awang A Z and Omar W 2018 Structural and material performance of geopolymer concrete: A review *Construction and Building Materials* **186** 90–102

[4] Provis J L 2018 Alkali activated materials *Cement and Concrete Research* **114** 40–48

[5] Provis J L and Deventer J S J 2014 Alkali Activated Materials *State of Art Report* RILEM TC 224-AAM Springer Publications

[6] Rakhimova N R and Rakhimov R Z 2019 Literature review of advances in materials used in development of alkali-activated mortars,concretes and composites *ASCE Journal of Materials in Civil Engineering* **31(11)** 1-13

[7] Shunxiang Wang and Min Wu 2020 Mechanical strength and durability properties of pervious concretes with blended steel slag and natural aggregate *Journal of Cleaner Production* **271** 1-29

[8] Choo H, Lim S, Lee W and Lee C 2016 Compressive strength of one-part alkali activated fly ash using red mud as alkali supplier *Construction and Building Materials* **125** 21–28

[9] Yuan B, Yuan S, Straub C and Chen W 2020 Activation of binary binder containing fly ash and portland cement using red mud as alkali source and its application in controlled low-strength materials *Journal of Materials in Civil Engineering* **32(2)** 1-11

[10] Jian Zhang, Shucai Li, Zhaofeng Li, Chao Liu and Yifan Gao 2020 Feasibility study of red mud for geopolymer preparation: effect of particle size fraction *Journal of Material Cycles and Waste Management* **22** 1328–1338

[11] El-Kholy, Hasan S A, El Deen H, Asran, Gomaa A, Abd Al-Kader and Al Sayed A 2018 Experimental studies on effect of alkaline solution type on behaviour of zeolite and red mud based geopolymer mortar *Life Science Journal*, **15(4)** 31-39

[12] Smita Singh, Basavanagowda S N, Aswath M U and Ranganath R V 2016 Durability of bricks coated with red mud based geopolymer paste *Materials Science and Engineering* **149** 1-9

IOP Conf. Series: Materials Science and Engineering **989** (2020) 012006 doi:10.1088/1757-899X/989/1/012006

Effects of crowd behavior on dynamic responses of permanent grandstand structure

Thasrifa M[1,2], Mohammed Thowsif[1] and Ramaswamy K P[1]

[1]Department of Civil Engineering, TKM College of Engineering, Kollam, Kerala, India.

[2]Email: thasrifa311@gmail.com

Abstract. Constructions industry's novel design and technological development have resulted in an enhanced use of slender structures having low natural frequency. The overwhelming happiness in concert, shows, sporting events etc., potentialize enthusiastic behaviour of people causes dynamic crowd load i.e., synchronized rhythmic movements may excite supporting slender structures such as stadium, grandstands, auditorium, bridges, malls and convention centre. The combination of slender structures with dynamic crowd load makes the structure easily vulnerable to excessive vibrations and dangerously affects the safety and comfort of occupants. Hence the vibration problems due to increased structural slenderness and dynamic crowd excitation have become more common. So studies regarding the behaviour of dynamic crowd load are mandatory. The present work aims to find out the behaviour of permanent grandstand structure under dynamic crowd load. Here we investigate the behaviour of grandstand structure considering load at different frequency of jumpers in the range of 1.5 Hz to 3.5Hz, different types of activity such as normal jumping, high aerobics and low aerobics and various percentage of active crowd with total crowd, also the effects of rake angle on the structure to reduce the vibration. Active dynamic crowd load was analytically developed as taken from BS 6399-Part 1 and generated as time history and applied to the structure. The passive live load as taken 5kN/m^2 specified in IS 875 (Part II). A three dimensional finite element model is used to model the grandstand structure. For modelling the grandstand structure and dynamic analysis finite element software, SAP2000 was used. The results obtained are compared in form of horizontal frequency, vertical frequency, acceleration, displacement, shear force and bending moment.

1. Introduction

The enhancement in the construction's novel design and technological development made modern public structures, such as exhibition halls, long-span floors, concert venues and grandstands, have low natural frequencies. Purposes of these structures are, to hold concerts, sports event, and aerobics like events that make people to make overwhelming happiness [7, 23]. This resulted the structure to be prone to external stimuli having frequency in range of dance type activities. So crowds generate periodic dynamic loads due to rhythmic activities, such as jumping, bobbing, and foot stamping, in response to music. As a result, the natural frequency of the structure will be equal or close to frequency of jumpers, resonant or near-resonant vibrations will happen. The resonance imparts serious structural problems [15]. Also the people feel uncomfortable, panic and even cause casualties as a result of these vibrations. Hence it is mandatory to avoid unnecessary vibrations at a design stage by understanding rhythmic crowd loads [5, 11].

Nowadays crowd jumping to beats are usual for spectators at concerts or sport games. But the jumping action is one of the most severe load from all individual activities. Major problematic levels of incidents reported due to crowd behavior are; during a pop concert held in London in 1994, a temporary grandstand collapsed under the audiences' rhythmic jumping with the beat of music, and more than 50 people were injured. A 39-story building shook vertically for about 10 min in Seoul, South Korea in 2011, where a group of people were exercising at an aerobics. The exciting frequency was almost equal to the building's natural frequency of 2.7 Hz. A part of a temporary grandstand collapsed before a football match in Bastia, Corsica in 1992 [1], killed 17 people and injured over 2500. The London Millennium Bridge had exhibited severe lateral sway on the day of its inauguration and had to be close down for renovation. Similar incident was reported for Pont-de-Solférino in Paris. Complaints of excessive structural vibrations were reported from the crowds of Manchester United's Old Trafford Stadium [20] and Morumbi Stadium in Brazil [25]. In Rajasthan's Sri Ganganagar area, 50 people injured in a mishap during a tractor-race event, structure collapsed apparently due to the weight of a large number of spectators sitting over it. To tackle this problem, existing codes and guidelines, including [2, 4] and [13, 14] says that a dynamic analysis should be done for stadiums with natural frequencies below certain threshold values. However, none of these codes and guidelines provides the tools that would allow a designer to analyze the performance expected of these structures.

The use of sports stadia increased, as it can accommodate large number of audience. Therefore, the grandstands at stadia bearing a more severe loading regime are usual. Hence a precise check for safety and serviceability of these structures was necessary. Studies investigating the behavior of grandstand structure considering load at different frequency of jumpers in the range of 1.5 Hz to 3.5Hz [12], different types of activity such as normal jumping, high aerobics and low aerobics and various percentage of active crowd with total crowd, also the effects of rake angle on the structure to reduce the vibration. Active dynamic crowd load was analytically developed as taken from BS 6399-Part 1 and generated as time history and applied to the structure. The passive live load as taken $5kN/m^2$ specified in IS 875 (Part II).

2. Dynamic crowd load

For the practical design and serviceability check for structures, an accurate analysis and precise investigation of grandstand structure subjected to dynamic loads are required. Moreover, it is desirable to measure and analyze the dynamic loads of spectator activities like jumping, bobbing, swaying etc. because these dynamic loads cannot be easily expressed in a numerical formula. Even though many kinds of literature [8, 9, 10, 11] and code [2] have adopted a numerical formula for calculation of load function due to individual jumping are mentioned below.

Synchronized dynamic loading caused by jumping and dancing are periodic. This mainly depend upon: static weight of the dancer(s) (G_s), period of the dancing load(s) (T_p), and contact ratio (α), i.e. the ratio of the duration within each cycle when the load is in contact with the floor and the period of the dancing. Mathematically the load at any instant (t) may be expressed as in Equation (1)

$$F_s(t) = G_s \left(1 + \sum_{n=1}^{\infty} r_n \sin\left(\frac{2n\pi t}{T_p} + \varphi_n\right)\right) \tag{1}$$

IOP Conf. Series: Materials Science and Engineering **989** (2020) 012006 doi:10.1088/1757-899X/989/1/012006

Where $F_s(t)$ is the dynamic load, r_n is the Fourier coefficient (or dynamic load factor) of the nth term, n is the number of Fourier terms, and φ_n is the phase lag of the nth term. The values of r_n and φ_n for a given period of dancing T_p or a jumping frequency ($1/T_p$) in Table 1.

Table 1. Fourier coefficients [8].

Human activity		n					
		1	2	3	4	5	6
Low aerobics	r_n	9/7	9/55	2/15	9/247	9/391	2/63
	φ_n	-π/6	-5π/6	-π/2	-π/6	-5π/6	-π/2
High aerobics	r_n	π/2	2/3	0	2/15	0	2/35
	φ_n	0	-π/2	0	-π/2	0	π/2
Normal jumping	r_n	9/5	9/7	2/3	9/55	9/91	2/15
	φ_n	π/6	-π/6	-π/2	-5π/6	-π/6	-π/2

From the above equation (1) for a normal jumping 2 Hz load was developed by choosing appropriate coefficients from Table 1 as following in Figure 1. The diagram shows two cycles of normal jumping with frequency equals 2 Hz and the contact ratio α = 1/3. Contact ratio α = 1/3 is equivalent to that only one-thirds of each cycle of the body has contact with the ground. In the figure the force is normalized so that the body weight is unity (G_s=1).

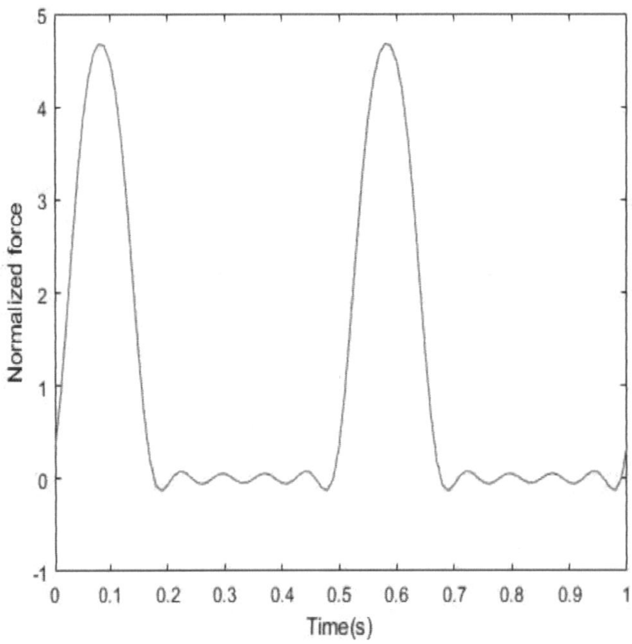

Figure 1. Dynamic crowd load (normalized) for normal jumping 2Hz.

Similarly load model for different range of frequency from 1.5 Hz to 3.5 Hz at an interval of 0.5 Hz for normal jumping was developed. The load taken in kN and static weight was taken as 78 kg. Also load model for different types of activity for 2 Hz frequency was developed as shown in below Figure 2. The coefficients are taken from the Table 1.

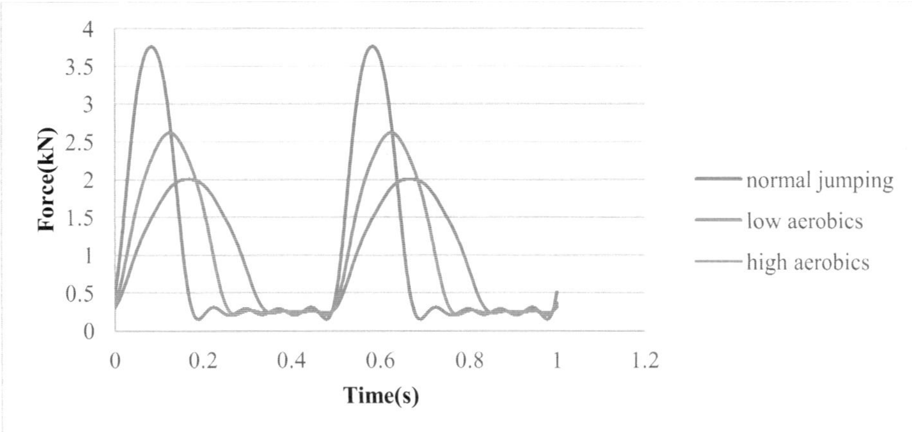

Figure 2. Load at different activity.

3. Modelling and dynamic analysis

3.1. Numerical modelling

A three dimensional finite element model is used to analyse the grandstand structure. A part of permanent grand stand structure was used for study. For modelling the grandstand structure and dynamic analysis finite element software, SAP2000 was used. For the three dimensional finite model of grandstand structure, beams and columns are modelled as frame elements (with six degree of freedom per node). Slabs are modelled with quadrilateral shell elements (with four node and six degrees per node).The dimensions selected for structure as; plan area 13.5x12m, Height of part – 6.2 m, Column size- 600x600mm, Beam size- 350x600mm, and Seating decks- Tread : 800mm and Rise : 312 mm. Figure 3 shows Plan and Sectional view of structure which considered for current work. The structure as modelled in software is shown in Figure 4.

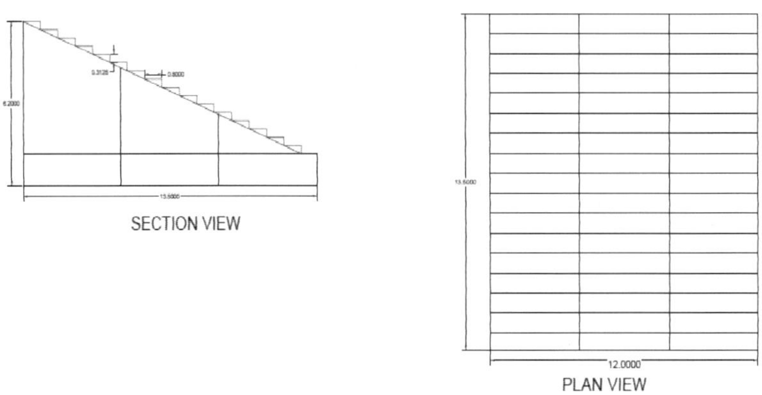

Figure 3. Plan and section view of structure.

Figure 4. Structural model in SAP2000.

3.2. Materials and section properties

Concrete material was used for the permanent grandstand structure. Concrete of M30 grade was used for columns, beams and slabs. For the reinforcement, HYSD415 bar was used for the longitudinal and mild250 was used for ties. Table 2 shows the material properties used for the members used in the structure. F_y and F_u respectively indicates the minimum yield strength and minimum tensile strength of steel.

Table 2. Material Properties of Members.

Material Type	Young's Modulus, E (MPa)	Poisson's Ratio, υ	Unit Weight (kN/m^3)	Design Strength (MPa)
Concrete, M30	27386.13	0.2	25	30
Longitudinal bar, HYSD415	200000	0.3	76.9729	$F_y = 415$ $F_u = 485$
Ties, Mild250	200000	0.3	76.9729	$F_y = 250$ $F_u = 410$

3.3. Load details

In this study dead load and crowd load only were considered for the analysis of the structures. The other loads such as live load, snow load, seismic load, etc. were not considered so as to realize the relevance of crowd load. The crowd load as already developed and explained above. For crowd load, the active load generated as time history and applied to the structure. The passive live load as taken $5kN/m^2$ specified in IS875 (Part II).To find the solution of problem on effects on the percentage of activeness on the total crowd load, there was used mixed combinations of dynamic active load and passive live load. The six combinations are used, such as the full passive load, 20%, 40%, 60%, 80% and full active load. The all model plan views are shown in Figure 5.

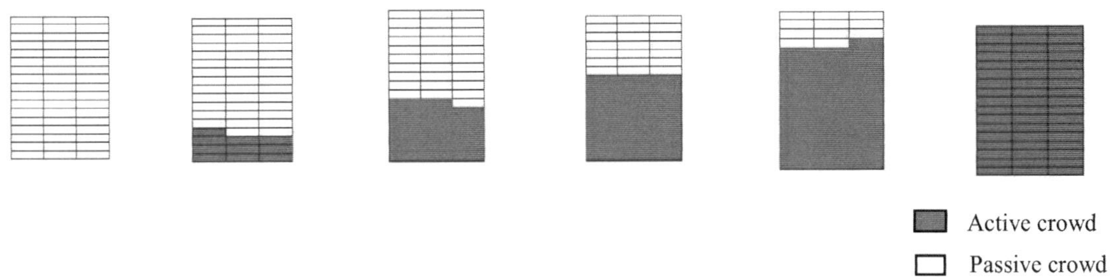

Figure 5. Combination of passive and active crowd load plan views.

3.4. Dynamic analysis
For dynamic analysis, SAP, the structural analysis software was used, which is user-friendly generalized analysis and design software. SAP2000 Version 15.1.0 has a powerful graphical interface with modelling, analytical, and design procedures. It is a full-featured program that can be used for the simplest problems or the most complex projects. Dynamic analysis of stadium structure has been carried out for different combinations of active and passive live load, different jumping frequencies, different activity and various rake angles. The results obtained are compared in form of horizontal frequency, vertical frequency, acceleration, displacement, shear force & bending moment.

From the results the peak acceleration response of the structure expressed as percentage of acceleration due to gravity (g). It can be related to the comfort level of occupants. The comfort level of an individual to the corresponding vibration level is calculated using the specifications listed in Table 3 as given by [10] for structures like grandstands with frequency less than 10 Hz.

Table 3. Crowd comfort level for various peak accelerations [10].

Vibration level	Comfort level
<5% g	Reasonable
<18% g	Disturbing
<35% g	Unacceptable
>35% g	Probably cause panic

4. Results and discussions
Dynamic analysis is carried out for problems of the stadium structure specified in this study using SAP2000. The results are in form horizontal frequency, vertical frequency, acceleration, displacement, shear force & bending moment values are taken from the software SAP2000. Each problem's results are shown in below. Earthquake force and any forces are not considered here because to understand the better effect of active live load.

4.1. Responses at different frequency
Analysis of structure for ranges of frequency from 1.5 Hz to 3.5 Hz was carried out and results in form of displacement, acceleration, shear force & bending moment are shown in the Figure 6. It is realized that when the frequency of load increases, the response of structure in form of displacement, internal member forces and moment increases.

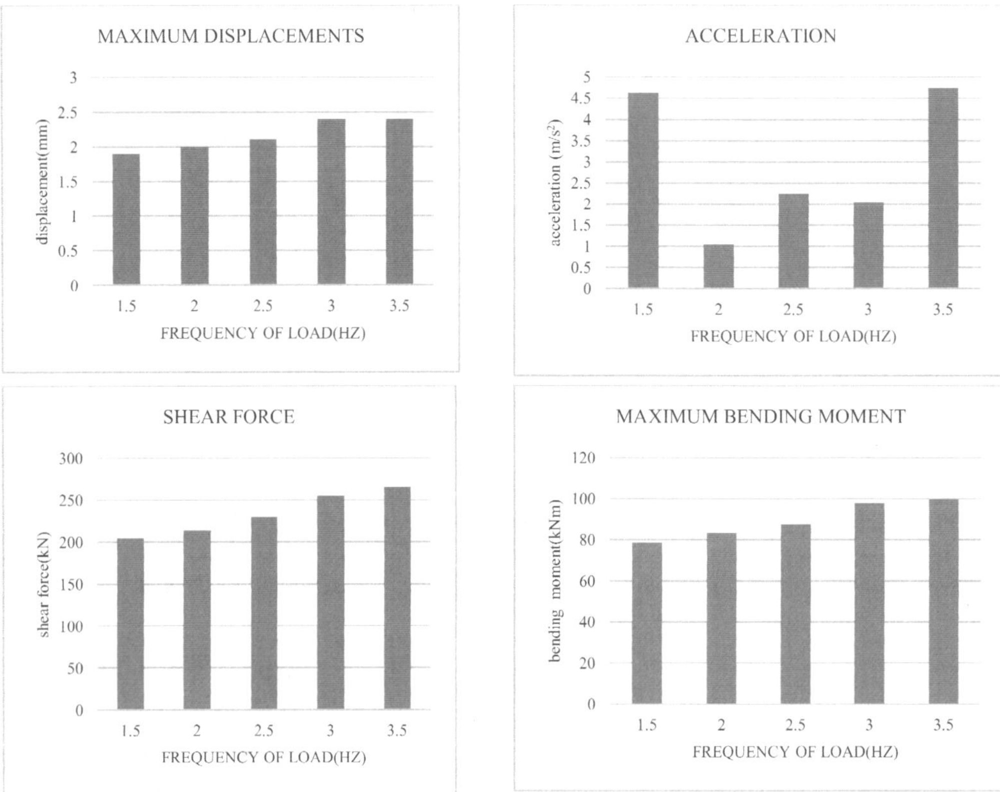

Figure 6. Responses at various frequency.

4.2 Responses at different activity

Analysis of structure for different activities such as normal jumping, high aerobics and low aerobics were carried out and results in form of displacement, acceleration, shear force & bending moment are shown in the Figure 7. From Figure 7, it is realized that responses are high for normal jumping, hence it is highly effect on the structure. For the low and high aerobics responses are comparatively less, these reduction may be due to synchronous of dancers.

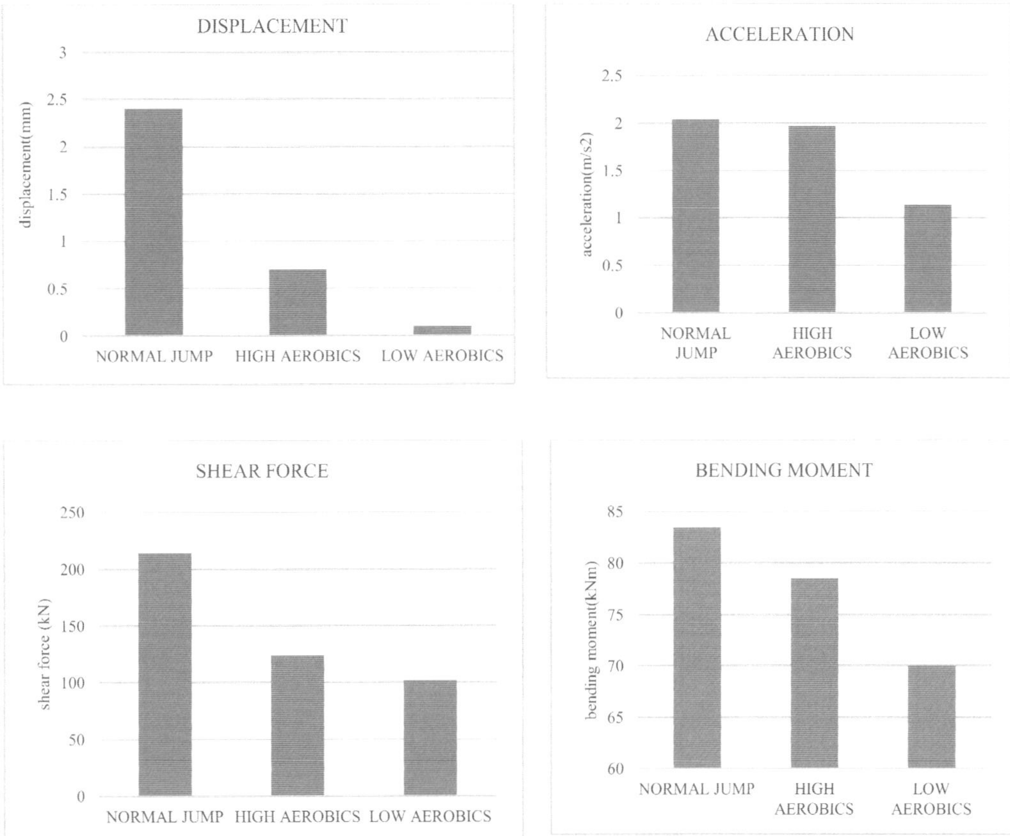

Figure 7. Responses at various activity.

4.3. Responses at various percentage of activeness

Dynamic analysis is carried out for all different load combination of active live load and passive live load of the structure. The responses of the structure such as horizontal frequency, vertical frequency, acceleration, displacement, shear force & bending moment values are shown below Figure 8. From the results, obtained that all the responses are increasing when increase the activeness of crowd. So the structure becomes risky for full active humans, safe for full passive humans. As per the responses, frequency of passive human was below the active crowd, but for better result frequency should be greater. Passive crowd will induce mass to the structure, hence the frequency was reduced. Horizontal frequencies for all the combinations greater than the 4.0 Hz means that it satisfies the code requirement of BS 6399: Part 1. Full active crowd generates higher maximum displacement, acceleration, shear force and bending moment, than all other cases while full passive crowd generates less bending moment among all cases. Acceleration is one of the criteria for serviceability. For better serviceability, the acceleration of structure should be less than 0.35g = 3.4335[3, 16]. Otherwise human comfort may become panic. Figure 8 is showing the comparison of acceleration. For all the combinations, it is below the desired limit. In passive case, structure acceleration is range of reasonable limit of 0.05g=0.49.

IOP Conf. Series: Materials Science and Engineering **989** (2020) 012006 doi:10.1088/1757-899X/989/1/012006

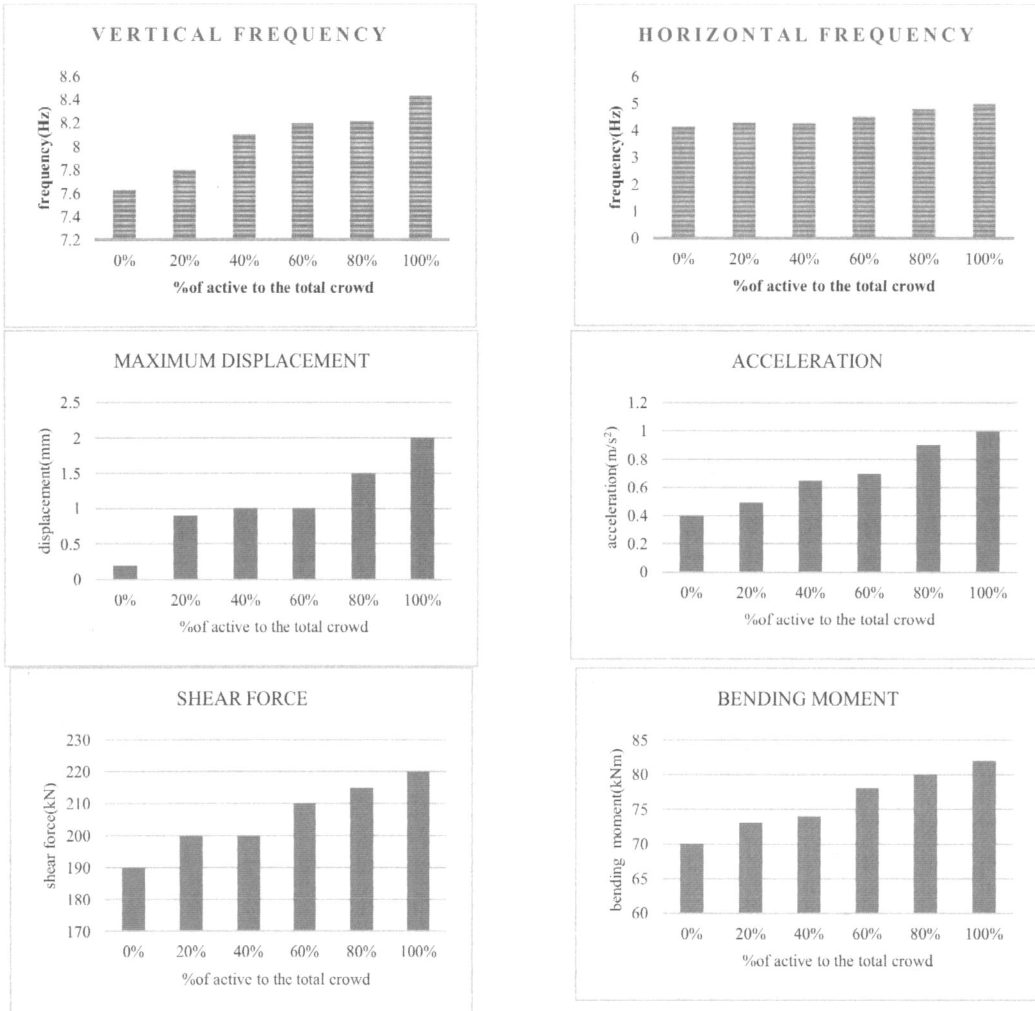

Figure 8. Responses at various percentage of activeness.

4.4. Responses at various rake angles

Analysis of structure for different rake angle for 21, 25, 30 and 35 degree was carried out and results of vertical frequency and horizontal frequency are shown in the below table. Frequency is an important parameter for any dynamic analysis and to prevent resonance condition. Frequency indicates cycles completed in unit time. Tables 4 and 5 gives the frequency of empty structure i.e., before the loading. Table 6 and 7 gives frequency values after applying the crowd load. From the results of vertical frequency, it is clear that for rake angle upto 30 degrees, there is no need of dynamic analysis, as the frequency is above the code specified value 8.2 Hz. While above 30 degree the structure needs dynamic analysis. But by the action of crowd load natural frequency was reduced, it is below the code recommendation. Hence for rake angle above 21 degree structure will be critical and need dynamic analysis.

While for four angles, horizontal frequencies of empty structure are greater than the 4.0 Hz, means it satisfies the code requirement of BS 6399 – Part 1. Horizontal frequency of Structure should be greater than 4.0 Hz as per code provision BS 6399 – Part 1. But by the action crowd load, above 25 degree frequency is below the recommended limit.

Table 4. Vertical frequency of empty structure.

Rake angle(degree)	Vertical frequency(Hz)
21	11.93
25	10.50
30	8.50
35	8.02

Table 5. Horizontal frequency of empty structure.

Rake angle(degree)	Horizontal frequency(Hz)
21	5.84
25	5.24
30	5.1
35	4.56

Table 6. Vertical frequency of structure under crowd load.

Rake angle(degree)	Vertical frequency(Hz)
21	8.43
25	7.42
30	6.01
35	5.67

Table 7. Horizontal frequency of structure under crowd load.

Rake angle(degree)	Horizontal frequency(Hz)
21	4.96
25	4.35
30	3.89
35	3.5

5. Conclusions

An analysis of dynamic loads induced by crowd movements was required for an accurate analysis of grandstand structure. Hence, the dynamic loads induced by spectators' jumping are calculated by numerical model and analyzed. This study investigated the behavior of grandstand structure considering load at different frequency of jumpers, different types of activity and various percentage of active crowd with total crowd, and the effects of rake angle. Based on the study, the following conclusions are drawn:

- Generally, human induced frequencies are in the range of 1.5 to 3.5 Hz. So, the horizontal frequency of structure should be higher than 4Hz. While the vertical frequency should be more than 8.4Hz.
- Static design for stadium structure is safe for loading but may fail in serviceability criteria so it is necessary to perform dynamic analysis considering human structure.
- In this study, as the frequency of load increases, the response of structure in form of displacement and internal member forces increases.
- Acceleration is the one of the criteria for serviceability. As per literature acceleration should be less than 0.35g (3.5m/s^2), in study for effects of frequency variation the excitation at 1.5 and 3.5 Hz was higher than limits.
- By the acts of dynamic crowd load the natural frequency of structure will decreases.
- For different types activity, normal jumping is highly effect on the structure due to all responses are less for the other activity. The reduction may be due to synchronous of dancers.
- In this study results for horizontal frequency, vertical frequency, maximum displacement, maximum bending moment, maximum shear force, and acceleration are measured higher for active live load in comparison with passive live load.
- When the percentage of activeness increases from full passive to active at an increase of 20% activeness with respect to total crowd, the responses like horizontal frequency, vertical frequency, acceleration, displacement, shear force & bending moment values are increases.
- All combinations of active and passive load, vibration level within the limit of 0.35g. So does not cause panic, but it may become disturbing one.
- In the study for effects of rake angle, it could be concluded that under crowd load above 21 degree rake angle, the vertical natural frequency will be below the limit, needed to be analyzed dynamically.
- While for the four angles, horizontal frequencies of empty structure are greater than the 4.0 Hz that means it satisfies the code requirement of BS 6399 – Part 1.
- The active crowd solely acts as external load to the structure while passive crowd was acting solely a mass, so the frequency of stand would decrease with the increasing size.

This study gave great importance in the dynamic analysis of long slender structures like grandstands when crowd load affected the structure. Indian standard codes may be modified to include guidelines for the same.

6. References

[1] Batista R C and Magluta C 1993 Spectator-induced vibration of Maracana Stadium *EURODYN '93* 985-992
[2] BS 6399-1:1996 (British Standards- *Code of Practice for Dead and Imposed Loads*)
[3] Campista D F, Santos da Silva J G, Fontela Sieira A C C and Burgose R B 2016 Dynamic analysis and human comfort evaluation of soccer stadium grandstands *Int. Journal of Engineering Research and Application* **6**(11) 23-31

[4] Commentaries on Part 4 of the *National Building Code of Canada* 1980
[5] Corner A J, Blakeborough A and Williams M S 2013 Rhythmic crowd bobbing on a grandstand simulator *Journal of Sound and Vibration* **332** 442–454
[6] Ebrahimpour A, Hamam A and Sack R L 1996 Measuring and modelling dynamic loads imposed by moving crowds *Journal of structural engineering* **122**(12) 1468-1474
[7] Ebrahimpour A, Pavic A, Racic V and Zivanovic S 2017 Effect of group walking traffic on dynamic properties of pedestrian structures *Journal of Sound and Vibration* **387** 207–225
[8] Ellis B R, Ji T and Littler J D 2000 The response of grandstand to dynamic crowd loads *Proceedings of the Institution of Civil Engineers: Structures and Buildings* **140** 355-365
[9] Ellis B R and Ji T 2004 Loads generated by jumping crowds: numerical modelling *The Structural Engineer* **82**(17) 35 - 40
[10] Ellis B R and Littler J D 2004 Response of cantilever grandstands to crowd loads Part 1: serviceability evaluation *Proceedings of the Institution of Civil Engineers: Structures and Buildings* **157** 235- 241
[11] Ginty D, Derwent J M and Ji T 2001 The frequency ranges of dance-type loads *The Structural Engineer* **79**(6) 27-31
[12] International Organization of Standardisation 1981 *Vibration and shock Mechanical driving point impedance of the human body ISO 5982: 1981* Geneva Switzerland
[13] IStructE/ODPM/DCMS working group 2001 *Dynamic performance requirements for permanent grandstands subject to crowd actions Interim guidance on assessment and design* London UK
[14] Li G, Ji T and Chen J 2018 Determination of the dynamic load factors for crowd jumping using motion capture technique *Engineering structures* **174** 1-9
[15] Littler 1999 The Dynamic response of a three tier cantilever grandstand *European Conference on Structural Dynamics EURODYN- 99* 623-628
[16] Patel A J, Jivani D and Dhmsaniya R G 2018 Effects of human structure interaction on behavior of stadium structure *International Journal of Advance Engineering and research Development* **5**(4) 952-959
[17] Pernica G 1983 Dynamic live loads at a rock concert *Canadian Journal of Civil Engineering* **10**(2) 185-191
[18] Racic V, Pavic A and Brownjohn J M W 2009 Experimental identification and analytical modelling of human walking forces: Literature review *Journal of Sound and Vibration* **326** 1- 49
[19] Reynolds P and Pavic A 2006 Vibration Performance of a Large Cantilever Grandstand during an International Football Match *Journal of performance of constructed facilities* **20**(3) 202-212
[20] Salyards K A 2013 Factors to consider in design for rhythmic crowd loading for vibration serviceability *Structures Congress* 2466 - 2477
[21] Sachse R, Pavic A and Reynolds P 2002 The influence of a group of humans on modal properties of a structure *5th European Conference on Dynamics EURODYN* 1241-1246
[22] Shahabpoor E, Pavic A and Racic V 2017 Effect of group walking traffic on dynamic properties of pedestrian structures *Journal of Sound and Vibration* **387** 207–225
[23] Sim J 2008 Statistical Model of Crowd Jumping Loads *Journal of structural engineering* **134**(12) 1852–1861
[24] Vasilatou V, Harrison R and Nikitas N 2017 Development of a human-structure dynamic interaction model for human sway for use in permanent grandstand design *Procedia Engineering* **199** 2895–2900

IOP Conf. Series: Materials Science and Engineering **989** (2020) 012007 doi:10.1088/1757-899X/989/1/012007

Degradation kinetics of alkali-activated mortar in aggressive citric acid environment

Nidhu M S[1], Sniktha V R[1], Libin Benny[1], Jerin Thomas[1], Athira Ajay[1,2] and Ramaswamy K P[1]

[1]Department of Civil Engineering, TKM College of Engineering, Kollam, Kerala, India.

[2]Email: athiajay96@gmail.com

Abstract. Acid attack is a complex phenomenon arising in construction industries worldwide as it is responsible for the deterioration of concrete in acidic environments resulting in premature degradation with regard to mass changes, weakening of mechanical properties, and increase in porosity due to calcium leaching. This paper investigates the citric acid resistance of geopolymer composites (GP) and its comparison with conventional concrete system (OPC) in lower and higher concentrations of acid medium. The GP samples displayed increased resistance to citric acid attack than OPC system in terms of mass and strength loss. It was also found that geopolymer mortar made with a blend of ground granulated blast furnace slag (GGBS) and red mud (RM) offered increased acid resistance based on the parameters studied in the degradation kinetics

1. Introduction

Concrete has been one of the most versatile materials adopted in the construction of key infrastructural components like sewerage pipes, industrial floors and foundations, water treatment facilities and many more. These infrastructural elements are easily deteriorated over the years when in contact with an acidic environment. Since high alkalinity is mandatory for the stability of the cementitious matrix, concrete is prone to attack by acidic media which affects the pH of the pore solution, that tends to crumble the hydrated cement paste to different levels based on the predominant exposure condition and some mixture design parameters of concrete. Numerous industries still rely on measures that create effluents or wastewater. Such effluents hold a wide variety of inorganic and organic acids whose activity may prompt extreme and quick deterioration of cementitious structures. Understanding the alteration kinetics and mechanisms of degradation of concrete by these aggressive media is a fundamental step towards making a durable concrete, which will then increase the service life of the concrete form.

Organic acids are naturally found in the farming products or as intermediary or final products of the metabolism of microorganisms growing within the effluents. Food and drink industries also produce large amount of organic acids. Alcoholic beverages contain lactic acid, while pharmaceutical industry uses citric acid. Concrete structures in cattle milk sheds degrade because of lactic acid attack combined with abrasive action of animals. Liquid manure also contributes to a wide variety of organic acids, including butyric, propionic, acetic, iso-butyric acids etc. and concrete structures used to store manure are prone to attack by these acids.

Among the agro-food effluents, whey water from the dairy industry contains acetic, lactic and citric acids. Molasses which has malic, lactic, acetic, oxalic, succinic and citric acids are produced by sugar industry. The fermenting business effluents mainly contain acetic, tartaric and succinic acid. The factors such as the binders used, acid type and test methods seem to influence the degradation of cementitious materials by acids [1]. Underlining the exact mechanism of deterioration is not yet inferred. Therefore, thorough investigations are required to understand the actual mechanisms involved in the deterioration process to propose a durable binder system.

Generally, cementitious materials are said to be in disequilibrium with an acidic environment due to the inherent alkaline nature of its pore solution. Concrete with OPC is reported to be less resistant to these acidic media. Moreover, the use of OPC is discouraged nowadays as its production contributes alarmingly to carbon dioxide emissions and thus global warming [2]. As we strive towards achieving sustainable development, replacement of cement-based systems with alkali activated/ geopolymer (GP) composites can be considered as a viable alternative for cementitious materials exposed to aggressive acidic environments. This also paves the way to utilize industrial wastes as a precursor binder material in the production of alkali activated binder systems. One such unused and untapped waste is red mud which is generated during the production of alumina in aluminium industry.

Around 35-40% of the processed bauxite mineral is deposited as an alkaline red mud slurry comprising 0.8-1.5 tons of red mud and 15-40% solids that are developed per ton of alumina produced [3]. It is assessed that every year, 70 million tons of red mud is manufactured worldwide [4]. Red mud is commonly disposed in mud lakes as slurry impoundment or stacked in ponds as dry mud close to alumina plants or straight headed off into the sea. The disposal of untreated red mud has caused genuine natural issues such as groundwater pollution, soil pollution and fine particles suspension in the sea [5,6]. Therefore, the incorporation of such waste materials as an alternative binder material in concrete could enhance the reduction of pollution level across the globe.

Geopolymer system is an innovative technology that utilizes industrial wastes such as red mud (RM), fly ash and ground granulated blast furnace slag (GGBS) as a binder material to produce a 3D system of cementitious materials under alkaline conditions through polymerization [7]. Individually, alkali activated GGBS and fly ash-based composites have been studied by many authors. However, the behaviour of such materials in aggressive acids such as citric acid has not been researched yet. Presently there are no codes or norms accessible for assessing the resistance of the binder constituents to acid attack. From the review of recent literatures on acid attack on concrete, enormous investigations have been done on the impact of inorganic acids on concrete but only a few investigations have concentrated on the impact of organic acids that could be detrimental to the matrix leading to premature deterioration of concrete structures. Therefore, this paper evaluates the degradation kinetics and the resistance of geopolymer composites in citric acid, an acid that is considered as one of the most aggressive organic acids that is detrimental to cementitious systems.

2. Action of organic acids

Concrete is highly susceptible to severe deterioration under acid attack. The essential mechanism behind the decay of cement by acid attack is acidolysis. When acid comes into contact with concrete surface, the hydrates of the concrete reacts with the acid protons, decomposing the hydrates. Other two different mechanisms are additionally discovered to be effective in organic acid attack. First is complexolysis, which happens when complex formation occurs by the reaction of acid ions and metal ions, allowing the dissolution of more of the solid phases. It is considered as a significant mechanism only where the metal ions are in solution because of acidolysis.

The subsequent mechanism is the precipitation of reaction products that are expansive in nature. It happens only when the salts formed by the reaction between metal ions (from hydrates) and the acidic species are of low solvency and are precipitated from the solution. Now, this doesn't have a harming impact, as precipitation may not upset the concrete framework on account of low molar volume salts. When the salt has a higher molar volume than the hydrates it replaces, its precipitation leads to the rupture and defragmentation of the cement matrix [8, 9]. Among the organic acids, the most aggressive and the least investigated acid is perhaps citric acid. Citric acid is a weak organic acid that exhibits poly acidity. Due to the citric acid's chemical nature, it can react with hydrates to produce calcium citrate salts of low solubility. As the acid is neutralized, the solubility decreases, and the calcium citrate precipitates out, which eventually delaminates from the surface owing to weakening in the acid solution. Moreover, as hydrates of conventional cementitious systems offer less resistance, the acid resistance of geopolymer (GP) composites is explored in the current study.

GP are amorphous in nature and are considered to be inorganic polymers. They are viewed as a subset of alkali activated binder systems [10,11]. Geopolymerization is an exothermic reaction where the reaction happens between silica (Si) - alumina (Al) under the alkaline condition which forms three-dimensional polymeric chain of Si-O-Al-O bonds. The soundness of a geopolymer framework depends on alumina-silicate instead of calcium silicate hydrate in conventional systems [12, 13]. On exposure to acid, the ion exchange reaction between the metal cations and hydronium ions are followed by an electrophilic attack of acid protons on Si-O-Al-O bonds, thereby leaching out the sodium ions that act as charge compensation cations of the aluminosilicate framework. Now, the Si-to-Al increases, highlighting the dissolvement of the tetrahedral Al from the aluminosilicate framework by breaking the Si-O-Al bond [12, 14, 15]. It has been established that the durability in acids increased with a decrease in CaO content within the hydrates i.e. N-A-S-H > C-A-S-H > C-S-H [16]. These are the proposed mechanisms and the actual resistance could vary with the changes in the chemical composition of binders and formulation of the mixes. Acid attack itself is still an infant area of research. Moreover, the literature investigations on citric acid resistance of geopolymer systems are rather limited and hence these aspects and findings are presented in this paper. New mixture formulations with red mud are also attempted and the citric acid resistance of mixes is evaluated.

3. Experimental program

3.1. Materials

The precursors used in the current investigation include GGBS and RM. Ordinary Portland Cement (OPC) 53 Grade with a specific gravity of 3.15 is taken to be the control mix. Slag having a specific gravity of 2.90 was used. RM was procured as wet sludge from HINDALCO, Belgaum. It was then oven dried at 100^0C for 24 hours. Later, the oven dried RM was put into Los-Angeles abrasion machine for 1000 revolutions to attain the required fineness. The specific gravity of red mud was 2.63. Alkaline activators, sodium hydroxide (10M) and sodium silicate solution were adopted with alkali silicate/alkali hydroxide ratio as 2.50. The admixture used is Sulphonated Napthalene Formaldehyde (SNF) superplasticizer with 3% by weight of binders for the preparation of geopolymer mixes. The mortar ratio for mix design is taken as 1:1.5 (binder: sand) by weight. Mortar specimen size adopted is cylindrical specimens having 25 mm in diameter and 50 mm in height. The control mix was done using distilled water with a water to cement ratio of 0.50. GP mixes were prepared with Alkali/Binder (a/b) ratio of 0.50. Manufactured sand (finer than 2 mm) was used as fine aggregates in the preparation of mortar specimens.

3.2. Mixing, casting, curing and testing procedures

The mixes adopted are OPC (control mix) and geopolymer mixes (SGP (100% GGBS), RSGP50 (50% GGBS + 50% RM) and RSGP25 (75% GGBS + 25% RM)). The mixing was done using Hobart mortar mixer. The plastic cylindrical moulds were oiled. Filling of mortar into the moulds were done manually with tamping. After casting, specimens were kept for 24 hours in atmospheric conditions for initial set before demoulding. For OPC mix, lime solution curing was done for 14 days to maintain the alkalinity and to prevent any leaching of metal ions from the matrix. A fourteen days air (ambient) curing was done for SGP and RSGP specimens. Chemical admixture was not used for the preparation of the OPC mix.

Mortar samples were submerged completely in citric acid solution of 0.15M and 0.31M concentrations for a period of 4 weeks to examine the resistance of OPC and GP based systems to citric acid attack. The volume of acid solution to the volume of specimens was kept as 5. The samples were observed visually for the changes, and their altered depth, mass change, pH variation of the solution and compressive strength was also recorded periodically. The acid solution was replaced once in two weeks with a fresh solution to maintain the aggressiveness of the solution and to sustain the reactions.

4. Results and discussions

4.1. Visual appearance

The extent of deterioration can be visually explored by taking digital photographs on a regular basis. Figure 1 showcases the photographs of specimens after 4 weeks of exposure in various concentrations of citric acid solution.

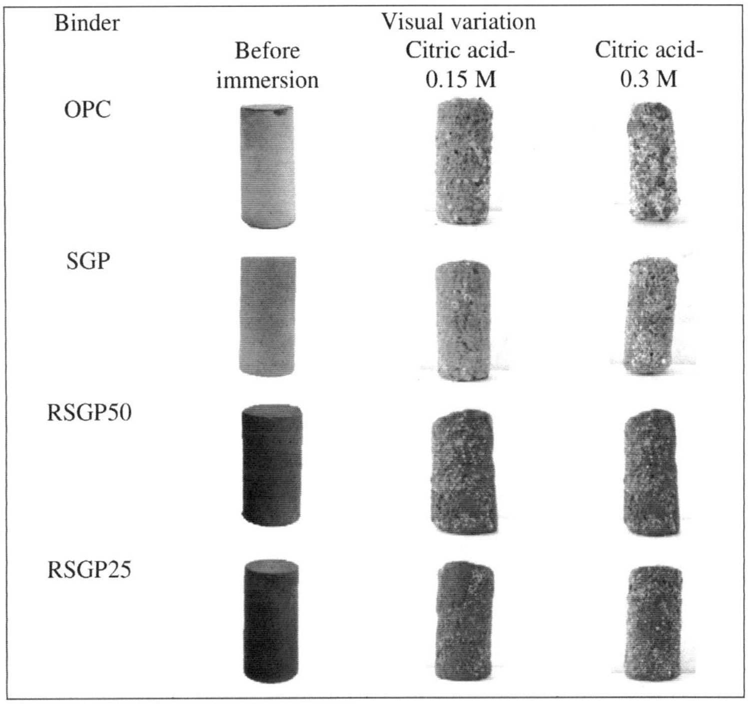

Figure 1. Visual observation of specimens immersed in various concentrations of citric acid after 4 weeks of exposure.

When specimens were closely observed, the formation of a white coloured salt in substantial quantities was observed. This salt was confirmed as calcium citrate tetrahydrate by X-ray diffraction in a similar study done by Ramaswamy and Santhanam [1]. The precipitated salt has low solubility (0.95 g/L), and observed to be non-protective, and loosely stuck to the specimen. On comparing the intensity of degradation with the concentrations of acids for all the binders, more damage was observed in the case of 0.31M citric acid than 0.15 M solution. The salt formed does not have a shielding effect on the matrix due to the high molar volume of the salt. More precipitate formation and damage effect was observed on OPC specimens when compared to all the geopolymer mixes. It was also noted that the precipitation formation was lesser at lower concentrations of exposure.

4.2 Mass changes

On a weekly basis, the mass of specimens (in saturated but surface dry condition) were noted with a precise digital weighing balance. The average of the mass of three specimens was then calculated. Mass loss is determined with respect to the initial saturated mass taken before the acid exposure and showed in %. The changes in mass of the specimens with age of exposure to different concentration of citric acid medium are shown in Figure 2.

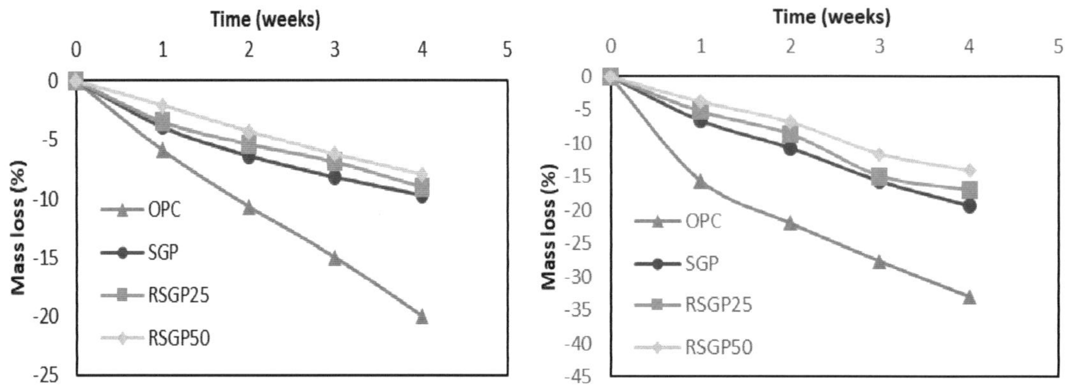

(a) 0.15M **(b) 0.31M**

Figure 2. Mass change of mixes in various concentrations of citric acid solution.

From the graph (Figure 2), it can be seen that there is a continuous decrease in the mass of all specimens in citric acid medium irrespective of the medium concentrations due to the dissolution of cementitious material to form the high molar volume precipitates. The rapid and severe mass loss prove that the intensity of degradation was high, deteriorating the inner layers on progressive exposure. Among the various mixes studied, the RSGP50 mix indicated lower mass loss while the OPC mix indicated the least resistance as the mass loss was higher. OPC specimens seems to have the highest mass loss in all acid medium with highest mass loss of about 35% in 0.31M citric acid medium. These observations are in alignment with the visual observations. On the contrary, RSGP50, RSGP25 and SGP mixes have lower mass loss in all acid medium, with a maximum of 13%, 15% and 18% mass loss in 0.31M medium. It was also observed that the mass loss in 0.15M citric acid was much lesser than 0.31M citric acid.

The severe mass loss and the aggressiveness of the attack may be ascribed to a combination of factors such as non-protective nature and high molar volume of the salt, poly acidic nature and buffer action exhibited by the acid and the high concentration being used for obtaining accelerated results in short span of time. Citric acid is polyacidic as it dissociates to give several protons at their corresponding pKa values of 3.14, 4.76 and 6.40 and this could lead to a sustained attack, if the pH of the solution is the range of pKa values.

4.3. Strength changes

The compressive strength was measured for all specimens on a weekly basis using a modified digital compression testing machine having a capacity of 50 kN. The test was conducted with a start load of 1 kN and stop load at 40%. The loading rate was maintained as 0.5 kN/sec. The initial compressive strength for OPC, SGP, RSGP50 and RSGP25 are found to be 44, 53, 36 and 48 N/mm^2. SGP have higher initial strength than the other mixes. Similar to mass loss, strength loss was computed based on the strength at particular age of exposure and the initial strength, and expressed as %. The strength loss of various mixes with the period of acid exposure is shown in Figure 3. All the mixes incurred a significant loss in strength with time. Notably, OPC specimens showed the highest strength loss in all acid medium in spite of having good initial strength. Loss of strength for OPC varies from 48% in 0.15M to 55% in 0.31M medium. The calcium hydroxide and CSH gel being least stable hydrates in OPC system might have been decalcified by the citric acid causing leaching and dissolution of calcium ions into the solution to effect the formation of expansive salts. The dissolution of calcium hydroxide and formation of silica gel due to decalcification of CSH gel brings about an increase in the porosity and adds to a rapid reduction in compressive strength [17].

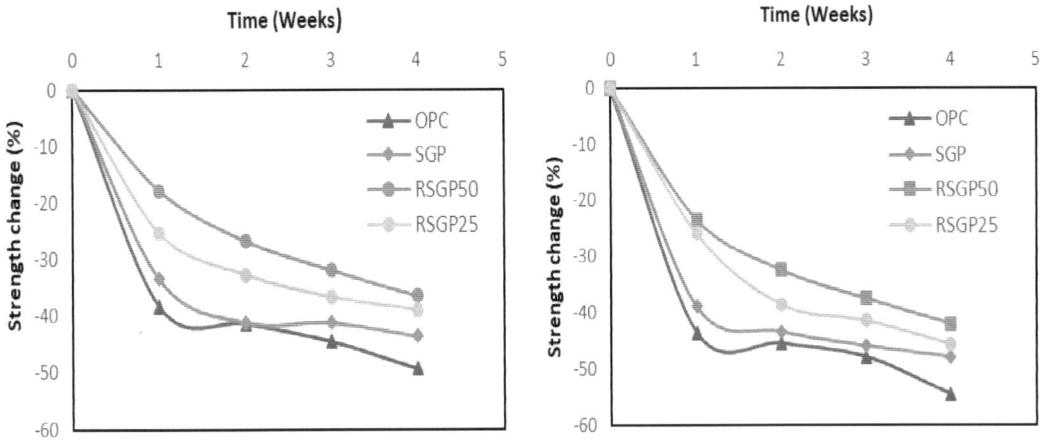

(a) 0.15M **(b) 0.31M**

Figure 3. Strength change of mixes in various concentration of citric acid solution.

The RSGP50 showed the lowest mass loss of about 34% in 0.15M and 40% in 0.31M citric acid medium. Although, the initial strength showcased by RSGP50 is the lowest, its retainment of strength with regard to time of exposure and acid concentration seems to be better than OPC, SGP and RSGP25. This is in concurrence with the mass loss where the OPC specimen shows the larger mass losses than all the GP mixes. It is noted that SGP mix also suffered strength loss. This is probably because the microstructure consists also of CASH and/or CSH gel which could be attacked by the acid ions at lower pH.

4.4. Altered depth

Altered depth is defined as the depth directly affected due to acid attack and is considered as a primary indicator of degradation kinetics. This can be estimated as the sum of depth of leaching/erosion and depth of reaction. The depth of reaction was determined by sprinkling phenolphthalein solution to a freshly cut surface of mortar specimens, and by measuring the depth (thickness) of colorless layer in the specimen. The depth of erosion is deduced from the difference of diameter of specimens before and after exposure, noted using a digital vernier caliper. Figure 4 shows the cross-section of specimens before and after citric acid attack. The pink shaded region shows the area which remain unaffected when exposed to acids after the application of phenolphthalein solution whereas the colourless or white regions show the deteriorated zone consisting of silica gel and/or calcium citrate salt.

For an elaborate understanding, altered depth of mortar specimens in various concentrations with respect to the exposure age is plotted and shown in Figure 5. The altered depth is seen to be higher for RSGP50 mix with a depth of 2.16mm when compared to other mixes. The altered depth of OPC in 0.15M medium was 1.78mm and is the least on comparing SGP and RSGP mixes. On the contrary, in the higher concentration i.e. 0.31M medium – OPC mix depicted a depth of 1.82mm. The lower depth in OPC may be ascribed to the additional calcium buffer available in OPC systems. In the case of GP systems, acid may be attacking directly the alumino-silicate gel and the penetration proceeds at a faster rate as measured by the application of phenolphthalein.

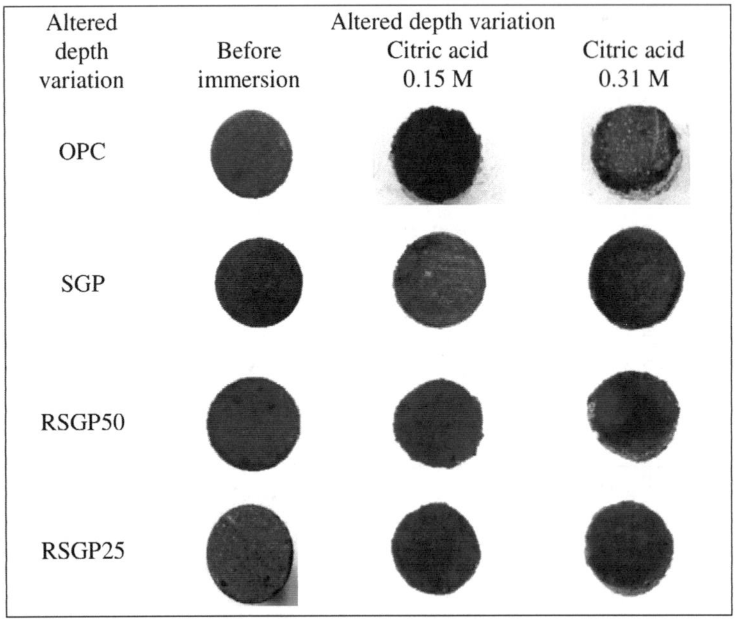

Figure 4. Visual examination of altered depth of specimens in different concentrations of citric acid.

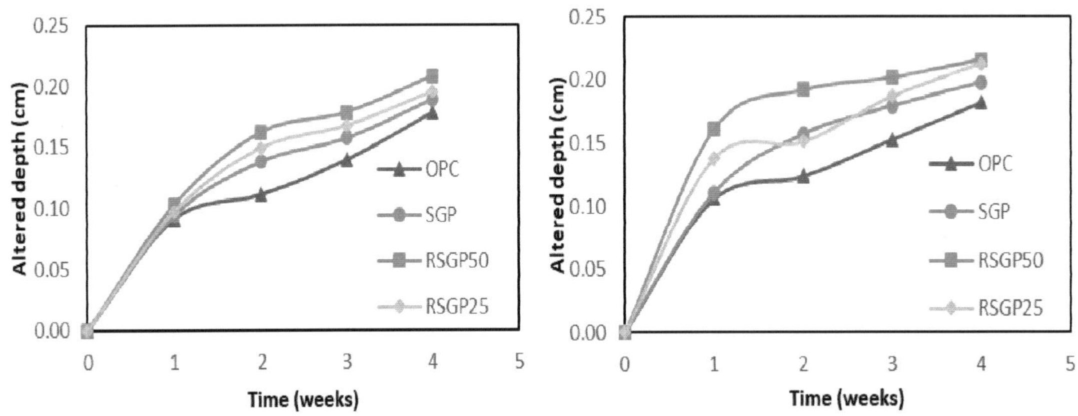

(a) 0.15M **(b) 0.31M**

Figure 5. Altered depth variation of mixes with age of exposure in different concentrations of citric acid.

4.5. Changes in pH

Acid attack is a classic example of acid - base reaction to form salt and water. In this neutralization process, ions such as calcium, aluminium from the cementitious framework gets into the reaction with the acid solution. Acid protons are consumed and to maintain the equilibrium, hydroxyl ions are released into the solution causing a rise in pH of the acid solution. Hence, estimating pH on consistent basis could be another measure of the degradation kinetics as it gives a reflection about the hydrogen ion utilization rate or in other words, the neutralization capacity of the system. Here, the initial pH of the 0.15M citric and 0.31M citric acid solutions were 2.06 and 1.97 respectively. The cyclic plots depicting increase in pH after every 14 days of exposure each time back to pH values of fresh acid solution are shown in Figure 6. There is a rapid increase in pH for the first week compared to the initial pH for the acid. This must be due to the rapid neutralization process when the specimen comes into contact with acid solution. The pH increase of acid medium of OPC mix is found to be higher on comparison with SGP and RSGP mixes. The increased pH could also be one of the reasons for the lowered altered depth in the case of OPC mix. The pH rise was observed to be lowermost for RSGP50 mix suggesting that the attack is possibly lower, and this connects well with the mass loss and visual observations.

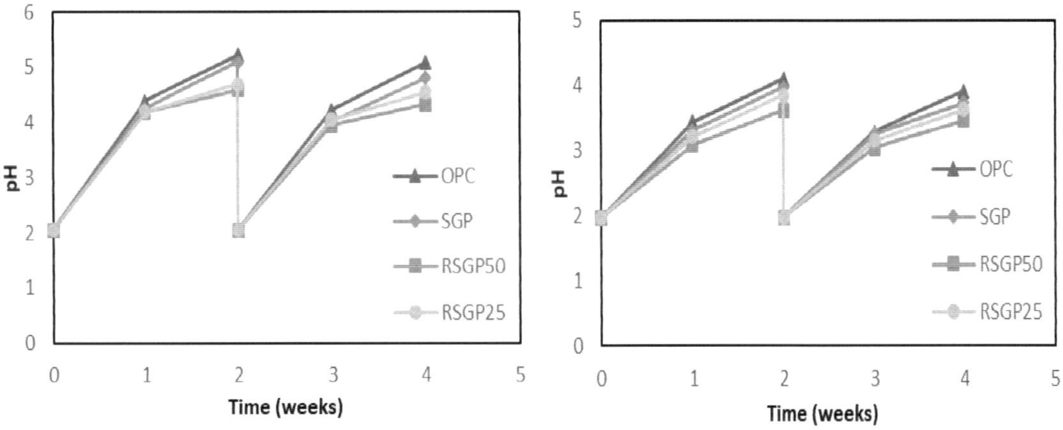

(a) 0.15M **(b) 0.31M**

Figure 6. Change in pH of different concentrations of citric acid solution with age of exposure.

5. Conclusions

Concrete structures are often exposed to aggressive environments due to growing production of acidic industrial effluents in urban areas. This has led to severe degradation of concrete structures made with OPC in contact with the acidic solutions. There is a need to think of a suitable alternative to OPC system and geopolymer technology utilizing industrial wastes is considered to be a potential solution. This paper has exhibited the results and discussion of an accelerated laboratory study conducted to examine and compare the resistance of geopolymer and OPC mortar systems to deterioration by citric acid. The mass loss for both slag and red mud blended slag geopolymer mixes is lesser than OPC during 4 weeks of exposure to different concentrations of citric acid.

The compressive strength of OPC specimens was found to reduce by 55% after 4 weeks of exposure to 0.31M citric acid solution when compared to 47%, 45% and 40% for the SGP, RSGP25 and RSGP50 samples. However, the altered depth (depth of penetration) was higher for SGP and RSGP mortar specimens in 0.15M and 0.31M acid medium unlike OPC which had the least altered depth. Based on the visual observations, mass loss, and strength loss, the performance of geopolymer mixes were better compared to OPC. Among the geopolymer mixes, the mortar mix containing 50% slag and 50% red mud exhibited superior acid resistance. However, when altered depth is considered, OPC showed greater resistance to attack compared to the geopolymer mixes. More studies with prolonged duration of exposure must be conducted to come up with geopolymer mixture formulations which will have better acid resistance with respect to mass changes, strength loss and altered depth.

6. References

[1] Ramaswamy K P and M Santhanam 2016 Degradation kinetics of cement-based materials in citric acid *10th Structural Engineering Convention* Chennai 1690-1695

[2] World Energy Council 1995 Efficient use of energy utilizing high technology: An assessment of energy use in industry and buildings *London: World Energy Council*

[3] Liu X and Zhang N 2011 Utilization of red mud in cement production: a review *Waste Management and Research,* **29** 1053-1063

[4] Vangelatos I, Angelopoulos G N and Boufounos D 2009 Utilization of ferro alumina as raw material in the production of ordinary portland cement *Journal of Hazardous Materials* **168** 473–478

[5] Agrawal A, Sahu K K and Pandy B D 2004 Solid waste management in non-ferrous industries in India *Resources Conservation and Recycling* **42** 99–120

[6] Wang S, Ang H M and Tade M O 2008 Novel application of red mud as coagulant, adsorbent and catalyst for environmentally benign processes *Chemosphere* **72** 1621– 1635

[7] Li Y, Min X, Ke Y, Liu D and Tang C 2019 Preparation of red mud-based geopolymer materials from MSWI fly ash and red mud by mechanical activation *Waste Management* **83** 202–208

[8] Dyer T 2016 Influence of cement type on resistance to organic acids *Magazine of Concrete Research* **69(4)** 175-200

[9] Dyer T 2017 Influence of cement type on resistance to attack from two carboxylic acids *Cement and Concrete Composite* **83** 20-35

[10] Provis J L 2018 Alkali-activated materials *Cement and Concrete Research* **114** 40-48

[11] Provis J L and Deventer S J V 2009 Geopolymers Structure, processing, properties and industrial applications *Woodhead Publishing Limited and CRC Press LLC*

[12] Stephan D, Firdous R, Noël J and Djobo Y 2018 Natural pozzolan based geopolymers: A review on mechanical, microstructural and durability characteristics *Construction and Building Materials* **190** 1251-1263

[13] Lodeiro G, Palomo A, Fernández-Jiménez A and Macphee D E 2011 Compatibility studies between N-A-S-H and C-A-S-H gels Study in the ternary diagram Na_2O–CaO–Al_2O_3–SiO_2–H_2O *Cement and Concrete Research* **41** 923–931

[14] Aiken T A, Sha W, Kwasny J and Soutsos M N 2017 Resistance of geopolymer and Portland cement-based systems to silage effluent attack *Cement and Concrete Research* **92** 52-65

[15] Gu L, Bennett T and Visintin P 2018 Evaluation of accelerated degradation test methods for cementitious composites subject to sulphuric acid attack; application to conventional and alkali activated concretes *Cement and Concrete Composites* **87** 187-204

[16] Gu L, Bennett T, and Visintin P 2019 Sulphuric acid exposure of conventional concrete and alkali activated concrete: Assessment of test methodologies *Construction and Building Materials* **197** 681-692

[17] Koenig A, Herrmann A, Overmann S and Dehn F 2017 Resistance of alkali activated binders to organic acid attack: Assessment of evaluation criteria and damage mechanisms *Concrete and Building Materials* **151** 405-413

Design comparison of conventional steel structure with pre- engineered structure

Bishbu Jacob[1,2] and Althaf M[1]

[1]Department of Civil Engineering, TKM College of Engineering, Kollam, Kerala, India

[2]Email:bishbujacobv@gmail.com

Abstract. The pre- engineered buildings are the steel buildings which are predesigned and prefabricated. For pre- engineered structure the member size is provided as per the requirement at the location and hence the members are tapered in nature. The study is on the design of the pre- engineered structure and its comparison with conventional structure. The objective of the study is to analyze and design a structure as a conventional steel structure and Pre-engineered structure and to compare the design of the structure in terms of structure weight. The use of cold-formed steel sections as secondary members reduces the self-weight of the pre- engineered structure. It was expected that the pre-engineered buildings will be more economical than conventional structure.

1. Introduction

The concept of pre-engineered building involves the steel buildings which are predesigned and prefabricated. The PEB has frame geometry which matches to the shape of the bending moment (internal stress) diagram of the member and thus optimizing material usage and reducing the total weight of the structure. Pre-engineered Buildings focuses on using pre-designed connections, design and pre-fabricated members for the building structures, thus significantly reduces the time for design, fabrication and installation. Pre-engineered buildings have custom made primary and secondary framing members for each building designed. Instead of standard uniform sections, tapered members can be used in the primary frame. Secondary members are steel Z and C shaped cold formed sections. All girts and purlins are lightweight steel as well.

2. Design of structures

The primary frame system of conventional steel structure and pre- engineered structure were designed in E-tabs software. The purlin was designed manually calculating the loads coming on them. The structures were designed considering dead load, collateral load, live load, wind load and earthquake load. The plan of the building is shown below in figure 1.

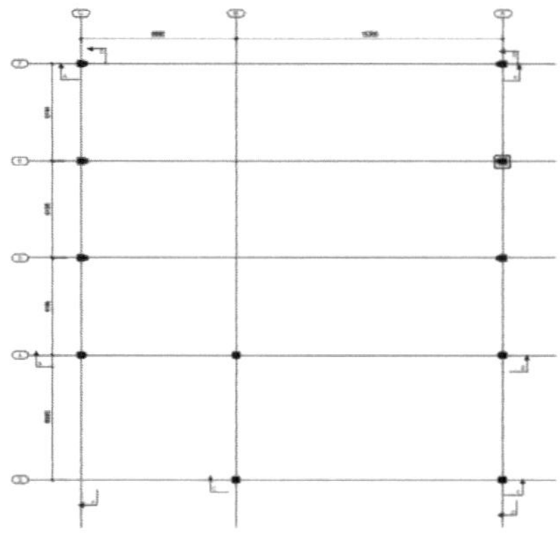

Figure 1. Plan of the structure.

2.1. Design results

2.1.1. Design results of conventional system

Table 1. Column dimensions of CSB.

Column	Column dimensions			
	Depth of web (mm)	Thickness of web (mm)	Width of flange (mm)	Thickness of flange (mm)
C1	550	10.5	250	17.6
C2	600	11.2	250	21.3

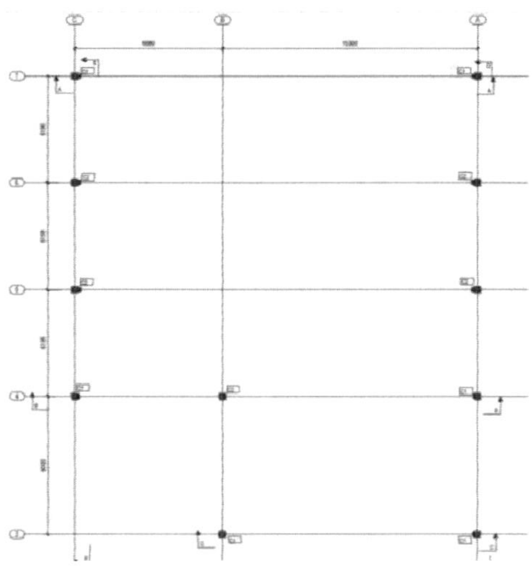

Figure 2. Column layout plan of CSB.

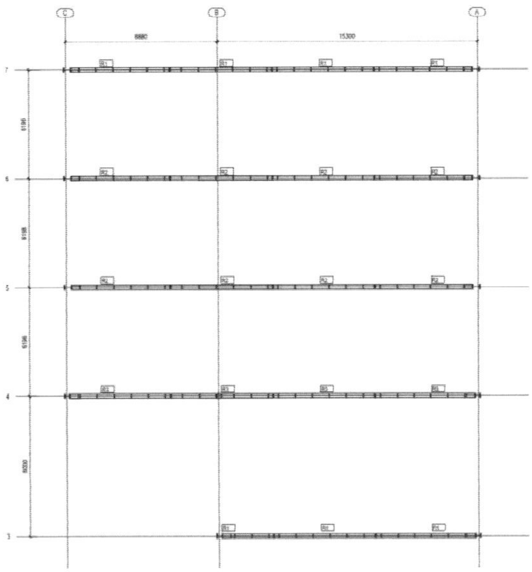

Figure 3. Rafter layout plan of CSB.

Table 2. Rafter dimensions of CSB.

Rafter	Rafter dimensions			
	Depth of web (mm)	Thickness of web (mm)	Width of flange (mm)	Thickness of flange (mm)
R1	550	9.9	190	15
R2	600	10.5	210	15.5
R3	450	8.6	170	13.4

Table 3. Steel weight take off for CSB.

Item	Weight/m(kg)	Length (m)	Weight (kg)
C1	112.5	28.14	3165.75
C2	133.7	23.64	3160.668
R1	86.3	40.4	3486.52
R2	99.5	48.48	4823.76
R3	65.3	24.24	1582.87
Purlins	20.6	448.944	9248.246
	Total weight		25467.82 kg

2.1.2. Design results of pre- engineered structure

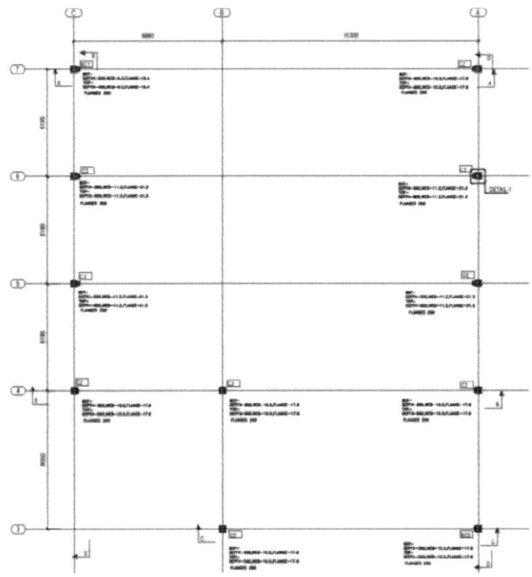

Figure 4. Column layout plan of PEB.

Table 4. Column dimensions of PEB.

Column	Start	End
C1	Depth- 300 Web Thickness- 9.2 Flange Width- 200 Flange Thickness- 15.4	Depth- 450 Web Thickness- 9.2 Flange Width- 200 Flange Thickness- 15.4
C2	Depth- 300 Web Thickness- 10.5 Flange Width- 250 Flange Thickness- 17.6	Depth- 550 Web Thickness- 10.5 Flange Width- 250 Flange Thickness- 17.6
C3	Depth- 300 Web Thickness- 11.2 Flange Width- 250 Flange Thickness- 21.3	Depth- 600 Web Thickness- 11.2 Flange Width- 250 Flange Thickness- 21.3

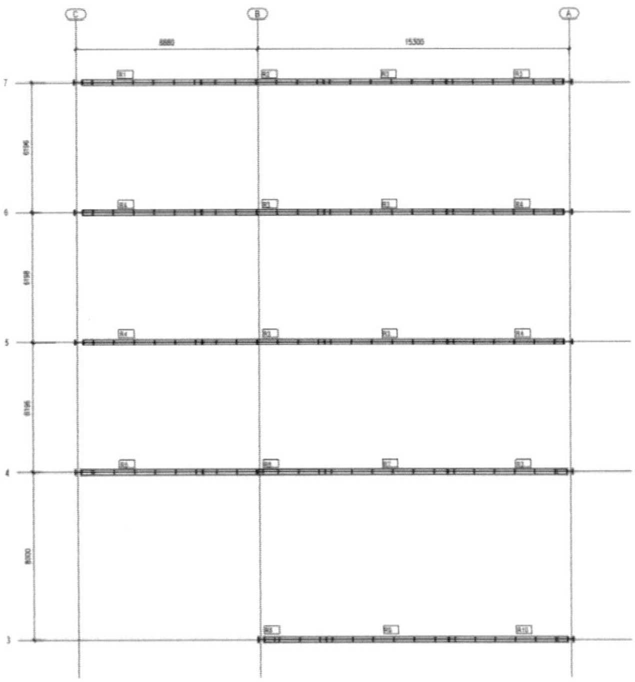

Figure 5. Rafter layout plan of PEB.

Table 5. Rafter dimensions of CSB.

Rafter	Start	End
R1	Depth- 450 Web Thickness- 8 Flange Width- 165 Flange Thickness- 12.5	Depth- 550 Web Thickness- 8 Flange Width- 165 Flange Thickness- 12.5
R2	Depth- 550 Web Thickness- 9.9 Flange Width- 190 Flange Thickness- 15	Depth- 550 Web Thickness- 9.9 Flange Width- 190 Flange Thickness- 15
R3	Depth- 550 Web Thickness- 9.9 Flange Width- 190 Flange Thickness- 15	Depth- 550 Web Thickness- 9.9 Flange Width- 190 Flange Thickness- 15
R4	Depth- 600 Web Thickness- 9.9 Flange Width- 190 Flange Thickness- 15	Depth- 550 Web Thickness- 9.9 Flange Width- 190 Flange Thickness- 15
R5	Depth- 550 Web Thickness- 9.9 Flange Width- 190 Flange Thickness- 15	Depth- 350 Web Thickness- 9.9 Flange Width- 190 Flange Thickness- 15
R6	Depth- 350 Web Thickness- 7.4 Flange Width- 165 Flange Thickness- 11.4	Depth- 350 Web Thickness- 7.4 Flange Width- 165 Flange Thickness- 11.4

R7	Depth- 550 Web Thickness- 9.9 Flange Width- 190 Flange Thickness- 15	Depth- 350 Web Thickness- 9.9 Flange Width- 190 Flange Thickness- 15
R8	Depth- 400 Web Thickness- 8 Flange Width- 165 Flange Thickness- 12.5	Depth- 275 Web Thickness- 8 Flange Width- 165 Flange Thickness- 12.5
R9	Depth- 550 Web Thickness- 9.9 Flange Width- 190 Flange Thickness- 15	Depth- 275 Web Thickness- 9.9 Flange Width- 190 Flange Thickness- 15
R10	Depth- 400 Web Thickness- 8 Flange Width- 165 Flange Thickness- 12.5	Depth- 400 Web Thickness- 8 Flange Width- 550 Flange Thickness- 12.5

Table 6. Steel weight take off for PEB.

Item	Length (m)	Weight(kg)
C1	4.5	219.82
C2	28.14	1868.9
C3	18	1391.82
R1	6.06	277.82
R2	6.06	379.02
R3	42.42	2653.132
R4	24.24	1562.991
R5	6.06	332.1
R6	6.06	238.95
R7	6.06	332.1
R8	4.04	144.14
R9	6.06	314.51
R10	6.06	268.34
Purlins	448.944	4471.48
Total Weight		14455.13 kg

3. Summary of results

The total weight of conventional structure was 25467.82 kg and pre- engineered structure had a weight of 14455.13 kg only. The steel take directly shows that the pre- engineered structure is light compared to conventional structure. There was a material saving about 43.24 %. The weight of primary structure (column and rafter) of conventional structure and pre-engineered structure was 16219.57 kg and 9983.65 kg respectively. There was a material saving of 38.44 %. While comparing the secondary member (purlin) pre –engineered building was 4776.766 kg lighter than conventional structure. The secondary members in conventional structure were almost 51.65 % heavier than secondary members in pre- engineered building.

4. Conclusion

A structure was designed as conventional portal frame structure and pre- engineered structure to study the design of both and to compare its design. The design was compared in terms of total steel weight take off. The conclusions of the study are as follows.

- The pre- engineered structure was 43.24% lighter than ordinary conventional portal frame structure.
- The secondary members in conventional structure were almost 51.65 % heavier than secondary members in pre- engineered building.
- The weight contribution by secondary members was 36.1% and 30.93% of total weight for conventional structure and pre- engineered structure respectively.
- The pre- engineered buildings are more economic in terms of steel take off.

5. References

[1] Aijaz A , Kulkarni A V and Aslam H 2013 Comparative study of analysis and design of pre-engineered buildings and conventional frames *Journal of Mechanical and Civil Engineering* **5** 32-43
[2] Anisha G and Tushar S 2018 Pre-engineered building design of an industrial warehouse *International Research Journal of Engineering and Technology* **5 (6)** 1484-1488
[3] Bhavikatti S S 2010 Design of steel structures by limit state method as per IS 800-2007 International publishing house Pvt. Ltd New Delhi
[4] Meera C M 2013 Pre-engineered building design of an industrial warehouse *International Journal of Engineering Sciences & Emerging Technologies* **5(2)** 75-82
[5] Hancock G J 2016 Cold-formed steel structures: Research review 2013–2014 *Advances in Structural Engineering* **19**(3) 1-16
[6] Christopher T and Benjamin W 1990 Behaviour of cold-formed steel purlins under gravity loading *Journal of structural engineering,* **116(8)** 2061-2069
[7] *Darshana P Z* 2012 Evaluation of pre-engineering structure design by IS-800 as against pre-engineering structure design by AISC *International Journal of Engineering Research &Technology* **1**(5) 1-9
[8] Duggal, S K *Earthquake Resistance Design of Structure* (2010) Fourth Edition Oxford University Press New Delhi
[9] Sai K, Kailasa R and Pradeep K 2014 Comparison of design procedures for pre-engineering buildings (PEB): A case study *International Journal of Civil, Architectural, Structural and Construction Engineering* **8(4)** 477- 481
[10] Gurusharan S Introduction to pre-engineered buildings *Civil Engineering Portal*
[11] Lip H Teh and Benoit P G 2014 Net section tension capacity of cold-reduced sheet steel angle braces bolted at one leg lip *Journal of Structural Engineering* **139** 328-337
[12] IS 1893 (Part I) Criteria for earthquake resistant design of structures part 1 general provisions and buildings *Bureau of Indian Standards* New Delhi
[13] IS 801(1975) Code of practice for use of cold formed light gauge steel structural members in general building construction *Bureau of Indian Standards* New Delhi
[14] IS 811(1987) Specification for cold formed light gauge structural steel sections *Bureau of Indian Standards* New Delhi
[15] IS 875 (Part II) Code of practice for design loads (other than earthquake) for buildings and structures part 2 imposed loads *Bureau of Indian Standards* New Delhi
[16] IS 800 (2007) General construction in steel- code of practice *Bureau of Indian Standards* New Delhi

[17] IS 4923 (1997) Hollow steel sections for structural use-specification *Bureau of Indian Standards* New Delhi

[18] IS 875 (Part I) Code of practice for design loads (other than earthquake) for buildings and structures part 1 dead loads - unit weights of building materials and stored materials *Bureau of Indian Standards* New Delhi

[19] IS 875 (Part III) Code of practice for design loads (other than earthquake) for buildings and structures part 3 wind loads *Bureau of Indian Standards* New Delhi

[20] Jatin D T and Patel P G 2013 Comparative study of pre- engineered steel structure by varying width of structure *International Journal of Advanced Engineering Technology* **4(3)** 56-62

[21] Jinsha M S and Linda A M 2013 Analysis of pre –engineered buildings *International Journal of Science and Research* **5 (7)** 1049-1051

[22] Monika N, Jigar Z and Darshan S 2017 Cost effectiveness of multi storey CSB & PEB industrial steel building *Journal of Emerging Technologies and Innovative Research* **4(4)** 264-268

[23] Kyvelou P, Kyprianou C, Gardner L and Nethercot D A 2019 Challenges and solutions associated with the simulation and design of cold formed steel structural systems *Thin-Walled Structures* **141** 526-539

[24] Salim P M, Seshadri S T and Mahima G 2017 Pre- engineered buildings - A review *i-manager's Journal on Structural Engineering* **6 (2)** 39-48

[25] Sigmund A F 2007 Response spectra as a useful design and analysis tool for practicing Structural Engineers *Journal of Earthquake Technology* **44 (1)** 25–37

[26] SP 38(1987), Handbook of typified designs for structures with steel roof trusses (with and without cranes) (based on is codes), Bureau of Indian Standards, New Delhi

[27] Syed F, Sarath C K B and Kanakambara Rao S 2012 Design concept of pre-engineered building *International Journal of Engineering Research and Applications* **2(2)** 267-272

[28] Mythili T D 2017 An overview of pre-engineered building systems *International Journal of Scientific & Engineering Research* **8(4)** 557-563

[29] Tejaswi K and Rajesh K 2016 Response spectra analysis of symmetric and asymmetric buildings with variation in natural period and soil strata *ELK Asia Pacific Journals – Special Issue* 1-6

[30] Vlad Cu, Daniel C and Dan-Ilie B 2015 Structural efficiency of cold-formed steel purlins, *International conference knowledge-based organization* **21(3)** 809-814

Mechanical Characteristics of Geopolymer Mortar in Ambient Air and Immersion Curing

S S Sneha[1] and K P Ramaswamy[2]

[1] M.Tech Student, Department of Civil Engineering, TKM College of Engineering, Kollam, Kerala, India
[2] Assistant Professor, Department of Civil Engineering, TKM College of Engineering, Kollam, Kerala, India
[1]snehasrss97@gmail.com, [2]ramaswamykp@tkmce.ac.in

Abstract: Geopolymer concrete is an environmentally sound alternative to the conventional concrete, which could assist in reducing the carbon footprint that augments the process of global warming. The current work presents a comparative study on the compressive strength and ultrasonic pulse velocity of geopolymer mortar mixes at activator modulus (SiO_2/Na_2O) of 1.25, 1.50 and 1.75 subjected to ambient air curing and immersion curing. The dynamic modulus of elasticity determined from the UPV results are analogized with the compressive strength. The study also aims at developing economic geopolymer mortar with acceptable mechanical properties. The three mixes considered for the study include geopolymer mortar with 100% slag, 50% slag and 50% fly ash and 50% red mud and 50% slag. The molarity of alkaline solution and sand-binder ratio is fixed as 8M and 1.50 respectively. The results indicate that the immersion cured specimens of the mix with 50% slag and 50% fly ash corresponding to the activator modulus 1.50 has satisfactory strength performance. The dynamic modulus of elasticity was found to be higher for the immersion cured specimens.

Keywords: Geopolymer, mortar, Activator modulus, Fly ash, Slag, Red mud, Strength, Ultrasonic Pulse Velocity

1. Introduction

The geopolymers are inorganic polymers that are capable of efficaciously reducing the emission of carbondioxide, significantly contributed from the construction industry. Geopolymers are produced by reacting the alkaline solution with the silicates and aluminates to produce binders. The waste products and industrial by products such as ricehusk ash, flyash, slag, metakaolin, redmud and volcanic ash are the source materials for the silicates and aluminates [1]. Geopolymer concrete is a cementless concrete that helps in achieving sustainability in the construction field. The effective utilization of the waste materials produced is possible with the systematic usage of geopolymer concrete.

The prime focus of the study is on the compressive strength development of the geopolymer mortar which is highly influenced by the parameters namely activator modulus (SiO_2/Na_2O), water to geopolymer solids (w/s) ratio, workability, curing conditions, silica content, alkali content, alkali to binder ratio (a/b), sand to binder ratio, temperature and duration of curing. Murali et al. [2] suggested that the compressive strength of fly ash based geopolymer mortar increases linearly with an increase in the alkali content. There is a remarkable decrease in the strength development with increase in the w/s ratio. It is well established in the study concluded by Islam et al. [3] that the early strength development in geopolymer mortar is significantly higher than the ordinary cement mortar. About 90 percent of compressive strength is achieved during the first 7 days. Razak et al. [4] reported that the strength attainment was greater when the specimens were subjected to curing in 1M NaOH compared to ambient curing. Ma et al. [5] reported that curing temperature considerably affected the strength enhancement of the flyash - slag blends. Temperature is responsible for the activation of aluminosilicates. The slag activation occurs at a lower temperature whereas higher temperature promote both the flyash and slag activation. Nuruddin et al. [6] observed that when the molarity of the solution exceeded 12M, the strength attainment was negatively affected.

The need of the hour is to optimise the activation condition so as to obtain environmental as well as economic sustainability by producing geopolymer with minimum activator modulus. This could be satisfied if researches are done by focussing on the activator modulus. Activator modulus (AM) is a vital parameter which governs the activation condition. It can be defined as the ratio of SiO_2 to Na_2O of the alkaline solution [7]. This paper presents a comparative study on the compressive strength of different geopolymer mortar mixes in ambient air and immersion curing condition by varying the activator modulus.

2. Experimental Programme

The objective of the paper is to conduct a comparative study of the mechanical property of the geopolymer mortar. Three different mixes were prepared using slag (GGBS), fly ash (FA) and red mud (RM). Mix 1 (M1-100% Slag), Mix 2 (M2-50% Slag and 50% FA) and Mix 3 (M3-50% Slag and 50% RM) were prepared with activator modulus of 1.25, 1.50 and 1.75. The specimens cast for each mix were subjected to both ambient and immersion curing. The work is also meant to assess the possibility of producing economic geopolymer mortar with minimum alkali content by varying the activator modulus. The mix design is done as per the standard codal provisions [7]. AM and w/s ratio was chosen based on the codal provisions which specified that AM is to be between 1 and 2 and w/s to be less than 0.50 respectively. The w/s ratio was fixed as 0.30 and the sand to binder ratio was fixed as 1.50 respectively. The ratio of Na_2SiO_3/NaOH was calculated corresponding to the fixed activator modulus of 1.25, 1.50 and 1.75. From the ratio obtained, the quantity of alkaline solution required for the each mix was calculated.

Materials procured for producing geopolymer mortar are as given below:

- Ground Granulated Blast Furnace Slag (GGBS)
- Fly ash (FA)
- Red Mud (RM)
- Sodium Hydroxide (NaOH) pellets
- Sodium Silicate (Na_2SiO_3) solution
- Graded M-Sand (passing 2 mm sieve)

Fly ash (Class F) and slag was obtained in powdered form. Red mud from an aluminium industry was sieved through IS 150 micron sieve and used. Measured quantity of NaOH pellets was dissolved in appropriate amount of water and then mixed with Na_2SiO_3 solution to form the 8M alkaline solution required to produce the geopolymer mortar mix. The 8M solution was prepared atleast 24 hours prior to mixing. Na_2SiO_3 was obtained in solution form from Minar Chemicals, Ernakulam, Kerala which contained 50.3 percent solids with an AM of 2.20. Graded M-Sand was used for the casting purpose. The sand passing 2 mm and retained in 1 mm, 500 micron and 90 micron sieve was taken in equal quantity for obtaining the standard gradation of sand. The mixes used are shown in the Table 1 given below:

Mix	Binder	AM	$(Na_2SiO_3)/(NaOH)$	Alkali/Binder
Control	100% OPC	-	-	-
		1.25	1.70	0.60
Mix 1 (M1)	100% Slag	1.50	2.77	0.67
		1.75	5.02	0.73
		1.25	1.70	0.60
Mix 2 (M2)	50% Slag + 50% FA	1.50	2.77	0.67
		1.75	5.02	0.73
		1.25	1.70	0.60
Mix 3 (M3)	50% Slag + 50% RM	1.50	2.77	0.67
		1.75	5.02	0.73

Binder	Specific Gravity
Cement	3.14
Slag	2.92
Fly ash	2.21
Red Mud	2.63

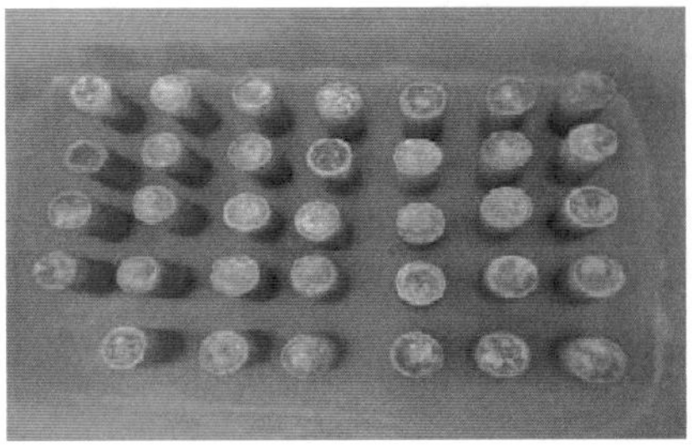

IOP Conf. Series: Materials Science and Engineering **989** (2020) 012009 doi:10.1088/1757-899X/989/1/012009

Pulse velocity by cross probing (km/sec)	Concrete Quality Grading
Above 4.5	Excellent
3.5 - 4.5	Good
3 - 3.5	Medium
Below 3	Doubtful

s of el , , in M
amic poisson's ratio (μ) a

$$E_d = \frac{\rho(1+\mu)(1-2\mu)}{(1-\mu)} V^2$$

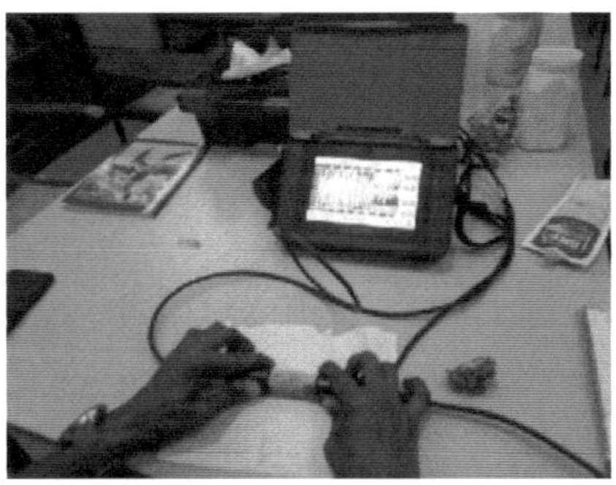

Mix	Activator modulus	7th day Compressive Strength (MPa)	
		Ambient curing	Immersion curing
M1(100% slag)	1.25	55.62	56.03
	1.50	47.63	55.30
	1.75	29.88	49.00
M2 (50% slag & 50% FA)	1.25	21.65	40.703
	1.50	20.36	36.05
	1.75	12.68	23.36
M3 (50% slag & 50% RM)	1.25	22.70	33.19
	1.50	20.30	32.44
	1.75	13.30	27.45
Control (100% OPC)	-	47.78 (immersion curing)	

Mix	Activator modulus	28th day Compressive Strength (MPa)	
		Ambient curing	Immersion curing
M1 (100% slag)	1.25	54.26	46.21
	1.50	40.21	46.79
	1.75	28.47	44.97
M2 (50% slag & 50% FA)	1.25	20.15	46.55
	1.50	18.47	46.39
	1.75	16.04	31.27
M3 (50% slag & 50% RM)	1.25	22.60	37.69
	1.50	19.44	36.89
	1.75	13.19	28.41
Control (100% OPC)	-	62.52 (immersion curing)	

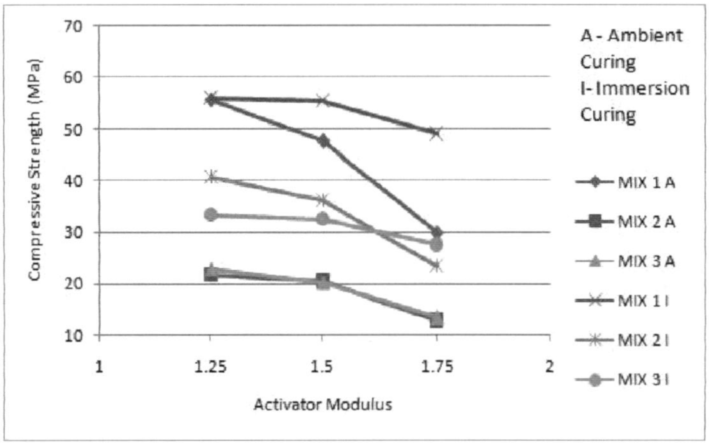

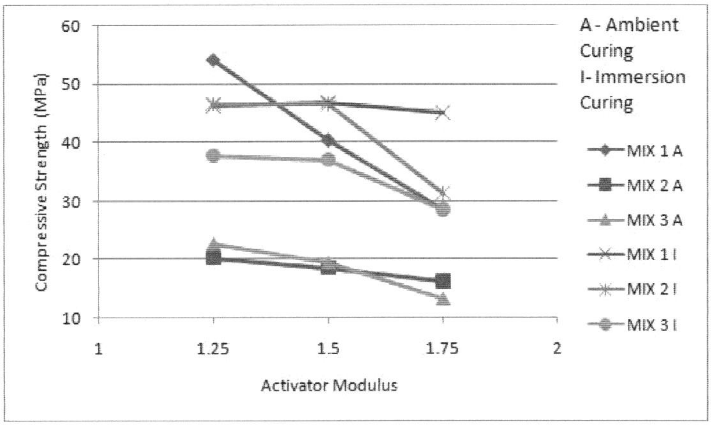
60
50
40
30
20
10
Compressive Strength (MPa)
1
1.25
1.5
1.75
2
Activator Modulus
A - Ambient Curing
I - Immersion Curing
MIX 1 A
MIX 2 A
MIX 3 A
MIX 1 I
MIX 2 I
MIX 3 I

Table 6. Ultrasonic Pulse Velocity Results of 7th day

Mix	Activator modulus	7th day UPV(m/s)	
		Ambient curing	Immersion curing
M1 (100% slag)	1.25	4439.50	4555.00
	1.50	3838.50	4309.50
	1.75	3106.50	4122.50
M2 (50% slag & 50% FA)	1.25	3000.00	4015.62
	1.50	2599.50	3762.50
	1.75	2460.50	4488.00
M3 (50% slag & 50% RM)	1.25	3213.50	4554.00
	1.50	2369.00	4619.50
	1.75	2249.50	4276.50
Control (100% OPC)	-	4975.45	

Table 7. Ultrasonic Pulse Velocity Results of 28th day

Mix	Activator modulus	28th day UPV(m/s)	
		Ambient curing	Immersion curing
M1 (100% slag)	1.25	5186.00	5593.00
	1.50	4019.00	4420.00
	1.75	3681.00	4204.00
M2 (50% slag & 50% FA)	1.25	3335.00	4388.00
	1.50	2913.00	4120.50
	1.75	2908.00	4595.50
M3 (50% slag & 50% RM)	1.25	3472.50	5057.00
	1.50	3027.00	5011.00
	1.75	2816.50	4342.50
Control (100% OPC)	-	6049.38	

Table 8. Density values of 7 day and 28 day cured specimens

Mix	AM	Density(kg/m³) 7th day		Density(kg/m³) 28th day	
		Air Curing	Immersion Curing	Air Curing	Immersion Curing
M1 (100% slag)	1.25	2307.64	2337.28	2399.16	2396.8
	1.50	2301.48	2311.68	2304.84	2310.04
	1.75	2155.20	2180.72	2287.04	2315.92
M2 (50% slag & 50% FA)	1.25	2148.92	2188.28	2186.16	2217.44
	1.50	2155.20	2180.96	2158.88	2191.52
	1.75	2124.52	2149.48	2089.96	2177.44
M3 (50% slag & 50% RM)	1.25	2327.72	2320.04	2308.64	2288.56
	1.50	2291.24	2319.00	2278.52	2287.64
	1.75	2277.80	2297.48	2237.96	2263.64
Control (100% OPC)	-	2336.38		2413.85	

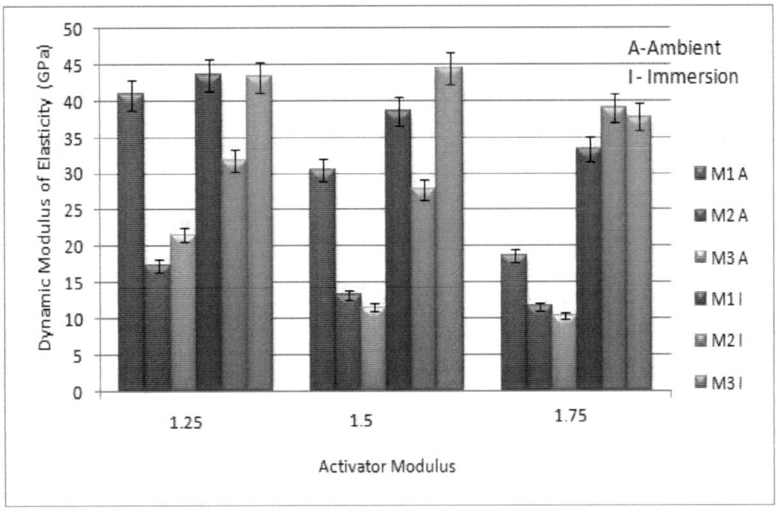
A-Ambient
I - Immersion
Dynamic Modulus of Elasticity (GPa)
50
45
40
35
30
25
20
15
10
5
0
1.25
1.5
1.75
Activator Modulus
M1 A
M2 A
M3 A
M1 I
M2 I
M3 I

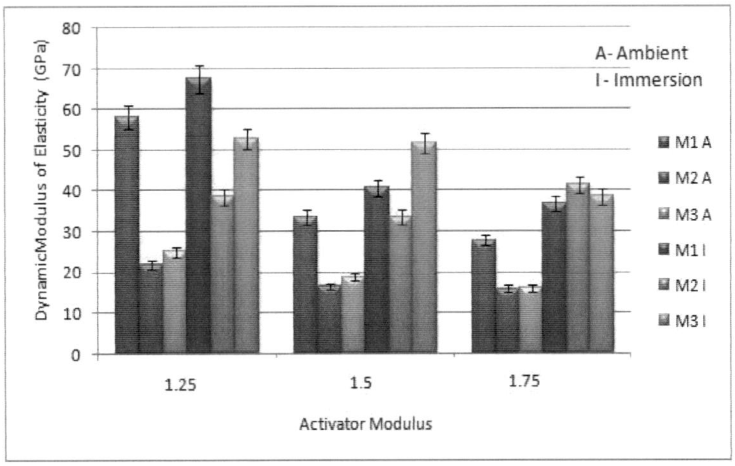
A- Ambient
I - Immersion
DynamicModulus of Elasticity (GPa)
80
70
60
50
40
30
20
10
0
1.25
1.5
1.75
Activator Modulus
M1 A
M2 A
M3 A
M1 I
M2 I
M3 I

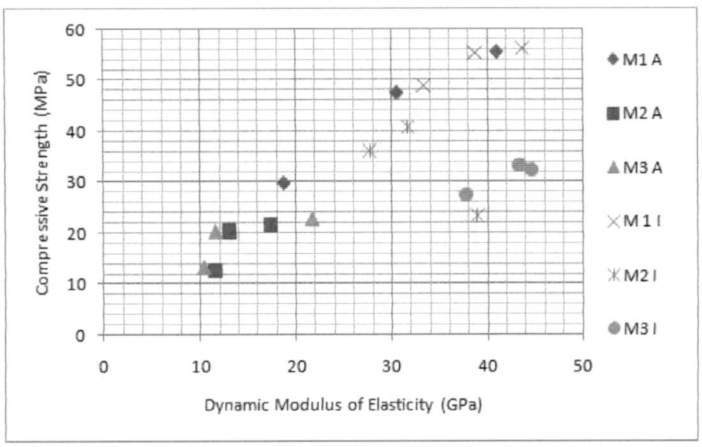
Compressive Strength (MPa)
60
50
40
30
20
10
0
0
10
20
30
40
50
Dynamic Modulus of Elasticity (GPa)
M1 A
M2 A
M3 A
M1 I
M2 I
M3 I

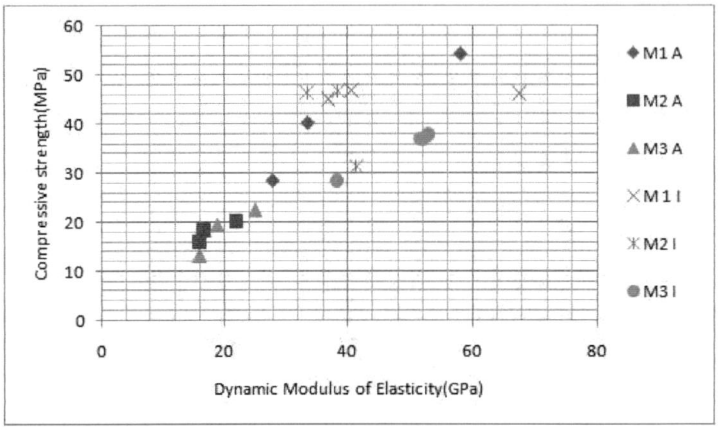
60
50
40
30
20
10
0
0
20
40
60
80
Compressive strength(MPa)
Dynamic Modulus of Elasticity(GPa)
M1 A
M2 A
M3 A
M1 I
M2 I
M3 I

IOP Conf. Series: Materials Science and Engineering **989** (2020) 012009 doi:10.1088/1757-899X/989/1/012009

5. References

[1] Davidovits J 1999 Chemistry of Geopolymeric System Terminology Proceedings of the Second International Conference, Geopolymere "99, *Geopolymer Institute*, Saint-Quentine, France **99** pp 9-93

[2] Murali K, Meena T, Sai P P and Chaitanya S T 2018 An Experimental Study on Factors Influencing the Compressive Strength of Geopolymer Mortar International Journal of Civil Engineering and Technology **9** pp 608- 616

[3] Islam A, Alengaram U J, Jumaat M Z and Bashar I I 2014 The development compressive strength of ground granulated blast furnace slag-palm oil fuel ash-fly ash based geopolymer mortar Materials and Design **56** pp 833-841

[4] Razak S N A and Nuruddin M F 2015 The Effects of Alkali Silica Reaction (ASR) towards FlyAsh Based Geopolymer Concrete Applied Mechanics and Materials **699** pp. 271-276

[5] Ma C K, Awang A Z and Omar W 2018 Structural and material performance of geopolymer concrete: A review Construction and Building Materials **186** pp 90-102

[6] Nuruddin M F, Demie S, Ahmed M F and Shafiq N 2011 Effect of Superplasticizer and NaOH Molarity on Workability, Compressive Strength and Microstructure Properties of Self-Compacting Geopolymer Concrete International Journal of Geological and Environmental Engineering **5** pp 187- 194

[7] CED 53- 14812 : 2019 Draft Indian Standard Guidelines for Use of Geopolymeric Concrete for Precast Products *Bureau of Indian Standards*, New Delhi

[8] IS 13311 Part 1 : 1992 Indian Standard Non-destructive testing of concrete- Methods of Test: Ultrasonic Pulse Velocity *Bureau of Indian Standards,* New Delhi

[9] Provis J L, Palomo A and Shi C 2015 Advances in understanding alkali-activated materials Cement and Concrete Research **78** pp 110-125

IOP Conf. Series: Materials Science and Engineering **989** (2020) 012010 doi:10.1088/1757-899X/989/1/012010

Degradation kinetics of alkali-activated pastes in acetic acid

Athira Ajay[1,2], Ramaswamy K P[1] and Anu V Thomas[1]

[1]Department of Civil Engineering, TKM College of Engineering, Kollam, Kerala, India.

[2]Email: athiajay96@gmail.com

Abstract. Growing urban and industrial activities globally are found to be the sources of rising acidic environment. These acidic environments ensue early degradation of concrete in terms of microstructural alteration of phases formulating to mass changes, weakening of mechanical properties, and increase in porosity due to calcium leaching. This paper probes into the acetic acid resistance of geopolymer composites (GP) and its comparison with conventional concrete system (OPC) in 0.125M and 0.25M concentrations of acetic acid medium. The GP samples displayed increased resistance to acetic acid attack in terms of mass and thickness loss. It was also found that blends of ground granulated blast furnace slag (GGBS) and fly ash (class F) (FA) with red mud (RM) offered increased acid resistance due to the nature of the main binder gel produced in GP systems.

1. Introduction

The never-ending hunger of the people for industrial products has directly or indirectly affected the durability aspect of concrete structures by exposing the latter to a vulnerable aggressive environment. Most industries worldwide are built upon processes that produce effluents or waste water. When untreated, the existence of complex acidic compositions within these effluents have led to severe degradation of structures involved in processes such as production, collection, storage and treatment. For example, about 40 billion litres of spent wash are annually generated in India with regard to the production from sugar mills and molasses-based distilleries [1, 2]. Spent wash is acidic in nature and contains a mixture of organic and inorganic acids. The direct disposal of untreated spent wash is considered hazardous as it arrests the growth of vegetation and damages the quality of ground water. Liquid manure, silage effluents, winery wastewaters from the distillery and fermentation industries contain predominant organic acids like acetic, lactic, citric, butyric and tartaric acids where the pH is around 4.5 generally [3]. Table 1 represents the predominant sources of organic acids that could endanger the service life of any reinforced concrete structures.

The acids present in silage effluents cause deterioration of concrete silos. Due to the increased horizontal pressure within the silos, larger silos are more prone to acid deterioration than smaller silos [4]. On placing moist plant material in a silo over a long period, the process releases silage acid mainly lactic and acetic. When the concrete silo walls come in contact with these silage acids, it causes a reduction in the strength, which leads to the buckling of walls and eventually causing failure. Most of the researches were carried out on the degradation of concrete caused by the presence of inorganic acids. Only limited studies are reported on the effect of organic acids. Regardless of organic acids being weak acids due to its partial dissociative nature, it could destruct the concrete matrix which is alkaline. These acids accelerate the corrosion of embedded reinforcements and disturb the production facilities.

Table 1. Predominant acids in agro-food industries [5].

Industry	Waste waters	Acids	Concentration (mmol/L)	pH
Wine industry	Vinasses	Tartaric	3.2–8.73	4-6
		Acetic	3.1–5.0	
		Lactic	2.2–5.6	
		Succinic	0.3–6.8	
		Malic	0.5–0.8	
		Total (max.)	26.9	
Sugar industries	Sugar cane vinasses	Lactic	18–80.4	4-7
		Glycolic	12–25	
		Citric	1.7–10.4	
		Trans-Aconitic	0.2–4.8	
		Cis-Aconitic	0.1–2.8	
		Oxalic	0.08–0.8	
		Fumaric	0.1–0.4	
		Total (max.)	124.2	
Dairy industries	Whey	Acetic	14.0	4–6
		Citric	9.2–9.6	
		Lactic	3.4–5.7	
		Propionic	1.0	
		Butyric	0.6	
		Total (max.)	30.9	
Animal rearing	Silage juice	Lactic	55.3–63.7	4–5
		Acetic	25.3–28.3	
		Butyric	43.4	
		Propionic	3.1	
		Total (max.)	99.1	
	Liquid manure	Acetic,	31.7–213	5-8
		Butyric	1.1–28.8	
		Propionic	5.4–37.8	
		Total (max.)	279.6	

Concrete with Ordinary Portland Cement (OPC) is reported to be less resistant in acidic media. Moreover, the use of OPC is discouraged nowadays as its production contributes alarmingly to carbon dioxide emissions and thus global warming. Contribution of CO_2 emissions worldwide by cement industry is approximated to be around 7% [6]. As we strive towards achieving sustainable development, the complete replacement of cement with alkali activated/ geopolymer composites can be considered as a viable alternative for cementitious materials exposed to aggressive acidic environments. Also, it has been estimated that 66 million tons of red mud is produced annually across the world and it is considered to be hazardous pollutant according to the Brazilian NBR 10004 standards [7, 8]. Therefore, the incorporation of such waste materials as an alternative binder material in alkali activated systems could lead to reduction of pollution level across the globe. GP system is an innovative technology that utilizes industrial solid wastes such as class F-fly ash (FA), ground granulated blast furnace slag (GGBS) and red mud (RM) as raw material to form a 3D network of cementitious materials under alkaline conditions through polymerization [9]. Individually, alkali activated GGBS and alkali activated FA has been studied by many authors. The organic acid resistance of these two alkali activated binder systems has not received much attention. Therefore, this paper investigates the resistance of GP materials in acetic acid solution by studying the degradation kinetics.

2. Mechanism of organic acid attack

Concrete, particularly the cement paste matrix is highly vulnerable under acid attack. The basic described mechanism of deterioration due to acid attack is acidolysis. In this process, the hydration products developed reacts with the incoming acid thereby releasing metal ions from the hydrates into the solution forming salts of varying characteristics and ultimately reducing the mass and integrity of the matrix. In the case of organic acid attack, complexolysis and precipitation of reaction products mechanisms are also involved. Complexolysis is a process where the metal ions extracted in the initial acidolysis phase reacts with acid ions and formulates complexes thereby dissolving the solid phase in the solution. The latter mechanism (i.e. precipitation) becomes significant based on the character of the salt formed i.e. when the salt formed has a higher molar volume than the originally formed hydration products, expansion and cracking of the matrix seems to accelerate then [2,10]. Table 2 represents the solubility data of salts formed as a result of the reaction of hardened cement paste with an organic acid. The belligerent nature of acids into the cementitious matrix depends on the acid dissociation constant, their solubility in water, mono/poly acidity of acids and the physical properties of salts they form.

Table 2. Calcium and aluminium salts solubility data of various organic acids [2, 10, 11].

Acid	Acid dissociation constant (25°C)	Aggressiveness	Major Calcium salts	Solubility (g/L)
Acetic	4.76	Moderate	Calcium acetate monohydrate {$Ca(CH_3COO)_2 \cdot H_2O$}	436 (0°C)/ Highly soluble
Citric	3.14/4.77/6.3	Very high	Calcium citrate tetrahydrate {$Ca_3(C_6H_5O_7)_2 .4H_2O$}	0.095 (25°C) / Slightly soluble
Tartaric	3.04/4.37	Moderately low	Calcium tartrate tetrahydrate {$CaC_4H_4O_6.4H_2O$}	Insoluble
Lactic	3.86	Moderate	Calcium lactate pentahydrate	0.97 (25°C) / Soluble
Succinic	4.16/5.61	High	Calcium succinate trihydrate	0.193 (0°C) / Slightly soluble
Oxalic	1.23/4.19	Very Low	Calcium oxalate monohydrate {$CaC_2O_4/CaC_2O. H_2O$}	Insoluble
Butyric	4.82	Moderate	Calcium butyrate monohydrate {$Ca(CH_3CH_2CH_2CO_2)2.H_2O$}	0.04 (25°C) / Slightly soluble

From the literatures, it is seen that acetic acid is an important organic acid predominantly present in waste waters. They are popularly known as weak acids but seem to be very aggressive when it comes into contact with cementitious systems. Acetic acid is a monocarboxylic acid that initiates the dissolution of calcium hydroxide present in the concrete and produces highly soluble calcium salts [11]. Since these calcium salts are very soluble, they are leached away exposing the fresh concrete surface to acid attack. Acetic acid also counters with the other hydrates like C-S-H gel and hydrated aluminates within the cement paste to form water soluble calcium and aluminium salts, enhancing the porosity and leads to reduction in the mechanical strength. As the salts formed are highly soluble, there are no protective cover formed on to the surface of the specimen to prevent further ingression of acid into the matrix and this quickens the degradation process. This reinstates the need to investigate the effect of acetic acid further. Moreover, as hydrates in conventional cementitious systems offer less resistance, the acid resistance of geopolymer (GP) composites is explored in the current study.

GP are inorganic polymers that are recognized as an alternative for conventional OPC system with regard to reducing CO_2 emissions. They are generally considered as a subset of alkali activated binder system and are amorphous in nature rather than crystalline. In alkali activated binder systems, alkali activation of a solid aluminosilicate produces a hardened binder which is based on a combination of hydrous alkali-aluminosilicate and/or alkali-alkali earth-aluminosilicate phases [12, 13]. On the other hand, Geopolymerization is the reaction taking place within the GP systems. It is an exothermic reaction between silica (Si)-alumina (Al) under alkaline condition thereby subsequently creating three-dimensional polymeric chain of Si-O-Al-O bonds. The structural integrity of GP systems relies on alumina-silicate rather than calcium silicate hydrate bond in traditional OPC based systems [14, 15]. Sodium aluminosilicate hydrate (N-A-S-H) (FA based system) or calcium aluminate hydrate silicate (C-A-S-H) (GGBS based system) or a combination sodium calcium aluminium silicate hydrate (N-C-A-S-H) gel are the main reaction products of the alkali-activated aluminosilicates [16]. When there is a disequilibrium in a geopolymer - acid environment, the ion exchange reaction proceeds in between the cations and hydronium ions and are then consecutively followed by an electrophilic attack of acid protons on Si-O-Al-O bonds. The sodium ions on exposure to acid, act as charge compensation cations for the aluminosilicate framework, are leached out and exchanged by hydronium ions from the solution. Also, the atomic ratio of Si-to-Al peaks up, designating the dissolvement of tetrahedral Al from the aluminosilicate framework by breaking the Si-O-Al bond [14, 16, 17]. Again, with a decrease in CaO content, the acid resistance of the binder gel is found to increase i.e. C-S-H < C-A-S-H < N-A-S-H [18]. Acid attack itself is still an infant area of research. Moreover, the literature investigations highlighting the post effects of organic acids such as acetic acid on geopolymer systems are rather limited and hence these aspects are explored in the findings reported in this paper.

3. Experimental program

3.1. Materials

The precursors utilized in the current investigation include Class F fly ash (FA), Slag (GGBS) and red mud (RM). Ordinary Portland Cement (OPC) 53 Grade is taken to be the control mix. RM was procured as wet sludge from HINDALCO, Belgaum. It was then oven dried at 100^0C for 24 hours. Later the oven dried RM was put into Los-Angeles abrasion machine for 1000 revolutions to attain the required fineness. Alkaline activators, sodium hydroxide (10M) and sodium silicate solution were adopted with Alkali silicate/Alkali hydroxide ratio as 2.5. The admixture utilized is Sulphonated Napthalene Formaldehyde (SNF) superplasticizer with 3% by weight of binders for the geopolymer mixes. Paste sample size adopted is cylindrical having 25 mm in diameter and 50 mm in height. The control mix was prepared using distilled water with water to cement ratio of 0.50. GP mixes were prepared with alkali/binder ratio of 0.50.

3.2. Mixing, casting, curing and testing procedures

The mixes adopted are OPC, FGP (100% FA), SGP (100% GGBS), FSGP (50% GGBS + 50% FA), RFGP (50% FA+ 50% RM) and RSGP (50% GGBS + 50% RM). The mixing was done using Hobart mixer. The plastic moulds were oiled. Filling of the paste into the moulds were done manually with tamping. After casting, specimens were kept for 24 hours in atmospheric conditions for initial set before demoulding. For OPC mix, lime solution curing was done for 14 days to maintain the alkalinity and to ensure that no degradation reaction takes place with an alteration in the pH level. A 14 days air (ambient) curing was done for SGP, FSGP and RSGP specimens. FGP and RFGP specimens were oven cured at 70^0C for 24 hours and then air cured for 14 days. No chemical admixture was used for the preparation of OPC mix investigated in the current study. To investigate the resistance of GP and OPC based systems to acetic acid attack, paste samples prepared are submerged completely in acetic acid solution of 0.125M and 0.25M concentrations for 8 weeks. The samples were visually monitored and their mass change, thickness change and pH variation of the solution was also recorded periodically. The acid solution was replaced once in two weeks to maintain the aggressiveness of the solution and to sustain the attack.

4. Results and discussions

4.1. Visual appearance

The extent of deterioration can be visually explored by taking digital photographs on a regular basis. Figure 1 showcases the photographs of specimens after 8 weeks of exposure in various concentrations of acetic acid solution. When specimens were closely observed, there is no formation or deposition of insoluble precipitate/products on the surface of the specimen. On comparing the intensity of degradation with the concentrations of acids for all the binders, more damage was observed in the case of 0.25M acetic acid than 0.125M acid solution.

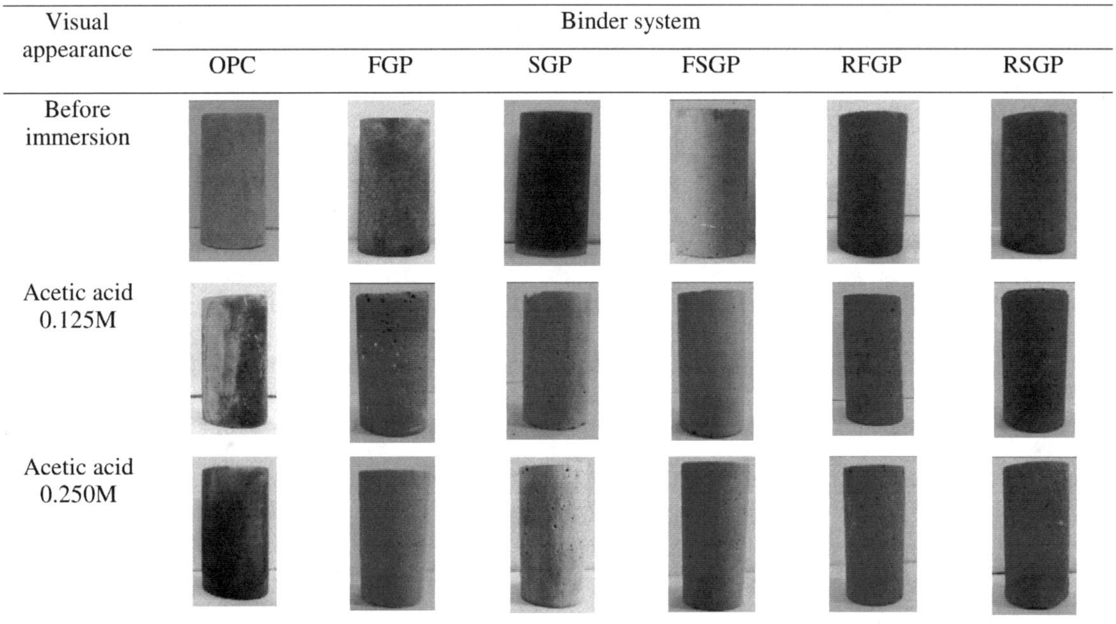

Figure 1. Visual observation of specimens immersed in various concentrations of acetic acid after 8 weeks of exposure.

This behaviour may be attributed to the fact that aggressiveness increases due to the increase in concentration. Again, on comparing the various mixes, extreme colour variation is found to be more predominant in OPC specimen on comparison with other mixes. This may be due to the reaction of acetic acid with Portlandite and C-S-H gel to form water soluble calcium acetate monohydrate which leaches to the solution leaving amorphous silica gel on the specimen, thereby causing disintegration of the internal microstructure and brown colour formation on the specimen. Visual examination of RSGP and RFGP specimens is difficult because these are originally rich red in colour even before the onset of exposure. Again, FGP and FSGP specimens did not depict extreme colour variations even after 8 weeks exposure.

4.2 Mass changes

On a weekly basis, mass changes (in saturated but surface dry condition) are calculated with respect to the initial saturated mass (noted before the acid exposure). The average of the mass changes of three specimens were calculated. The change in mass of the specimens with age of exposure in acid solution is shown in Figure 2. From the graph, it can be seen that there is a continuous decrease in the mass of all specimen in acetic acid medium irrespective of the medium concentrations. The continuous mass loss may be due to leaching of soluble salts formed as the reaction product. The mass loss prove that the intensity of degradation was high, deteriorating the inner layers by penetration of acid through the pores of the specimens.

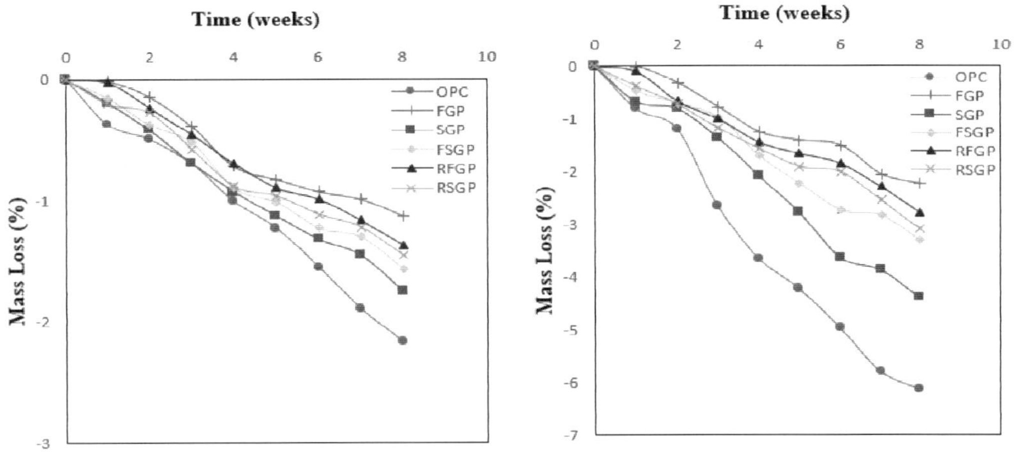

(a) 0.125M (b) 0.25M

Figure 2. Mass change of mixes in various concentrations of acetic acid solution.

The decrement in mass is due to the formation of water-soluble calcium acetate which leaches to outside solution, leading to increase in the inherent porosity within the specimen and thus leading to mass loss. OPC specimens appears to have the highest mass loss in both acid medium with highest loss of about 2.17% in 0.125M and 6.12% in 0.25M acetic acid medium. Clearly RFGP mix showed improved resistance than RSGP and FSGP mix too. This is due to the presence of enhanced N-A-S-H gel and low calcium content within the RFGP mix. GP systems exhibited lower mass loss than OPC mix in acetic acid medium for both concentrations. This is as a result of the difference in the chemistry of GP system, as there are no less stable hydrates such as Portlandite and CSH gel in GP systems. Mass loss has a profound increase with regard to higher concentrations of acid medium i.e. from 0.125M to 0.25M. This is due to the increased concentration gradient that exists between the acid solution and the pore solution of the matrix causing higher rate of diffusion of acid ions into the matrix at higher concentrations of acid.

4.3. Thickness changes

The diameter changes were recorded with a digital Vernier caliper to understand the kinetics of degradation. Readings were taken at every 7 days interval. Figure 3 shows the thickness loss of mixes with respect to time in 0.125M and 0.25M acetic acid solution.

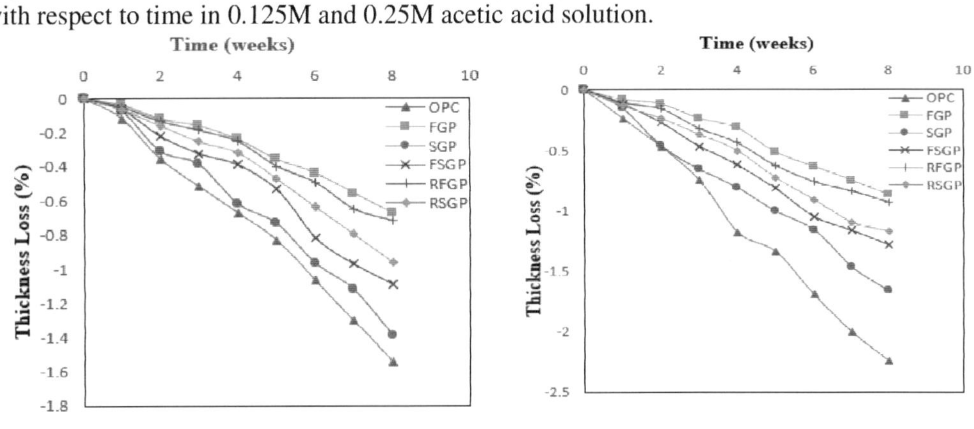

(a) 0.125M (b) 0.25M

Figure 3. Thickness loss of mixes in various concentration of acetic acid solution.

There is a continuous decrease in the diameter of all specimens in acetic acid medium irrespective of the medium concentrations. This coincides with the mass loss findings. OPC specimens seems to have the highest thickness loss in both acid medium with highest loss of about 1.54% in 0.125M and 2.24% in 0.25M acetic acid medium. Furthermore, FGP mixes have lower loss in both acid media i.e. 0.68% in 0.125M and 0.87% in 0.25M acetic acid medium. Among the GP mixes SGP is found to have the highest loss due to the presence of larger calcium content in its C-A-S-H gel unlike the other GP systems that consist of N-A-S-H or N-C-A-S-H gel that are found to be more stable in lower pH environment.

4.4. Changes in pH

Measuring pH of the solution gives an idea about the hydrogen ion consumption of a particular binder system. The reaction of acid solution with various phases of cementitious system tends to release Ca^{2+} and OH^- ions to the solution, thereby increasing the pH of the acid solution with age. Here, the initial pH of 0.125M and 0.25M acetic acid medium are 3.26 and 2.72 respectively. The cyclic plots depicting increase in pH after every 14 days of exposure each time back to pH values of fresh acid solution are shown in Figure 4. There is a rapid increase in pH for the first week on contrast to the initial pH for all the acid medium due to the rapid neutralization process. The pH increase of acid medium of OPC mix is found to be higher on comparison with the GP mixes. This is due to leaching of calcium from the hydrates and subsequent release of hydroxyl ions to the solution. The pH rise was found to be lowest for FGP mix and this is in alignment with the mass loss measurements. As concentration of acid increases, the kinetics of degradation was found to increase in a rapid rate.

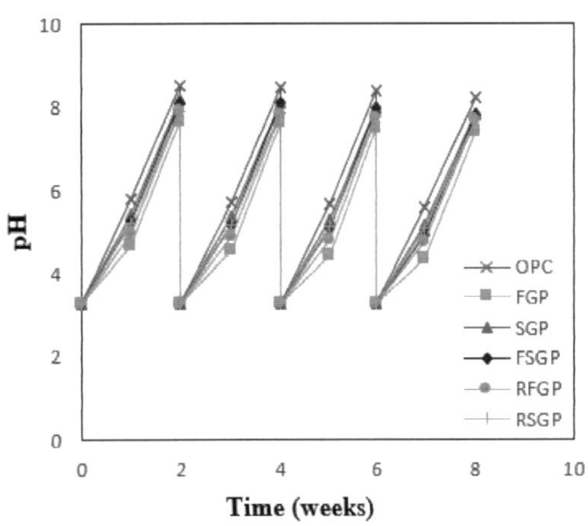

(a) 0.125M

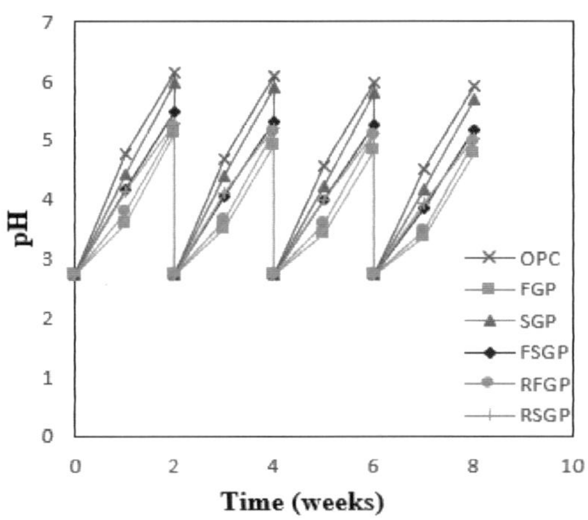

(b) 0.25M

Figure 4. Changes in pH of acetic acid solution of different concentrations with age of exposure.

5. Conclusions

The increased proliferation of industries has led to the large-scale generation of untreated effluents being acidic in nature. This has led to severe degradation of concrete structures made with OPC on contact with the acidic solutions and this demands a study on the acid resistance of alternative materials and composites. This paper has presented the results and discussion of an accelerated leaching study carried out to investigate and compare the resistance of GP and OPC paste systems to silage effluent attack caused primarily by acetic acid for an exposure period of 8 weeks.

OPC paste specimens appears to have the highest mass loss (6.12% in 0.25M) and thickness loss (2.24% in 0.25M) even in higher concentration of acetic acid medium. RFGP mix manifested improved acetic acid resistance than RSGP and FSGP mix. Accordingly, red mud (RM) material can be adopted as a binder or partial replacement for OPC as it promises both durability and sustainability from the preliminary results. The leaching of soluble salts formed as the reaction product are held accountable for the continuous mass loss and thickness loss in all the mixes. FGP delineated enhanced performance on resisting acetic acid attack. Thereby, from the overall assessment GP paste samples have increased acetic acid resistance when compared with OPC paste samples. Nevertheless, evaluating degradation kinetics based on only mass change/thickness change parameter are not admissible. Microstructure studies are recommended to explore the actual degradation mechanisms to understand the phenomena better.

The kinetics of degradation was more profound in the case of higher concentration of acids and the long-term effects can be observed from it. The degradation rate seems to rely on the concentration of acid, chemical composition of binders, solubility and characteristics of the salts formed. Again, this study is recommended to be carried on mortar and concrete specimens for a superior understanding of the real-life scenario. Evolution of test methods, studying the mechanism of acid attack, modelling of kinetics and examining the out turn of mixture of acids are some of the further research works that needs to be delved for the broader understanding of acid attack.

6. References

[1] Ramaswamy K P and Santhanam M 2017 Durability of cementitious materials in acidic environments: Evaluation of degradation kinetics *14th International Conference on Durability of Building Materials and Components (DBMC)*, University of Ghent, Belgium, 1–14

[2] Ramaswamy K P, Bertron A and Santhanam M 2017 Additional insights on the influencing factors and mechanism of degradation due to acid attack: Special case of acids forming soluble salts Proceedings of International Conference on Advances in Construction Materials and S ystems (ICACMS), Chennai **4** 279–290

[3] Bertron A, Alexander M G and De Belie N 2013 Performance of Cement-Based Materials in Aggressive Aqueous Environments *State-of-the-Art Report* RILEM TC 211 – PAE 131-173

[4] Dogangun A, Karaca Z, Durmus A and S ezen H 2009 Cause of Damage and Failures in Silo Structures *Journal of Performance of Constructed Facilities* **23(2)** 65-71

[5] Larreur-Cayol S, Bertron A and Escadeillas G 2011 Degradation of cement-based materials by various organic acids in agroindustrial waste-waters *Cement and Concrete Research* **41** 882-892

[6] Pacheco-Torgal F, Abdollahnejad Z, Camoes A F, Jamshidi M and Ding Y 2012 Durability of alkali-activated binders: a clear advantage over Portland cement or an unproven issue? *Constr. Build. Mater.* **30** 400–405

[7] Arhina D D, Nuamaha R A, Tuffoura B A, Obadac D O and Awaso A Y 2017 Bauxite red mud-Cement based composites: Characterisation for pavement applications *Case Studies in Construction Materials* **7** 45–55

[8] Deb P S, Nath P, Sarker P K 2014 The effects of ground granulated blast-furnace slag blending with fly ash and activator content on the workability and strength properties of geopolymer concrete cured at ambient temperature *Mater. Des.* **62**, 32–39

[9] Li Y, Min X, Ke Y, Liu D and Tang C 2019 Preparation of red mud-based geopolymer materials from MSWI fly ash and red mud by mechanical activation *Waste Management* **83** 202–208

[10] Dyer T 2016 Influence of cement type on resistance to organic acids *Magazine of Concrete Research* **69(4)** 175-200

[11] Bertron A, Larreur-Cayol S and Escadeillas G 2009 Degradation of cementitious materials by some organic acids found in agro industrial effluents *Proceedings of the RILEM TC 211-PAE Final Conference* Toulouse France **1** 96-107

[12] Provis J L 2018 Alkali-activated materials *Cement and Concrete Research* **114** 40-48

[13] Provis J L and Deventer S J V 2009 Geopolymers Structure, processing, properties and industrial applications *Woodhead Publishing Limited and CRC Press LLC.*

[14] Stephan D, Firdous R, Noël J and Djobo Y 2018 Natural pozzolan based geopolymers: A review on mechanical, microstructural and durability characteristics *Construction and Building Materials* **190** 1251-1263

[15] Lodeiro G, Palomo A, Fernández-Jiménez A and Macphee D E 2011 Compatibility studies between N-A-S-H and C-A-S-H gels. Study in the ternary diagram Na_2O–CaO–Al_2O_3–SiO_2–H_2O *Cement and Concrete Research* **41** 923–931

[16] Aiken T A, Sha W, Kwasny J and Soutsos M N 2017 Resistance of geopolymer and Portland cement-based systems to silage effluent attack *Cement and Concrete Research* **92** 52-65

[17] Gu L, Bennett T and Visintin P 2018 Evaluation of accelerated degradation test methods for cementitious composites subject to sulphuric acid attack; application to conventional and alkali activated concretes *Cement and Concrete Composites* **87** 187-204

[18] Gu L, Bennett T and Visintin P 2019 Sulphuric acid exposure of conventional concrete and alkali activated concrete: Assessment of test methodologies *Construction and Building Materials* **197** 681-692

IOP Conf. Series: Materials Science and Engineering **989** (2020) 012011 doi:10.1088/1757-899X/989/1/012011

Comparison of the energy consumption in the production of natural and recycled concrete aggregate: A case study in Kerala, India

Alan V Ittyeipe[1,2], Anu V Thomas[1] and Ramaswamy K P[1]

[1]Department of Civil Engineering, TKM College of Engineering, Kollam, Kerala, India

[2]E mail: alanittyeipe@gmail.com

Abstract. The growth of construction sector in India is gaining pace to cater to the overwhelming demands of the rising urban population and this has in turn resulted in overexploitation of natural resources for production of natural aggregates (NA). The prospects of recycling demolition debris to produce recycled concrete aggregate (RCA) is little explored, due to the underestimation of actual data on the construction and demolition (C&D) waste generated in India. The different stages involved in the production of NA and RCA were investigated by field study in an aggregate quarry and a demolition site in Kerala, India. The energy consumption involved in the recycling of the demolition debris from a reinforced concrete (RC) structure and the production of NA was determined. The energy consumed to produce 1 tonne of NA and RCA was determined as 21112 KJ/tonne and 16178 KJ/tonne respectively. The production of RCA showed a 30.5% savings in the energy consumption compared to that of NA. A specific sensitivity analysis suggested that the energy consumed to produce RCA was less than that to produce NA for a maximum distance of 13.5 km. As the energy consumption was evaluated based on the diesel consumption of the machinery involved, the reduced diesel consumption and its subsequent cost savings could also add to the prospects of using RCA. Although the results are site specific, such local case studies could contribute to the development of C&D waste recycling infrastructure in the country.

1. Introduction

The rising population and rapid urbanisation has significantly contributed to the growth of construction market in India. The urban population in India as per the Census 2011 is estimated to be 31.16% [1] and is likely to become 50% of the total population by 2050 [2]. In order to accommodate the increasing urban population, government initiatives such as Pradhan Mantri Awas Yojana (Urban) have been introduced to ensure housing for everyone by 2022. However, 7.07 million houses are yet to be constructed under this project [3]. Thus, India is set to become one among the flag-bearers of the global growth of construction market by 2030, thereby meeting the demands of its overwhelming urban population [4].

The expansion of construction sector has been witnessed by huge increase in the number of reinforced concrete (RC) structures and a steep increase in the demolition of existing buildings to pave way for newer construction. This had further strained the natural resources for the extraction of raw materials such as aggregates, as well as, stressed the importance of construction and demolition (C&D) waste management. The coarse and fine aggregates are indispensable in the construction of RC structures as it accounts for 60-75% of the total volume of concrete. The present annual consumption of sand and crushed stone in India is estimated to be about 1.4 billion tonnes and 2 billion tonnes respectively [5]. Also, river sand is now being replaced by manufactured sand (M sand), further aggravating the pressure on quarries. The indiscriminate mining to meet the overwhelming demands has led to exhaustion of natural aggregates and stress the need for secondary materials to replace it. Demolition of buildings holds the major contribution in the generation of C&D waste. As per the government figures, the annual C&D waste generation in India is about 12-15 million tonnes [6]. However, many researchers have reported that there is gross underestimation of the actual C&D waste generated in India. Ram and Kalidindi [7] estimated that the C&D waste generated was about 36% of the municipal solid waste in the Chennai city for the year 2013 and found that it was comparable to the C&D waste generation in other developed countries/regions. Jain et al. [8] estimated that the C&D waste generation in India was between 112 and 431 million tonnes for the year 2016. Figure 1 shows the composition of C&D waste in India, reported in different studies [6] [9] [10]. Ram et al. [10] reported the composition of C&D debris that gets dumped in two landfills in Chennai where materials such as metal and wood are excluded due to their recyclability. It can be observed that the concrete and masonry waste constitute 53% to 60% of the total C&D waste composition in India and are often directly landfilled. Thus, there exists a huge potential in the recycling of C&D waste in India, which is little explored. Recycling of C&D waste by introducing them in new concrete as recycled aggregates has become an effective solution to compensate for the depletion and exploitation of natural resources, as well as efficient waste management system for C&D waste. Provisions are introduced in IS 383: 2016 for the replacement of 20% natural coarse and fine aggregates with recycled concrete aggregates (RCA) in the reinforced concrete [11]. Thus, the introduction of RCA into the supply system can result in a sustainable construction practice.

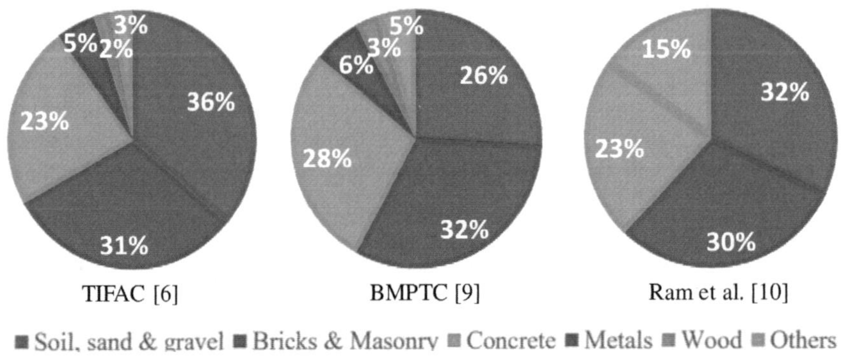

Figure 1. Composition of C&D waste.

However, the prospects of C&D waste recycling need to be assessed and quantified. Several studies have reported the environmental benefits of recycling C&D waste in comparison to landfilling by carrying out Life Cycle Assessment (LCA) [12] [13]. Blengini [14] reported that the recycling of steel could significantly contribute to the environmental benefits obtained in recycling demolition waste. In most of the studies, the environmental benefits obtained were mainly dominated by the recycling of materials such as steel and wood. Hence the environmental benefits of C&D debris alone, which forms the quantum of C&D waste in India, needs to be evaluated. Ram et al. [10] reported that recycling of C&D debris showed environmental benefits in all 15 impact categories compared to landfilling. Comparative LCA studies on the production of recycled aggregates (RA) and natural aggregates (NA) have also been carried out by the authors. Estanqueiro et al. [15] performed a comparative LCA of NA and coarse recycled aggregate (RA) used in concrete production either using a fixed or mobile plant. The environmental benefits of coarse RA when compared to NA were observed only in the land use and respiratory inorganics impact categories and it was suggested that the benefits can be improved by using fine RA also in the concrete production, instead of landfilling. Rosado et al. [16] performed a life cycle assessment of NA and mixed RA production in Brazil. The mixed recycled aggregate was a better option for all the environmental impact categories if the transportation distance of RA was 20 km longer than that of NA. Significant number of LCA studies on C&D waste, were from European countries whereas LCA studies based on Indian scenario is very limited. A cost reduction of 19-22% was estimated for the paving blocks made with C&D waste from Ahmedabad and Bangalore, than NA [17]. Thus, the environmental and economic aspects involved in the production of both natural and recycled concrete aggregate needs to be quantified and compared with the primary material savings and C&D waste management.

2. Research significance

Kerala is one of the densely populated states in India with a population density of 859 per sq. km [18]. Ernakulam district has the highest urban population in the state [19]. It has been reported that Ernakulam has the highest number of dwelling houses constructed in the urban sector of the state, for the year 2016-17 [20]. Ernakulam also tops in the number of active quarries in the state [21]. This gives a clear evidence on the dependence of the increasing construction activities on the natural aggregate quarries, to cater to the demands of the urban population. Ernakulam has witnessed the demolition of the 4 multistoried framed RC structures in the Maradu municipality in January,2020. The RC structures were demolished before the end of their service life, under the court order for violation of Coastal Regulation Zone. The debris generated from the demolition was to be recycled and reused, emphasizing the importance of C&D waste management. The Ministry of Urban Development had instructed all states to set up environment friendly C&D waste recycling facilities in all cities/towns with population of over 1 million. But, India with a population over a billion has only 5-6 operational recycling plants. Although, the Construction and Demolition Waste Management Rules was introduced in 2016 to address the roles and responsibilities of waste generators and various stakeholders in C&D waste management, its implementation remains in infancy [22]. Thus, local case studies on the strategy adopted in recycling of demolition debris could further accelerate the development of C&D waste recycling infrastructure in India. Thus, a study on the production of RCA is carried out with respect to the recycling strategy adopted on the demolition waste generated in Maradu. The present case study investigates the various stages and compares the energy consumption involved in the production of 1 tonne natural and recycled concrete aggregates.

3. Methodology

The nearest aggregate quarries to the Maradu demolition site were identified and were approached to conduct a field study. However, only one active aggregate quarry, located in the Mazhuvannur village, had granted permission to carry out the field study. The quarry engaged in the production of aggregates by extraction and crushing of granite rocks was visited. Among the 4 multistoried structures (Alfa Serene (A), Holy Faith H2O (H), Jain Coral Cove (J) and Golden Kayaloram (G)) demolished in Maradu municipality, the demolition sites of (A) and (H) were visited. Upon visual inspection, it was observed the debris in (A) was mostly of concrete chunks and bricks, whereas the debris in (H) was mostly of concrete chunks and solid blocks.

Discussions were carried out with the officials of the contracting firm which had been awarded the tender for the removal of debris and was informed that the interior walls of (A) and (G) were built using bricks, which makes segregation and downsizing of the mixed brick and concrete debris difficult. Whereas, the interior walls of (H) and (J) were built using solid blocks, which makes the debris more of concrete waste. Thus, the segregation of steel and concrete debris and subsequent downsizing becomes easy. Hence, the recycling of debris from demolition site of (H) was considered for the study. Figure 2 shows the location and photograph of the NA quarry and demolition site (H).

Figure 2. Location and photograph of the natural aggregate quarry and demolition site (H).

3.1. *Natural aggregate production*

The quarry has its own crushing facility within its area. The average production of NA from the quarry is estimated to be 1500 tonnes per day. Figure 3 shows the process flow in the production of NA.

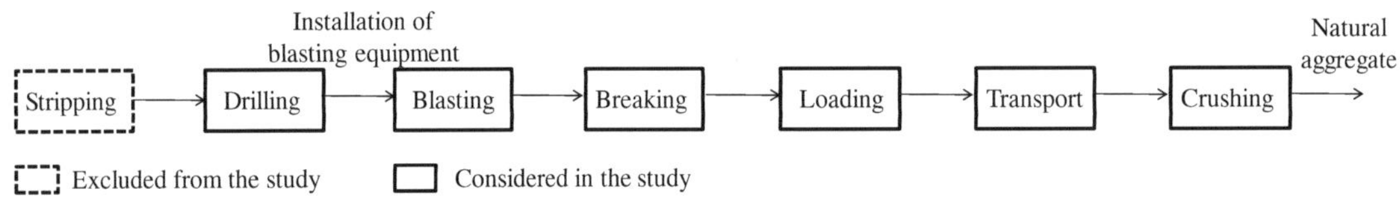

Figure 3. Process flow of natural aggregate production.

The overburden on the top of the quarry was stripped to expose the underlying ore body. Drilling involves the creation of holes in the ore body by using jack hammers, with the help of a compressor connected to the tractor providing sufficient pressure for its operation. Figure 4 shows the drilling process carried out using jack hammer. Drilling is done prior to blasting for the placement of the explosives. Figure 5 shows the various blasting equipment used. As per the drilling-blasting design adopted, gelatine sticks (see figure 5 (a)) were placed in each hole and were connected to a non electric detonator (see figure 5(c)). The holes were then filled with rock powder for one- third of its depth, known as stemming. The non electric detonators from each hole were connected in series and then connected to an electric detonator (see figure 5(b)). The electric detonator was connected to the initiation system (see figure 5 (d)) which initiates the blast. Breaking involves the size reduction of the rock boulders produced from blasting by using hydraulic breakers. Also, rock surface after blasting, was again fragmented using a hydraulic breaker so as to remove remaining loose rocks and create a hard surface for next drilling and blasting to be carried out. This process involves additional production of rock fragments. The fragmented rock was loaded into the trucks by an excavator. The truck carries the fragmented rock from the extraction site to the crusher unit located within the quarry. Figure 6 shows the process of breaking, loading, haulage and unloading at the crusher unit respectively. The size of the rocks obtained from the quarry may vary from 350 mm boulders to fine particles. Once the rocks are brought to the crusher unit, they are fed into the dump hopper. Aggregates of the required sizes of 40 mm, 20 mm, 12mm, 6 mm, 4.75mm and less than 3mm are obtained using vibrator screens of required size. Fine particles of both 4.75mm and less than 3mm are subsequently carried by the belt conveyors for washing and classifying. Fine aggregates of 4.75mm are used as M sand and those less than 4.75 mm are used as P sand.

Figure 4. Drilling.

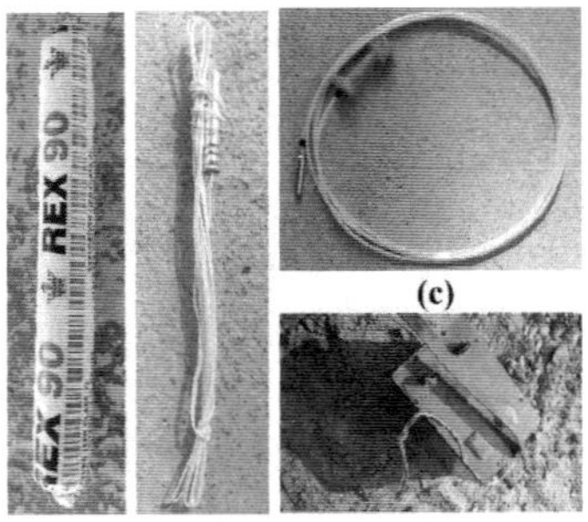

Figure 5. Blasting equipment.

Figure 6. (a) Breaking (b) Loading (c) Hauling and unloading.

3.2. Recycled concrete aggregate production
The demolition waste from the site (H) includes large sized concrete chunks, solid block and steel. Figure 7 shows the process flow in the production of RCA. The separation of the steel and concrete from the structural elements of the demolished RC building (H) was done using hydraulic breakers. It was used to break the concrete adhering to the steel reinforcements into smaller fractions, thereby separating out the steel. Thus, on site segregation was carried out. Oxy-fuel cutting process was used then to cut the steel, thereby removing the smaller concrete blocks adhering to the steel surface. Depending on the diameter of the steel obtained, it is separated out for further processing, either rolling or melting, which is beyond the scope of study. Figure 8 shows the segregation process carried out by (a) hydraulic breaker and (b) oxy-fuel cutting. The segregation process was carried out for 51 days. The segregated concrete debris was then loaded into trucks using excavator. The loaded trucks were then transported to crusher unit. Figure 9 shows the recycled concrete aggregate produced after crushing. The concrete debris was also transported to various godowns and industrial sites for road formation and basement filling activities. The consolidated data on the weekly transport of concrete debris was obtained from the contracting firm involved in the debris removal. The actual tonnage of debris removed from the demolition site (H) was approximately 20450 tonnes. The properties of the recycled concrete debris also need to be studied. The contracting firm had manufactured concrete blocks using the recycled fine and concrete aggregate after crushing.

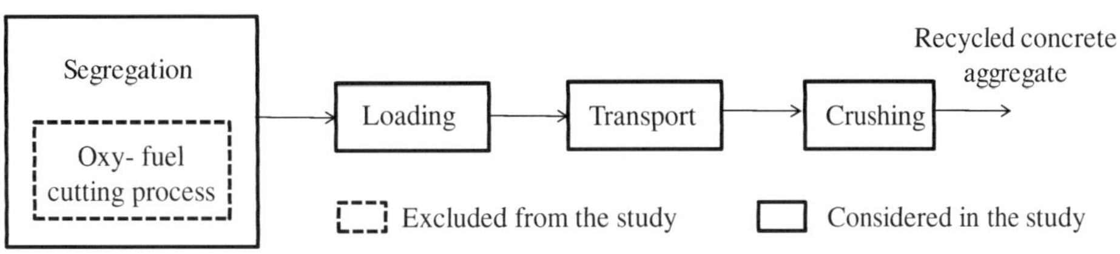

Figure 7. Process flow of recycled concrete aggregate production.

(a) (b)

Figure 8. Segregation process by (a) hydraulic breaker (b) oxy-fuel cutting.

Figure 9. Recycled concrete debris.

3.3. Computation of energy consumption

The process involved in the production of NA was thoroughly studied. The machinery involved in the different stages of production were identified and its average diesel consumption was obtained by verifying the diesel statement from the store. As the crusher unit was within the quarry, the transportation distance between the extraction site to crusher unit was assumed to be 1 km. Thus, the energy consumption involved in the different stages of production of 1 tonne of NA was formulated based on the diesel consumption of machinery. A questionnaire was formulated to obtain similar data on aggregate production from different quarries so as to compare and check for variability, if any, in the production process. Similarly, the different stages of production and the machinery involved in the production of RCA was studied with respect to the recycling strategy adopted on the demolition debris from (H). The diesel consumption of the machinery was collected from field. The total quantity of concrete debris generated from the site (H) was assumed to be converted to RCA in the crusher unit. The nearest crusher unit to the demolition site (H) was identified from KOMPAS [21] and the distance was measured to be approximately 7km. Hence, the transportation distance from the demolition site to crusher was assumed to be 7km. The energy consumption involved in the different stages of production of 1 tonne of RCA was formulated based on the diesel consumption of machinery. Since the crushing process was common to both the production, the energy consumption involved was assumed to be same and hence neglected in the calculation. The comparison between the energy consumption involved in the production of 1 tonne of NA and RCA was carried out based on the assumption that 1 tonne of RCA could significantly replace 1 tonne of NA in its usage.

4. Results and discussion

The energy consumption involved in the different stages of production of 1 tonne of NA and RCA were calculated and are summarized in Table 1. The energy consumed to produce 1 tonne of NA and RCA was determined as 21112 KJ/tonne and 16178 KJ/tonne respectively. It was found that there would be 30.5% savings in the energy consumption, if the production of RCA could replace that of the NA in the current scenario.

Table 1. Energy consumption in different stages of production of NA and RCA.

NA			RCA		
Stages of production	Energy consumption (KJ/tonne)	Total energy consumption (KJ/tonne)	Stages of production	Energy consumption (KJ/tonne)	Total energy consumption (KJ/tonne)
Drilling	3397	21112	Segregation	8715	16178
Breaking	8251		Loading	2367	
Loading	8736		Transport	5096	
Transport	728				

Figure 10 compares the energy consumption involved in the different stages of production of 1 tonne of NA and RCA. The energy consumption of the hydraulic breaker employed in breaking the rocks into fragments and segregating the concrete debris from steel was found comparable. The variation in the loading stage of both rock fragments and concrete debris may be due to the difference in the specifications of excavator employed in both job conditions. The excavator used in the quarry was of larger bucket capacity and power, to load the heavy boulders of rock, which involves higher fuel consumption and hence increasing the energy consumption. When comparing the effect of transportation in the savings for energy consumption, based on the assumptions it is found that the energy consumption was highly sensitive to the distance between the quarry/ demolition site and the crusher unit. In order to find the distance to which the energy consumption involved in the production of RCA is favourable than the production of NA, a sensitivity analysis is performed. The analysis has been performed by keeping the distance from the quarry to the crusher unit constant at 1 km, i.e., the crusher unit is within the quarry area. The distance from the demolition site to the crusher unit was gradually increased in steps of 1 km. Figure 11 shows the sensitivity analysis for the energy consumption, affected by the transportation distance. It was observed that the energy consumed to produce RCA was less than that to produce NA for a maximum distance of 13.5 km.

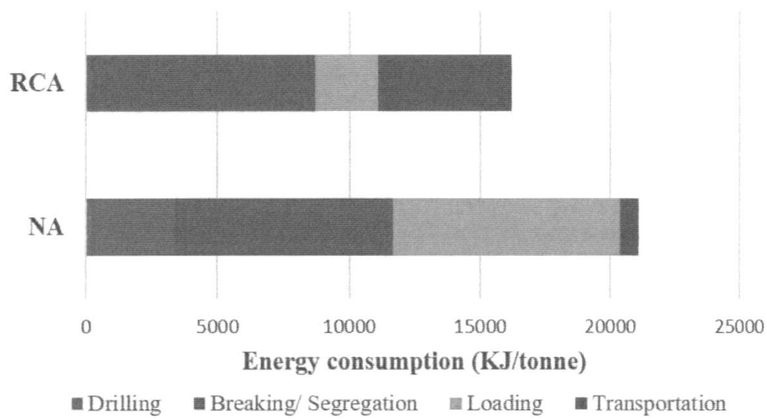

Figure 10. Energy consumption involved in the different stages of production of 1 tonne of natural and recycled concrete aggregate.

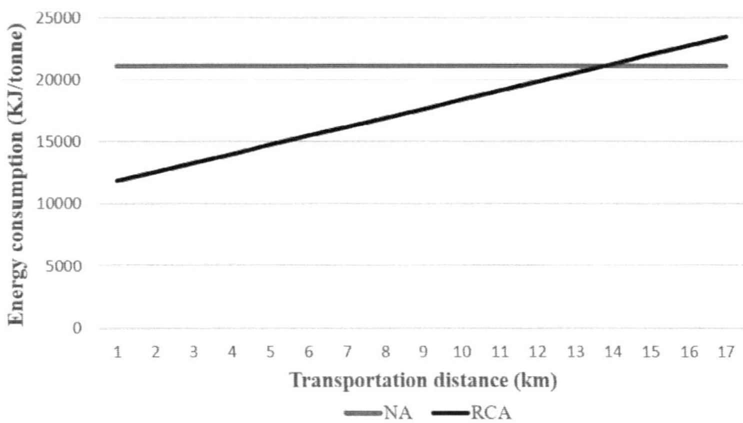

Figure 11. Sensitivity analysis on the energy consumption and transport distance of natural and recycled concrete aggregate.

The top five regions where maximum number of construction activities occurred in the district of Ernakulam were identified from the Building Statistics 2016-2017 conducted by the Government of Kerala [20]. The top five regions in the urban sector and the number of buildings constructed in the year 2016-2017 are as follows: Cochin corporation (7104), municipalities such as Kalamassey (2241), Thrikkakkara (1942), Trippunithura (1788) and Maradu (928). Figure 12 shows the location of top five regions where the highest number of construction activities were reported in the district for the year 2016-2017. Based on the sensitivity analysis for the energy consumption, it was known that the RCA production was favourable than NA only up to a distance of 13.5 km. Keeping the demolition site as the centre, a red coloured circle has been drawn with a radius of 13.5 km which indicates the regions where RCA production can be favourable than the NA production. From the figure it can be observed that the top five regions with the highest number of construction activities were much closer to the demolition site and falls within the red circle. Thus, recycling of the concrete debris from the demolition site to produce RCA could cater the demands of the construction activities in the district, as a replacement to NA with the prospects of reduced energy consumption. As the energy consumption was evaluated based on the diesel consumption of the machinery involved, the reduced diesel consumption and its subsequent cost savings could also add to the prospects of using RCA.

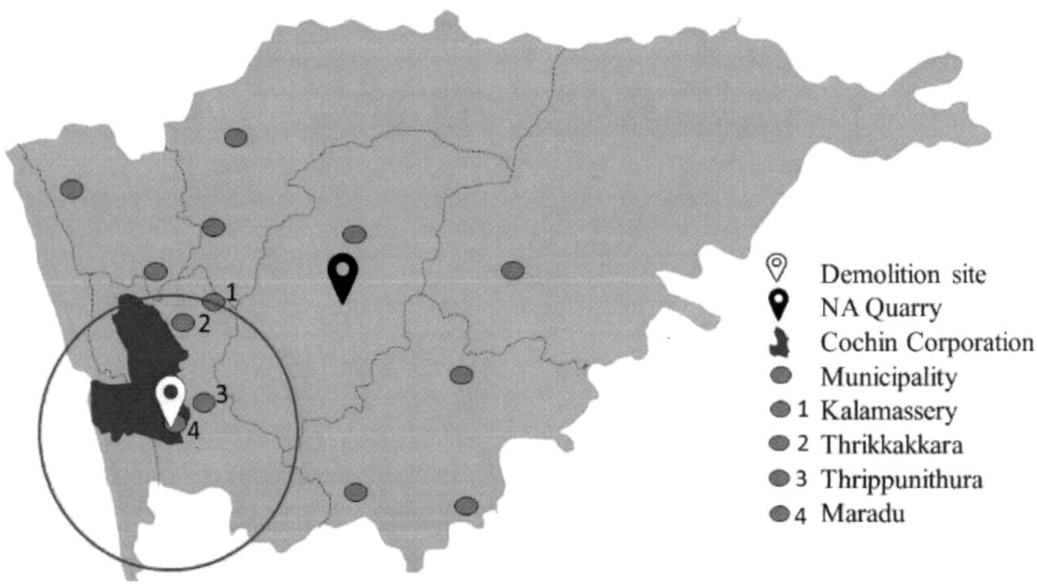

Figure 12. Location of the top 5 regions where the highest number of construction activities reported during the year 2016-2017.

Figure 13 shows the location and number of active quarries in the district engaged in the mining of granite and production of building stone as per the KOMPAS [21]. It can be observed that the quarries are in the eastern side of the district, whereas, majority of the construction activities are happening in the western side of the district. Hence the aggregates need to be transported to longer distances from the quarries located far away from the construction sites, which involves higher transportation cost. This gives an advantage to recycled aggregates in terms of lower transport distances and points to the prospects of setting up recycling units. However proper supply, quantity and quality of recycling C&D waste into RCA needs to be ensured. Experts state that the indiscriminate and rampant quarrying has significantly contributed to the heavy landslips in the northern districts of Kerala during the floods in 2019. Considering the number of active quarries in the central part of the state, which includes Ernakulam, there exists a possible risk of danger. Also, the illegal disposal of C&D debris was found to be a major reason for choking the sewers further leading to the floods in Chennai in 2015. Thus, proper C&D waste management with production of recycled aggregates can be the best solution to the environmental implications stated above. The study can be further extended to address the environmental benefits/impacts of recycling by carrying out LCA.

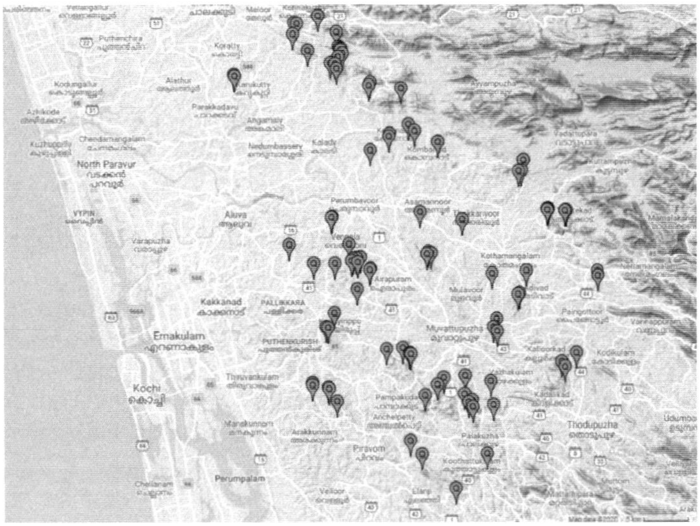

Figure 13. Location of active quarries in Ernakulam.

5. Conclusion

The actual potential of C&D waste recycling in India is least explored. The perception of takers for RCA in congruence with the NA can be ensured only if the quality and benefits of recycling can be quantified. Local case studies on recycling demolition waste can be used to quantify the impacts and further facilitate to the development of C&D waste recycling infrastructure. The study investigated the different stages of production of RCA and NA with respect to the recycling strategy adopted for the demolition debris generated from RC structure in Ernakulam, Kerala. About 30.5% savings was observed in the energy consumption to produce 1 tonne of RCA over NA. The sensitivity analysis on the energy consumption indicated the production of RCA was favourable compared to that of NA for a maximum distance of 13.5 km. It was observed that majority of the construction activities in the urban sector of the district was within the 13.5 km. Hence the recycled concrete debris from the demolition site could cater the requirements of NA with reduced energy consumption. The energy consumption was calculated based on the diesel consumption of the machinery involved. Thus, the reduction in energy consumption could attribute to savings in diesel consumption. The study could be further extended to compare the environmental impacts of the production of NA and RCA and also compare the environmental prospects of recycling over landfilling, by performing LCA.

6. References

[1] Census of India 2011 *Provisional population totals- Urban agglomerations and cities* Ministry of Home Affairs- Office of the Registrar General & Census Commissioner India

[2] Make in India 2020 Construction/ Reasons to invest

[3] Ministry of Housing and Urban Affairs 2020 Pradhan Mantri Awas Yojana (Urban) – *Housing for All* (HFA) State wise Progress (since 2014)

[4] Global Construction Perspectives & Oxford Economics 2015 *Global Construction 2030: A global forecast for the construction industry to 2030* (London: Global Construction Perspectives & Oxford Economics) 8

[5] GIZ 2016 *Material consumption patterns in India: A baseline study of the automotive and construction Sectors* (New Delhi: GIZ) 97

[6] TIFAC 2001 *Utilization of waste from construction industry*(New Delhi: Technology

Information, Forecasting and Assessment Council Publications)
[7] Ram V G, and Kalidindi S N 2017 Estimation of construction and demolition waste using waste generation rates in Chennai India *Waste Manag. Res.* **35** 529–51
[8] Jain S, Singhal S and Jain N K 2019 Construction and demolition waste generation in cities in India: an integrated approach *Int. J. Sustain. Eng.* **12** 333-340
[9] BMTPC 2018 *Utilisation of Recycled Produce of Construction & Demolition Waste: A Ready Reckoner* (New Delhi: Building Materials & Technology Promotion Council) 3
[10] Ram V G, Kishore K C and Kalidindi S N 2020 Environmental benefits of construction and demolition debris recycling: Evidence from an Indian case study using life cycle assessment *J. Clean. Prod.* **255**
[11] IS: 383 2016 *Coarse and Fine Aggregate for Concrete – Specification* Bureau of Indian Standards
[12] Ortiz O, Pasqualino J C and Castells F 2010 Environmental performance of construction waste: Comparing three scenarios from a case study in Catalonia Spain *Waste Manag.* **30** 646–54
[13] Blengini G A and Garbarino E 2010 Resources and waste management in Turin (Italy): The role of recycled aggregates in the sustainable supply mix *J. Clean. Prod.* **18** 1021-30
[14] Blengini G A 2009 Life cycle of buildings, demolition and recycling potential: A case study in Turin Italy *Build. Environ.* **44** 319-30
[15] Estanqueiro B, Silvestre J D, De Brito J and Pinheiro M D 2016 Environmental life cycle assessment of coarse natural and recycled aggregates for concrete *Eur. J. Environ. Civ. Eng.* **22** 429-49
[16] Rosado L P,Vitale P, Penteado C S G and Arena U 2017 Life cycle assessment of natural and mixed recycled aggregate production in Brazil *J. Clean. Prod.* **151** 634-42
[17] GIZ 2015 *Resource Efficiency in the Indian construction sector: Market evaluation of the use of secondary raw materials from construction and demolition waste*(New Delhi: GIZ)
[18] Census of India 2011 Density of population Ministry of Home Affairs- Office of the Registrar General & Census Commissioner India
[19] Census 2011 District/ Ernakulam District: Census 2011-2020 data
[20] Department of Economics & Statistics 2019*Building Statistics 2016-2017*(Thiruvananthapuram: Government of Kerala)
[21] KOMPAS- *Kerala Online Mining Permit Awarding Services* Quarry/Crusher/ Depot Location Map
[22] Ministry of Environment Forest and Climate Change 2016 *Construction and demolition waste management rules* G.S.R. 317 (E) The Gazette of India Part-II Section-3 Sub-section (ii)

IOP Conf. Series: Materials Science and Engineering **989** (2020) 012012 doi:10.1088/1757-899X/989/1/012012

Effect of dimensional parameters on the seismic performance of circular CFST columns

Reshma S[1,2] and Bushra M A[1]

[1] Department of Civil Engineering, T K M College of Engineering, Kollam, Kerala, India

[2] E-mail: reshmasraveendran333@gmail.com

Abstract. Due to the expansion of cities and increase in population, the need for construction of high rise buildings becomes more essential in the society. As earthquakes are of the greatest damaging natural hazards to the buildings, the design and construction of structures which are capable of resisting the adverse effects of earthquakes become necessary. Concrete Filled Steel Tubes (CFSTs) are the steel tubular sections in-filled with concrete and with a composite action. CFSTs are becoming a popular solution for structures located in seismic areas, due to their high strength, excellent ductility and large energy absorption capacity. This study was mainly concentrated to find out the effect of dimensional parameters on the seismic performance of CFSTs. Width- thickness ratio (D/t) and slenderness ratio (L/D) were taken as two parameters and the study was carried out using circular CFST buildings. It was observed that, there will be a particular limit for D/t and L/D ratio upto which the thickness and diameter of the CFSTs can be decreased. At that point the building will show a better seismic performance.

1. Introduction

Concrete-filled steel tubes (CFST) are composite structures of steel tube and in-filled concrete. Concrete filled steel tubular member uses the advantages of both steel and concrete. They comprise of two main parts, an outer steel hollow section of circular or rectangular shape and an inner filler material of plain or reinforced concrete. A CFST member consisting of a steel tube which is filled with concrete material realizes the importance of steel reinforcement to provide confinement for the concrete and to increase the load-carrying capacity of the composite member. From the structural point of view, the inner concrete material not only prevents the occurrence of inward buckling of the outer steel tube but also enhances the ductility of the CFST member up to the ultimate load. CFST member performs under composite action, ie both the steel and concrete will resist the external loading by interacting together by bond and friction action. In a CFST member, concrete and steel are combined together in such a fashion that the advantages from both the materials should be effectively utilized.

First research on CFST members was started by 20th century which mainly concentrated on strength capacity. Application of CFST concept can lead to the reduction in total consumption of steel and thereby can reduce the overall construction cost for the building compared to the conventional structural steel system. CFST members can be used in multi-storey buildings as both columns and beams and also as bridge piers. Concrete core enhances higher compressive strength, stiffness and damping whereas tensile strength is provided by the outer steel tube. Also high strength CFST columns require only smaller cross-sections compared to the conventional reinforced cement concrete columns under the same loading conditions. The behavior of CFST members in bending, shear, compression and fatigue resistance are also superior over reinforced members under cyclic seismic loading. The steel concrete composite columns have several advantages such as high axial load carrying capacity, good ductility performance, energy absorption capacity and high fire resistance as well as fast construction [1].

In this study the effect of dimensional parameters on the seismic performance of CFSTs were evaluated. Width- thickness ratio (D/t) and slenderness ratio (L/D) were considered as two parameters for the study. ETABS- 2018 software was used for the design and analysis and the columns of the building were chosen to be circular CFSTs.

2. Modeling and analysis

The plan of the building was fixed to be in rectangular shape having dimension of 25m x 20m with the longest side along the X- axis. The spacing of each column was fixed as 5m and the number of bays provided along X and Y directions were 5 and 4 respectively. G+12 storey buildings were used by fixing each storey height as 3m, obtaining an overall height of 39m. Usually pile foundations are used for high rise buildings and here also the support conditions are chosen to be fixed. The detailed plan of the building was given in Figure 1. The material needed for the CFST members were fixed to be Fe345, for the outer structural steel and M30, for the inside concrete. Floors and roofs were modeled using RCC slab with a fixed thickness of 150mm.

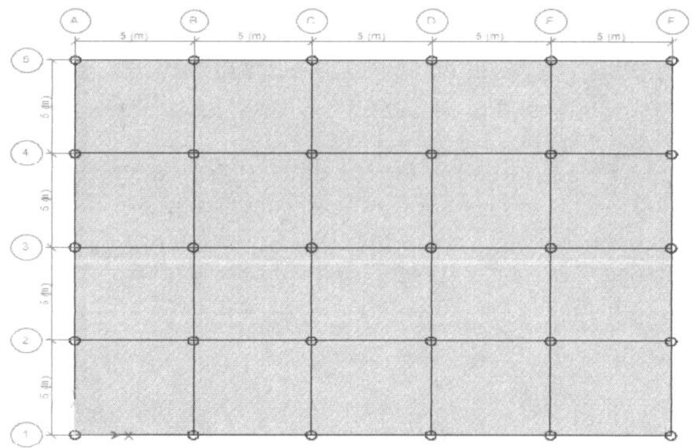

Figure 1. Plan of the building.

Three types of loads were considered for the analysis. They are dead loads, live loads and seismic loads. For the roofs and floors, the dead and live loads were applied separately. Dead loads were applied as frame loads with a value of 6 kN/m and 12 kN/m on roofs and floors respectively. Live load of 1.5 kN/m^2 was applied on roofs and 3 kN/m^2 was assigned for floors, by considering them as the shell loads. Seismic loads on the building were calculated according to the Indian standard code, IS 1893: 2016. Loads were applied both in X and Y directions without considering eccentricities. The details of the seismic characteristics used for the study were shown in the Table 1.

Since there is no particular code for the design of CFSTs in India, the American Standard code, AISC 360-10 was used for the design of the building. The beams of the buildings were modeled using the steel I-section, ISMB 300. With all the details, building was modeled using ETABS-2018 software and seismic analysis was carried out using linear dynamic analysis. Circular CFST columns having dimensions 450mm x 12mm was obtained from the design, for which the thickness and diameter were varied and analysis were carried out separately to find out the effect of dimensional parameters under seismic loads.

IOP Conf. Series: Materials Science and Engineering **989** (2020) 012012 doi:10.1088/1757-899X/989/1/012012

Table 1. Details of seismic load characteristics.

Seismic characteristics	
Design Code	IS 1893: 2016
Soil Type	II (Medium Soil)
Seismic Zone	III
Seismic Zone Factor	0.16
Importance Factor	1.5
Response Reduction Factor	5
Damping Ratio	0.05
Modal Combination	CQC
Directional Combination Type	SRSS

2.1. Width- thickness ratio (D/t)

Width- thickness ratio is one of the main parameters to understand the behavior of CFSTs. Here the study of this parameter was conducted by varying the thickness of CFSTs for that already designed circular column (450mm x 12mm) building as 6mm, 8mm, 10mm and 14mm. The diameter for each circular CFSTs were fixed as 450mm itself. The dimension of CFST columns used for each building and their D/t ratios are shown in the Table 2.

Table 2. Dimensions of circular CFSTs for each D/t ratios.

Building No	Dimensions of circular CFSTs	D/t ratio
1	450mm x 6mm	75
2	450mm x 8mm	56.25
3	450mm x 10mm	45
4	450mm x 12mm	37.5
5	450mm x 14mm	32.14

2.2. Slenderness ratio (L/D)

Slenderness ratio is calculated as the length to width ratio. In this study circular CFST columns were considered and hence length to diameter ratio (L/D) was taken as the parameter by varying the diameters of the columns. For this to be done, the designed circular column CFST building with column size 450mm x 12mm was considered. Length and thickness of CFSTs were remained as same and only the diameter of the columns were varied from 350mm to 550mm. Table III shows the dimensions of the CFST columns used for the five buildings which were taken for the study and the corresponding L/D ratios.

Table 3. Dimensions of circular CFSTs for each L/D ratios.

Building No	Dimensions of CFST columns	Length of columns	L/D ratio
1	350mm x 12mm	3000mm	8.57
2	400mm x 12mm	3000mm	7.5
3	450mm x 12mm	3000mm	6.67
4	500mm x 12mm	3000mm	6
5	550mm x 12mm	3000mm	5.45

3. Results and discussions

The thickness and diameter of the CFST were varied and for each building seismic analysis was carried out by using linear dynamic method. Fundamental time periods, storey displacements, storey drifts and base shears were the seismic parameters considered for the comparative study. The results obtained from the study are discussed here.

3.1. Effect of width- thickness ratio (D/t)

The time period values obtained for each building with different D/t ratios were illustrated in the Figure 2. Maximum value of time period was shown by the lowest thick CFST building ie, by the 450mm x 6mm CFST building and also it was observed that while increasing the thickness of the CFSTs, the time periods start decreasing upto 12mm thickness and after that gets increasing. Total 6 modes were considered for the study and for every mode the same pattern of variation in time periods was observed.

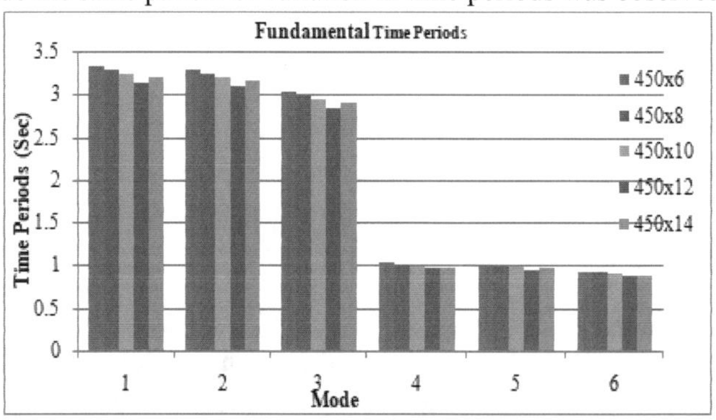

Figure 2. Fundamental time periods (Seconds).

Maximum displacements obtained by the response spectrum analysis for each floors for the different CFST buildings along X and Y directions were shown in the Figure 3 (a) and (b) respectively. From the pattern of variation of the graphs, it is found that the displacements at each floor starts decreases from the 6mm thick CFST to the 12mm thick CFST and thereafter increases for the 14mm thick CFST building.

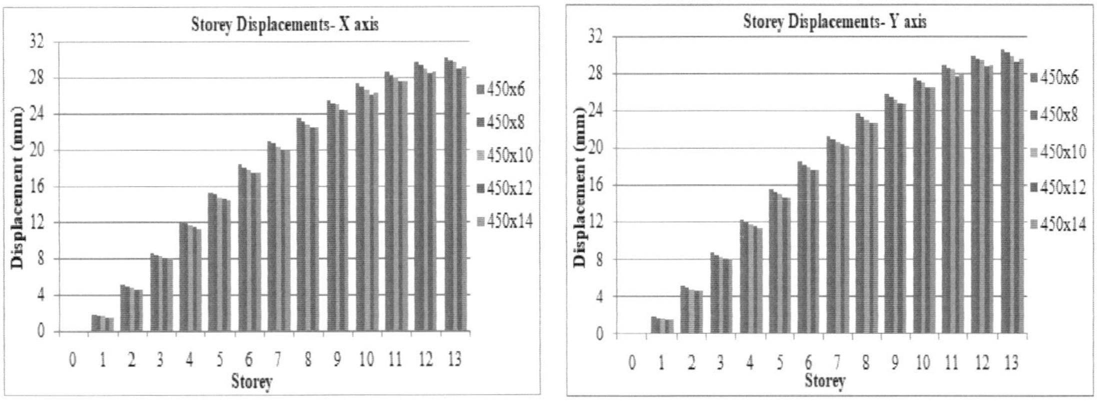

(a) (b)

Figure 3. Maximum storey displacements along X-axis (a) and Y-axis (b).

The graphs obtained for storey drifts for the different CFST buildings were shown in Figure 4. From the graph it can observe that for the bottom most stories 6mm thick CFST building shows higher storey drifts and continues the decreasing pattern giving lowest value for 14mm thick CFST building. But while moving to the top stories the pattern of variation starts changing by giving deceasing values of storey drifts from 6mm to 12mm and after that a higher value for 14mm thick CFST. The same variation was observed for both the axes. From this graphs it is clear that for the lower storey buildings higher thick CFST will cause low storey drifts whereas when increasing the number of stories in the building there will be a limit for the thickness of CFST beyond which can lead to high storey drifts.

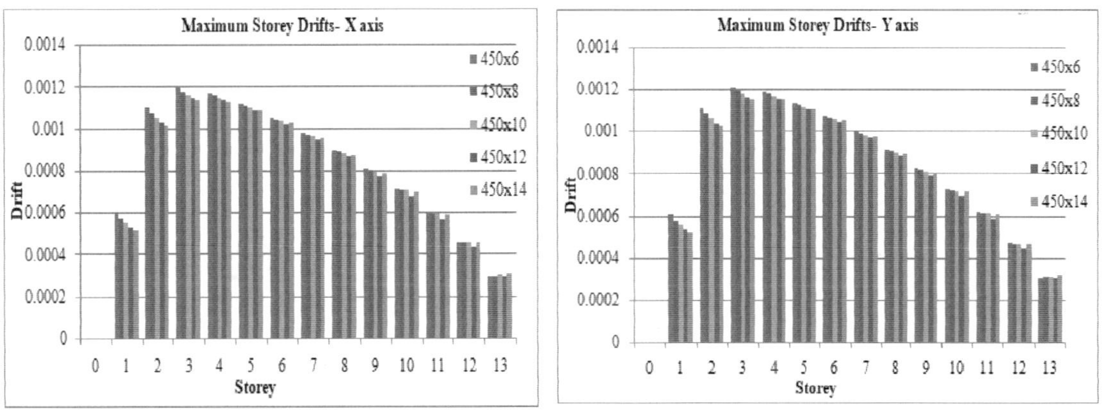

(a) (b)

Figure 4. Maximum storey drifts along X-axis (a) and along Y-axis (b).

Table 4 shows the base shears obtained for the buildings with different D/t ratios. For along both the X and Y directions it is observed that increase in thickness of CFST causes increase in base shears also, ie here the building with 6mm thick CFST gave lowest base shears and with 14mm thick CFST showed highest base shears for both the axes.

Table 4. Base shears along X and Y axes.

Building Type	Base shear (kN)	
	X- axis	Y- axis
450X6	646.4724	639.6522
450X8	654.2699	647.5332
450X10	661.0347	654.4106
450X12	666.2838	660.1835
450X14	672.3353	666.0142

3.2. *Effect of slenderness ratio (L/D)*

Time periods obtained for the buildings upto 6 modes were shown in the Figure 5. It was observed that when diameter of the CFST increases time periods start decreasing, ie highest value of time period was shown by the 350mm diameter CFST building whereas lowest value was given by the 550mm diameter CFST building. The building with higher L/D ratio gave the higher time period and lower L/D ratio showed the lower time period value for every mode.

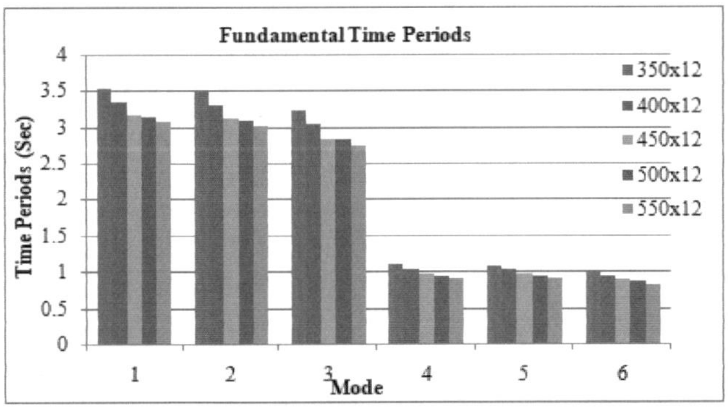

Figure 5. Fundamental time periods (Seconds).

Storey displacements were found to be higher for the building with lower diameter CFST columns, ie increase in diameter of CFST leads to decrease in displacements. This can be clearly observed from the Figure 6, which represents the storey displacements obtained for each building having different L/D ratios. The building with higher L/D ratio gave higher displacements and with lower L/D ratio showed lower displacements. The same pattern of variation was observed for both the directions.

For bottom stories upto 7th floor 350mm diameter CFST building showed highest storey drifts and gave a decreasing pattern of variation from 350mm diameter CFST to 550mm diameter CFST building. Whereas from the 8th floor to 12th floor, the decreasing pattern was shown upto 450mm diameter CFST and after that increase in storey drifts were observed upto 550mm diameter CFST building. But for the top most storey, the condition was totally reversed compared to bottom most stories. ie the highest storey drift was given by the 550mm diameter CFST and lowest value was shown by the 350mm diameter CFST building. The same pattern of variation was observed for both the directions. Figure 7 illustrates the variation of storey drifts observed for the buildings having different L/D ratios along both the X and Y axes. From the study it can be given that for low storey buildings, increase in diameter of CFST, ie low values of L/D ratios can decrease the storey drifts. But when the number of stories increases that can give a reverse result.

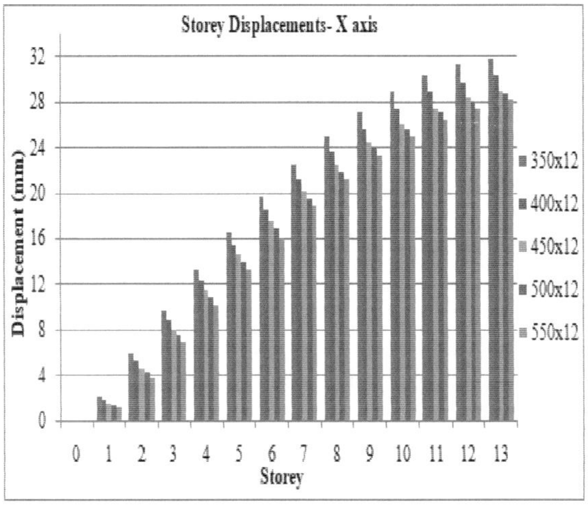

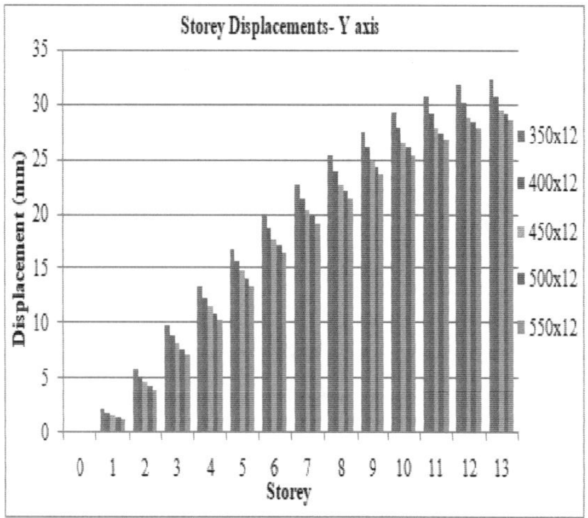

(a) (b)

Figure 6. Maximum storey displacements along X-axis (a) and along Y-axis (b).

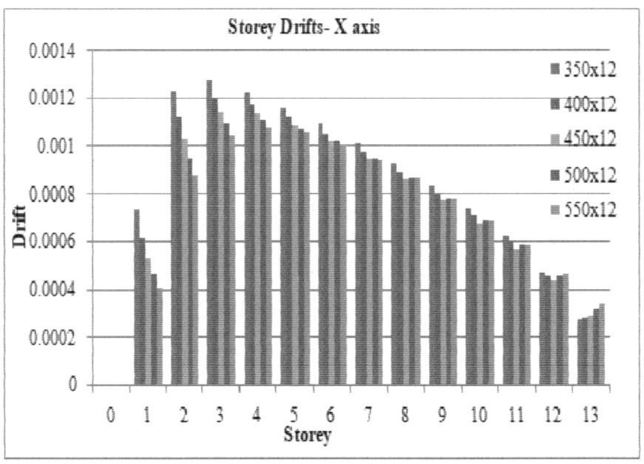

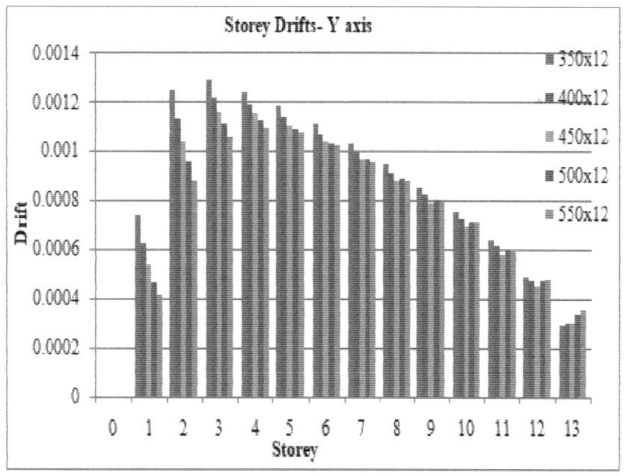

(a) (b)

Figure 7. Maximum storey drifts along X-axis (a) and along Y-axis (b).

Table 5 represents the maximum storey shear obtained at the base for each building. It is clear from the table that, as the diameter of the CFST increases the base shear of the building also increases, ie the highest base shear was given by the 550mm diameter CFST building whereas lowest base shear was observed for the 350mm diameter CFST building. In terms of L/D ratio, it can be stated that as the L/D ratio decreases, base shear starts increasing.

Table 5. Base shears along X and Y axes.

Building Type	Base shear (kN)	
	X- axis	Y- axis
350x12	604.7761	599.2981
400x12	638.9634	632.2069
450x12	666.2838	660.1835
500x12	692.1746	686.0971
550x12	716.2297	710.4688

4. Summary

When considering the D/t ratios, it was observed that time periods and storey displacements of the buildings give higher values for 6mm thick CFST and then starts decreasing upto 12mm thick CFST and thereafter again increases for the 14mm thick CFST building. So it can be given that there will be a particular limit for D/t ratio at which the time periods and storey displacements will be minimum. The increase in thickness of CFST can improve the behavior of steel in the building and can lead to high lateral displacements due to the increase in flexibility. So there have to keep a limit for thickness upto which thickness can be reduced to get a better performance. For bottom stories, storey drifts were found to be higher for 6mm thick CFST building or the building with higher D/t ratio and then showed a decreasing pattern upto the 14mm thick CFST building or the building with lower D/t ratio. But when moving to the upper stories, decrease in storey drifts were found from 6mm thick CFST to 12mm thick CFST and thereafter increased for 14mm thick CFST building. Therefore the storey drifts will be decreasing upto a particular limit of D/t ratio. In this study D/t ratio of 37.5 was found to be the limiting value of D/t ratio at which the time periods, storey displacements and storey drifts were found to be minimum. Base shears were observed to be increasing with the increase in thickness of CFSTs, ie with the decrease in D/t ratios.

As with the decrease in L/D ratio, it was observed that time periods and storey displacements starts also decreasing, ie when the diameter changes from 350mm to 550mm the time periods and displacements starts decreasing giving lowest value for the 550mm diameter CFST building. When the diameter of CFST increases, the concrete content in the CFST also starts increasing thereby building behaves like an RCC building, ie stiffness of the building also increases giving low lateral displacements. For bottom stories storey drifts were found to be decreasing as with the decrease in L/D ratios. But when considering the top stories, there shows a particular limit of L/D ratio upto which the storey drifts decreases and after that increases. This was observed only upto 12th floor, whereas for the top most storey, ie the 13th storey, a continuous increasing pattern was observed from 350mm diameter to 550mm diameter CFST building. So it can be said that for higher storey buildings there has to keep a limit for L/D ratio below or beyond which structure can cause least performance under seismic loads. From this study, it was observed that at an L/D ratio of 6.67, the building shows better seismic performance. Base shears were found to be increasing with the decrease in L/D ratios due to the increase in weight of the buildings.

5. Conclusions

The effect of the dimensional parameters on the performance of CFSTs was evaluated by using the width-thickness ratio (D/t) and slenderness ratio (L/D). The study was carried out by varying the thickness and diameters of the circular CFSTs. It was observed that there will be a particular limit for D/t ratio, at which the building will show a good seismic performance. Decrease in D/t ratio causes decrease in time periods, storey displacements and storey drifts and after a particular limit of D/t ratio, they starts again increasing, ie increase in thickness of CFST can only be provided upto a particular limit. D/t ratio of 37.5 was obtained as the limiting value from the study. Above this ratio, there will be less confinement effect for the concrete due to less thickness of outer steel tube and below that ratio higher thickness of CFST will cause the steel behavior of the building to become more prominent. Both the conditions lead to less seismic performance and so there has to keep a limiting thickness or D/t ratio for attaining better performance of the building under seismic loads. Decrease in slenderness ratio, L/D ratio or the increase in diameter of CFST leads to the decrease in time periods and storey displacements. Storey drifts were found to be decreasing upto a limit of L/D ratio and after that starts again increasing. The building having L/D ratio of 6.67 was found to be better under seismic loads. Increase in thickness and diameter of CFSTs can cause the improvement in the behavior of CFSTs as steel and RCC respectively. So when considering the dimensions of CFSTs, there has to keep a limit for thickness and diameter of CFSTs or for the D/t and L/D ratios, below or beyond which can cause least performance under seismic loads.

6. References

[1] Kalingarani, Shanmugavalli K B and Sundarraja M C 2014 Axial compressive behaviour of slender CFST members–analytical investigation *Int. J. Innovative Research in Science Engineering and Technology* **3** 22- 25
[2] Bjorhovde R, Han L H and Li W 2014 Developments and advanced applications of concrete-filled steel tubular (CFST) structures: members *J. Constr. Steel Research* **100** 211- 228
[3] Chaoui K and Zeghiche J 2005 An experimental behaviour of concrete-filled steel tubular columns *J. Constr. Steel Research* **61** 53– 66
[4] Merool D V, Patel M N and Shah D K 2014 Behavior of concrete filled steel tube column *Int. J. Engg. Development and Research* **2** 325- 328
[5] Liew J Y R., Xiong D and Xiong M 2016 Design of concrete filled tubular beam-columns with high strength steel and concrete *J. Structures* **8** 213- 226
[6] Tailor A, Dalal S P and Desai A K 2017 Comparative performance evaluation of steel column building & concrete filled tube column building under static and dynamic loading *11th Int. Symp. on Plasticity and Impact Mechanics* **173** 1847 – 1853
[7] Jingbo L and Yangbing L 2008 Seismic behavior analysis of steel-concrete composite frame structure systems *14th World Conference on Earthquake Engg.* **6** 183- 191
[8] Sulke M S and Kewate S 2018 Seismic performance of concrete filled steel tubular column building *Int. J. Engg. Science Invention* **7** 80- 85
[9] Vidhya P and Sukumaran A 2017 Comparative study on seismic analysis of multi storied buildings with composite columns *Int. J. Engg. Research & Technology* **6** 555- 561
[10] Sachin S and Prasad H R S 2018 Study of seismic behavior on multi-storied buildings with composite columns *Int. J. Research in Applied Science & Engg. Technology* **6** 4779- 4785
[11] Lu X and Lu W 2000 Seismic behavior of concrete and steel composite columns under cyclic loading *12th World Conference on Earthquake Engg.* **1416** 1- 8
[12] Jiang Y, Silva A, Macedo L, Castro J M and Monteiro R 2018 Seismic performance of composite structures made with concrete-filled steel tubular members *16th European Conf. on Earthquake Engg.* **18** 1-12
[13] Evirgen B, Tuncan A and Taskin K 2014Structural behavior of concrete filled steel tubular sections (CFST) under axial compression *J. thin walled structures* **80** 46- 56
[14] Alireza K, Farshid H and Ali T M 2011 Seismic performance of structures with CFST columns and steel beams *1st Int. Conf. on Urban Construction in the Vicinity of Active Faults 1-9*
[15] Mohod M V and Wagh M M 2016 Behaviour of concrete filled steel tube with reference to different shape of column *Int. J. Innovative and Emerging Research in Engg.* **3** 290- 296

IOP Conf. Series: Materials Science and Engineering **989** (2020) 012013 doi:10.1088/1757-899X/989/1/012013

A study on partial and full replacement of fine aggregate in high-performance concrete with granulated blast furnace slag

Dodda Srinivas [1,4], Hardik Chitirala[2], Srinivasarao C H[3] and Dibyendu Adak[1]

[1]Departmant of Civil Engineering, National Institute of Technology Meghalaya, India.
[2]Departmant of Civil Engineering, V N R Vignana Jyothi Institute of Engineering and Technology, India.
[3]Department of Civil Engineering, CMR Technical Campus, India.

[4]Email: dsrinivas709@gmail.com

Abstract. The influence of the partial / full replacement of fine aggregates with granulated blast furnace slag (GBS) and its effect on mechanical strength and durability properties (acid attack) of high-performance concrete of grade (M60) had been studied in the present work. The high-performance concrete mix had been designed in such a way that one group with natural sand and another with manufactured sand (Quarry dust). Each group of concrete mix with natural and manufactured sand consisted of five sub-groups, having 0%, 30%, 50%, 80% and 100% of GBS content. Other pozzolanic materials such as fly ash and silica fume were also employed to replace ordinary Portland cement (OPC) at 20% and 8% respectively, with a constant w/c ratio of 0.32 for the desired slump of 100 mm. According to the results, all the replacement levels of fine aggregate by GBS proved to enhance the mechanical and durability performance of concrete. It is concluded that the high-performance concrete can be produced by utilizing 100% GBS as fine aggregate, with improved mechanical and durability performance.

1. Introduction

Due to the increase in urbanization and infrastructural development all over the world, especially in developing countries like India, it requires a huge amount of natural resources used to make the concrete. One such resource is the fine aggregate (river sand). As the deposits of natural sands have slowly been depleted and posing many serious problems to both environment and humankind, it has resulted in a search for other suitable alternatives in the form of fine aggregate. From the literature [1]–[4], it was observed that the manufactured sand (Msand), granulated blast furnace slag (GBS), waste foundry sand, glass waste, etc. were used as a fine aggregate alternative to natural river sand. Among these, though Msand is a by-product of the stone crushing industry, due to the increased demand and the intentional manufacturing of Msand further causes the depletion of rock sources which takes millions of years to form.

A non-metallic Granulated blast furnace slag is the by-product of steel and iron industries, consisting essentially of non-crystalline calcium-aluminosilicates [5], GBS is produced by rapid quenching of molten slag with water jets, which resulted in a granular glassy aggregate. This granular aggregate with little fines used for fine aggregate replacement in the present study.

High-performance concrete (HPC) has led to a decrease in section size of structural elements due to its high compressive strength, toughness and durability when exposed to severe environmental conditions [6]. But, to produce HPC it requires higher paste content. If we use OPC alone to produce HPC, it will lead to increased cost, the heat of hydration and drying shrinkage. Generally, HPC is produced by utilizing other supplementary cementitious materials (SCM's) such as fly ash (FA), silica fume (SF), rice husk ash, ground granulated blast furnace slag (GGBS) [7]–[9]. HPC demands a lower w/c ratio to give higher strength and low permeable concrete. So, water-reducing admixtures such as naphthalene based or ether-based are employed to increase the workability. Utilization of SCM's further led to reduced cost of concrete and at the same time made it more sustainable. Although some studies have been reported on utilizing the GBS as fine aggregate in normal concrete [2], [10] and self-compacting concrete [1], there has been little research on the influence of GBS as fine aggregate in HPC.

For the present study two different groups of concrete were developed, one with river sand (RS) as fine aggregate and other with manufactured sand (Msand) as fine aggregate. Compressive, split tensile, flexural strength and acid attack tests were conducted to evaluate the performance of concrete containing GBS as fine aggregate, in place of RS and Msand.

2. Experimental details

2.1. Materials

Ordinary Portland cement (OPC) of 53 grade confirming to IS 269: 2015 having the physical properties as given in table 1 was used for making the high-performance concrete (HPC). Pozzolanic materials such as fly ash (FA) having a specific gravity of 2.5 confirming to IS 3812 (Part 1): 2013 and silica fume (SF) having average particle size of 150 nm and specific gravity of 2.2 confirming to IS 15388: 2003 were employed to replace total cementitious material by 20% and 8% respectively, by weight based on the trial mixes.

Locally available coarse aggregate (CA) passing 20 mm and 12.5 mm, manufactured sand (MS), river sand (RS) and granulated blast furnace slag (from JSW Steel Ltd.) confirming to IS 12089: 1987 were used for making the concrete. The standard physical properties of these materials were shown in table 2. To improve the workability of HPC, PCE (poly carboxylic ether) based superplasticizer (SP) (ECMAS HP-904) having a specific gravity of 1.13 and tap water confirming to IS 456:2000 was used.

2.2. Mix design and methodology

Trial mix designs were carried and finalized the material proportions per cubic meter volume as given in table 3 according to IS 10262: 2009 to achieve the predetermined compressive strength and workability.

The methodology of the present study is to compare the mechanical and resistance to acid attack properties of the control sub-groups in each group and concrete in which fine aggregate is replaced with 30, 50, 80, and 100% of GBS at a constant w/c of 0.32.

2.3. Test procedure

- Workability test on fresh concrete was conducted with the slump cone apparatus in accordance with IS 1199: 1991.
- Compressive strength on 150 mm size cubes after 7 and 28 days of curing and flexural strength on simple beam (100 mm × 100 mm × 500 mm) with fourth point loading after 28 days of curing was conducted in accordance with IS 516: 1959.
- Split tensile test was conducted on cylindrical specimens (150 mm dia. and 300 mm height) after 28days of curing in accordance with IS 5816: 1970.
- Acid attack test was conducted in accordance with ASTM C267 - 01 (2012) on 70.6 mm concrete cubes and residual strength was found after exposure to 5% concentrated sulfuric acid solution for 28 days.

Table 1. Properties of cement.

Property	Result
Specific gravity	3.16
Fineness (m^2/kg)	311.85
Soundness (mm)	0.6
Initial setting time (min)	145
Final setting time (min)	215
Compressive strength (MPa)	
3 days	42.8
7 days	48.2
28 days	62.1

Table 2. Properties of aggregate.

Property	RS	Msand	GBS	CA
Specific gravity	2.65	2.80	2.85	2.80
Bulk density (kg/m^3)	1630	1850	1220	1560
Moisture content (%)	0.8	1.5	1.0	-
Water absorption (%)	1.0	4.0	1.0	0.04
Fineness modulus	2.63	2.67	2.86	2.86
Impact value (%)	-	-	-	9.22
Crushing value (%)	-	-	-	0.82

Table 3. Mix proportions of concrete specimens (kg/m^3).

	RS					Msand				
Material	100% RS	30% GBS	50% GBS	80% GBS	100% GBS	100% Msand	30% GBS	50% GBS	80% GBS	100% GBS
Cement	333.0	333.0	333.0	333.0	333.0	333.0	333.0	333.0	333.0	333.0
Fly ash	92.5	92.5	92.5	92.5	92.5	92.5	92.5	92.5	92.5	92.5
Silica fume	37.0	37.0	37.0	37.0	37.0	37.0	37.0	37.0	37.0	37.0
RS/Msand	767.4	541.1	386.5	154.6	0	806.2	569.0	403.1	162.5	0
GBS	0	257.2	428.6	687.8	858.0	0	257.2	429.0	686.4	858.0
CA	1078	1078	1078	1078	1078	1078	1078	1078	1078	1078
Chemical admixture	2.08	2.08	2.08	2.08	2.08	2.08	2.08	2.08	2.08	2.08
Water	166.9	166.3	163.9	162.5	160.9	184.2	177.4	172.7	165.7	160.9
w/c ratio	0.32	0.32	0.32	0.32	0.32	0.32	0.32	0.32	0.32	0.32

3. Results and discussion

3.1. Workability

From table 4 it can be observed that with the increase in GBS content, the super plasticizer requirement increased to achieve the desired slump of 100 mm, irrespective of the concrete series. Due to more angular structure and water absorption, Msand utilized more super plasticizer (SP) than river sand. The ability to reduce the workability of fine aggregate was in the order as RS < Msand < GBS.

Table 4. Variation in dosage of superplasticizer.

GBS content (%)	Design SP in gm	RS		Msand	
		Total added SP in gm	Extra dosage in %	Total added SP in gm	Extra dosage in %
0	2081	2111	1.4	2151	3.3
30	2081	2143	2.9	2175	4.5
50	2081	2161	3.8	2195	5.4
80	2081	2176	4.5	2205	5.9
100	2081	2193	5.4	2193	5.4

3.2. Compressive strength

Average compressive strength results at the end of 7 and 28 days of curing of both the groups (RS and Msand) are shown in figure 1. It is observed that the target mean strength i.e., 46.58 MPa for 7 days cured specimens and 68.25 MPa for 28 days cured specimens has been achieved by all mix proportions. Also, compressive strength of mixes having GBS as fine aggregate is higher (26 – 33% at 7 days and 2 – 13% at 28 days) and (25 – 47% at 7 days and 11 – 34% at 28 days) when compared to mix having only Msand and RS as fine aggregate respectively.

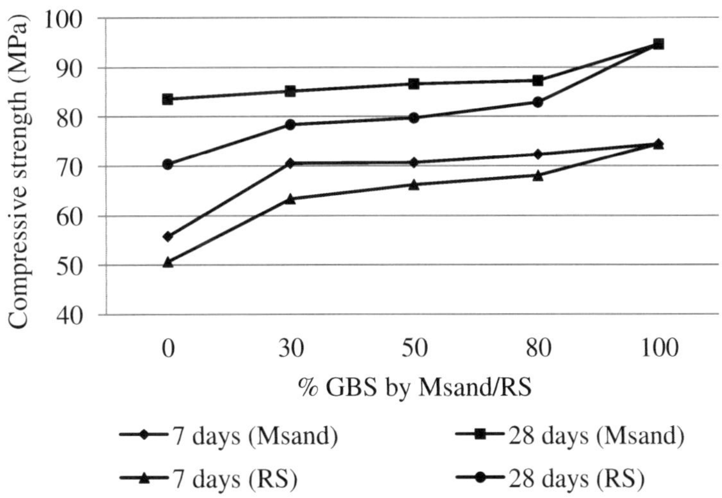

Figure 1. Variation in average compressive strength.

Whereas this increment is lower in case of normal and standard grade concrete [11], [2], this higher increment in compressive strength can be attributed to the inclusion of fly ash and silica fume as supplementary cementitious materials [12]. The maximum compressive strength is achieved at 100% GBS as fine aggregate. It is clear that both Msand and GBS not only helped in achieving the early age strength but also higher strength at later ages.

3.3. Flexural strength

Figure 2 represents variation in the average flexural strength of 28 days cured specimens. It is observed that the flexural strength of concrete mixes is increased with the increase in the incorporation of GBS. The flexural strength value of control mix with only Msand as fine aggregate is 6.89 MPa and this value further increased to 6.93, 6.96, 7.33 and 7.45MPa which is 0.58%, 1.01%, 6.38% and 8.12%

higher than control mix for 30%, 50%, 80% and 100% replacement with GBS respectively. Similar trend is also observed in case of concrete with RS as fine aggregate and these values are 0.73%, 1.31%, 4.1% and 9.23% higher than control mix having only RS as fine aggregate. Highest flexural strength is observed at 100% replacement levels for both the series. This is due to the lower w/c ratio and pozzolanic action of supplementary cementitious materials and GBS [3].

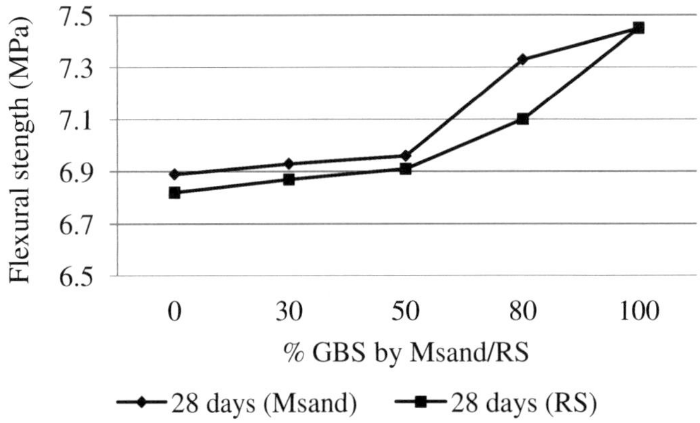

Figure 2. Variation in average flexural strength.

3.4. Split tensile strength

Variation in the average split tensile strength of 28 days cured specimens of all mixes is graphically represented in figure 3. Msand series mixes have 0.55%, 4.42%, 5.53%, 5.90% and 9.04% higher split tensile strength than target strength of 5.42 MPa at 0%, 30%, 50%,80% and 100% replacement with GBS respectively. Mix with 100% RS has 2.4% lesser split tensile than what it is designed for. On the other hand, remaining mixes of RS series have 0.73%, 2.58%, 4.24% and 9.04% higher split tensile strength than target mean strength at 30%, 50%, 80% and 100% replacement with RS with GBS. It is clear from the above results that the GBS as a fine aggregate along with lower w/c ratio and pozzolanic materials enhanced the tensile properties of HPC.

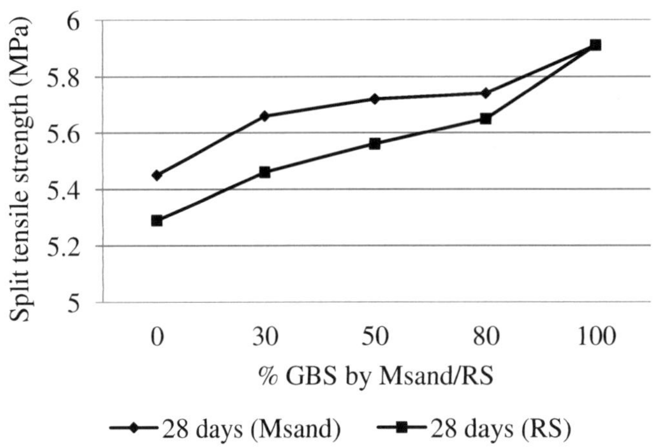

Figure 3. Variation in average split tensile strength.

3.5. Residual compressive strength after acid attack

The acid attack test was conducted to measure the durability of the concrete. The figure 4 indicates the variation of Residual compressive strength for different percentages of GBFS replacement. The percentage decrease in compressive strength due to acid attack are 53.88%, 31.73%, 29.64%, 21.48% and 31.97% for 0 %, 30%, 50%, 80% and 100% replacement with GBS respectively in case of Msand. Similarly, the percentage decrease in compressive strength due to acid attack in case of river sand are 48.87%, 33%, 31%, 32% for 0 %, 30%, 50%, 80% and 100% replacement with GBS respectively. Though the mix with 80% GBS performed well compared to all other mixes in both the series, the mix with 100% replacement with GBS also has better resistance towards acid attack, which is consistent with mechanical properties.

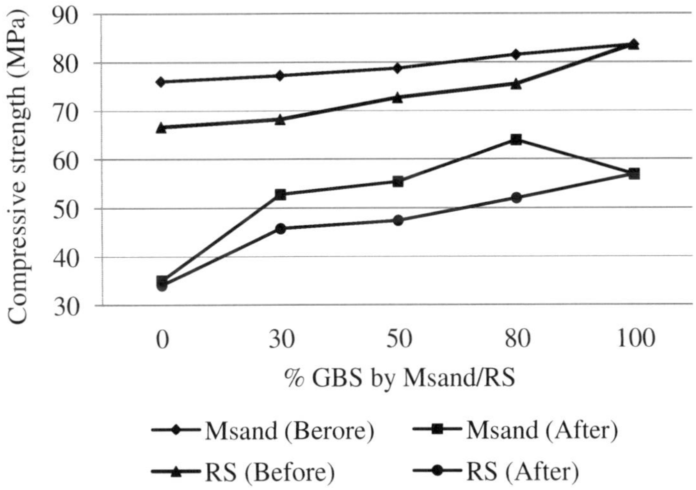

Figure 4. Variation in average residual compressive strength.

4. Conclusions

From the tests conducted, following conclusions can be drawn:

- Replacement of fine aggregate with GBS resulted in decrease in workability of both RS and Msand based concrete and increases with the increase in GBS content.
- The rate of increase in compressive strength is higher at lower replacement levels, whereas the rate of increase in tensile strength is higher at higher replacement levels.
- Though 80% replacement with GBS shows better performance against acid attack. 100% replacement can be done when compared with the control mix.
- Combination of GBS and Msand performs well compared to that of GBS and RS combination.
- Inclusion of pozzolanic materials to the concrete improves the mechanical and durability performance relatively at higher replacement percentages of fine aggregate with GBS.

5. References

[1] Nanthagopalan P and Santhanam M 2011 Fresh and hardened properties of self-compacting Concrete produced with manufactured sand *Cem. Concr. Compos.* **33** 353–358

[2] Srinivasarao C and Vijaya Bhaskar Reddy C Study of standard grade concrete consisting of granulated blast furnace slag as a fine aggregate *Materials Today: Proceedings* https://doi.org/ 10.1016/j.matpr.2020.01.024

[3] Patra R K and Mukharjee B B 2017 Influence of incorporation of granulated blast furnace slag as replacement of fine aggregate on properties of concrete J. Clean. Prod. 165 468- 476

[4] Tiwari A, Singh S and Nagar R 2016 Feasibility assessment for partial replacement of fine Aggregate to attain cleaner production perspective in concrete: A review J. Clean. Prod. 135 490–507

[5] Siddique R and Iqbal Khan M 2011 Ground Granulated Blast Furnace Slag : Supplementary Cementing Materials in Engineering Materials Springer Berlin, Heidelberg 37

[6] Gaurav Singha, Souvik Dasa, Abdulaziz Abdullahi Ahmeda, Showmen Sahab and Somnath Karmakar 2015 Study of Granulated Blast Furnace Slag as Fine Aggregates in Concrete for Sustainable Infrastructure *Procedia - Social and Behavioral Sciences* **195** 2272 – 2279

[7] Mostafa Jalal, Alireza Pouladkhan, Omid Fasihi Harandi and Davoud Jafari 2015 Comparative study on effects of Class F fly ash, nano silica and silica fume on properties of high performance self-compacting concrete *Construction and Building Materials* **94** 90-104

[8] Singh Ashita, Bhadauria Sudhir, Kushwah Suresh and Tare Kshitij 2019 The study on optimization of cement and fine aggregate by blast furnace slag in concrete, AIP Conference Proceedings 2158

[9] Malagavelli, Venu and Rao P 2010 High performance concrete with GGBS and ROBO sand *International Journal of Engineering Science and Technology* **2** 5107-5113

[10] Yüksel I, Bilir T and Özkan O 2007 Durability of concrete incorporating non-ground blast furnace slag and bottom ash as fine aggregate *Build. Environ.* **42** 2651–2659

[11] Reddy B V V 2013 Alternatives to river sand a sustainable approach Indian Concrete Institute 30–41

[12] Li G and Zhao X 2003 Properties of concrete incorporating fly ash and ground granulated blast-furnace slag *Cem. Concr. Compos.* **25** 293–299

IOP Conf. Series: Materials Science and Engineering **989** (2020) 012014 doi:10.1088/1757-899X/989/1/012014

Mechanical properties of rubber aggregates based geopolymer concrete- A review

Sesha Choudary Yeluri[1,2] and Nishant Yadav[1]

[1]School of Civil Engineering, Lovely Professional University, Punjab, India

[2]Email: yeluriseshachoudary@gmail.com

Abstract. Aggregates are the furthermost vital components of concrete, but there is a need for replacement of these natural resources due to the insufficient availability of the materials. To overcome the difficulty of materials and to develop sustainable concrete, much research has been conducting to date. The replacement of aggregates wholly or partially is one of the best alternatives to this problem. To develop sustainability in concrete waste materials that are causing depletion to the environment are used as alternative materials. Rubber tire waste is one such material which is causing a lot of greenhouse gas emission. This paper reviews the use of rubber tire waste as aggregates in geopolymer concrete which shows the effect of various sizes of rubber, untreated and treated rubber aggregate on various properties such as Fresh Properties where flow values of mortar reduced with escalation in the rubber and workability of the mortar and concrete increased. Nevertheless, the Mechanical Properties shows a decrement in compressive strength, and tension properties show an increment. There is a significant amount of increment in the possessions of mortar and concrete after the crumbed rubber treated. The loss of strength for the small amount of replacement of rubber, which is 5-10% shows the strength higher than the standard concrete. The outcomes show that the custom of the rubber in concrete primes to a substantial amount of strength loss, but the addition of mineral admixture and superplasticiser helps us to achieve the geopolymer mortar and concrete applications in the field of construction.

1. Introduction

Sustainability means assembly the requirements of the current without conceding the ability of the forthcoming generation for their use [1]. Sustainability can be obtained by recycled and reuse of waste material as the replacements of natural resources. The automobile industry is a rapidly growing industry, which results in an abundance of tire production, and, subsequently, wastes tire is generated [2]. Annually more than 500 million tonnes of tires disposed of [3]. Its believed that around 30 million cars are produced alone in India by 2019. Rubber tires of 24 million tonnes weight of the rubber used on average. Millions of rubber tires are dumped illegally without secondary usage. This waste tire disposal requires a massive amount of landfills which amasses the risk of combustion which results in pollution due to high toxin emissions and are non-biodegradable, which leads to poses deadly diseases and a threat to surrounding environment [4]. Degradation of rubber is a prolonged process. It almost takes 50-80 years for itself. Mostly tires are dumped in the typical dump yard and are lit with other waste, which leads to black pollution. Therefore, recycled and reuse of this rubber tire waste in other forms can reduce the adverse effect on the environment.

IOP Conf. Series: Materials Science and Engineering **989** (2020) 012014 doi:10.1088/1757-899X/989/1/012014

Rubber tire waste incorporated in concrete by shredding the waste tire into desirable sizes of natural aggregates and is replaced partially or entirely as aggregates in conventional concrete. Rubber addition in concrete shall enhance the Impact resistance, Abrasion resistance, Freeze-thaw cycles, Durability performance of concrete [4]. The rubber-based concrete used in places where the impact resistance, vibrational damping, heat isolation is required such as the Building bunkers, Jersey barriers, Foundation for machinery, Railway stations, Lightweight concrete blocks, Tiles, sea dense, Armor units, floor screeds, trench filling and pile heads end to end with paving slabs and pipes bedding [2]. Rubber-based concrete also helps in reducing the noise, which is caused by various sources such as vehicles, aircraft, power plants, and machinery. The sound caused by sources not only causes health problems but also leads to serious health issues [5]. The sound waves impeded materials with notable porosity, which results in sound waves into solid and energy transforms into a form of heat and shall be dissipated [5]. Nevertheless, the usage of the crumb rubber still used as both coarse and fine aggregates incorporated in Ordinary Portland Cement-based concrete, which results in a large amount of CO_2.

Concrete is a significant material in the field of construction, which consists of components such as cement, aggregates, water. The demand for the components of concrete can never be satisfied and expected to increase in the coming future. However, there is a various environmental issue associated with the components—for instance, the cement industry-acclaimed to be the second-largest contributor to greenhouse gas emission [6]. Moreover, cement production expected to reach 4380 million tons by 2050 [7]. One ton of cement production leads to 0.85-1 tons of CO2 [8]. With the urgency to calm down the global crisis due to the greenhouse gas emission, there is a need for replacing these environment unfriendly materials and consider some other replacement materials in place of cement, which is the motive of this research to present a supplementary cementitious material.

Geopolymer concrete is the alkali-activated aluminosilicate material wherein alkaline chemicals used as binding material. Depending on material castoff in curing conditions for preparing the geopolymer concrete, the geo-polymerisation is different for those produced in hydration of OPC [9]. For this, the calcium content should also be shallow to qualify the construction of the three-dimensional polymeric chain, and the ring construction consists of a Si-O-Al-O bond [10]. The geopolymer concrete consists of the SiO_4 and AlO_4, which has a tetrahedral bond interchangeably sharing all the oxygen with a monovalent alkali metal. The principal empirical of the geopolymer shown in equation (1).

$$M_n [-(SiO_2)_Z - AlO_2]_n.wH_2O \quad (1)$$

Where M is a monovalent alkali metal, n is the degree of the polycondensation of polymerisation, and z is the variable, which is 1,2,3 or higher.

2. Materials

The material component constitutes of binder materials such as aluminosilicate material such as the Fly ash, GGBS, burnt Rice husk ash, various quarry slag which forms geo-polymeric reaction when combined with alkali activator solution such as the sodium silicate and sodium hydroxide, potassium silicate and potassium hydroxide. Along with natural aggregates such as coarse aggregates and fine aggregate.

2.1Production of pozzolanic materials

The explosive volcanic eruption is the primary source for the natural pozzolans in the pyroclastic rocks. This explosion results in small atoms of melted magma to expose to an atmosphere where the rapid compression decreases results in the microporous structures by the emission of the gases and quenching lead to their glassy state. This deposits of volcanic tuff by further exposed to the weathering leads to zeolitisation results in zeolitic materials or clay minerals [11]. The artificial pozzolans such as fly ash, GGBS are the industrial produced during their production. This aluminosilicate material used for the geopolymer formation having a size of 20-40 µm.

2.2 *Reactivity of pozzolanic material*

Alkali activator solution (AAS) helps us to achieve the reactivity of the geopolymer binder. Reactivity mostly depends on the chemical composition, specific surface area, and mineralogical composition of materials. The chemical composition of fly ash used for geopolymer formation shown in Table 1. The significant portion of silica and alumina presence in the materials helps us to achieve the geopolymer blend. The sophisticated quantity of the loss of ignition in materials shows the presence of the CO_2 and water in the glassy form. It is difficult to differentiate between the high and low reactivity of material based on the chemical and mineral composition because only the glassy phase of mineral chemical composition used to determine the reactivity [12].

2.3 *Production of the rubber aggregates*

The production of the rubber aggregates starts with collecting rubber tire from the local sources. First, the tires inserted to cut the thickness of the tire in a circle, and then it sent into a strip cutter. The strip cutter helps us to obtain the rubber in strip form and then send it to the slice cutter. The slice cutter helps us to obtain the rubber blocks, and then rubber is crushed in the rubber crusher. The fiber obtained by crushing the rubber separated by using the fiber separator and remaining rubber waste sent for magnetic separator. The magnetic separator bifurcates the rubber powder and steel wires.

Treatment of the crumb rubber: The pre-treatment of the rubber helps us to obtained better results compared to untreated. The treatment of rubber is achieved by oxidation followed by sulphonation. In this treatment process, the rubber is immersed in the desired chemical, as mentioned for a stipulated amount of time, then cleaned and sundried. For the simple and cost efficiency purpose, the hydroxide treatment used. [13]

Table 1. Chemical composition of the binder materials.

Sources	SiO_2	Al_2O_3	Fe_2O_3	K_2O	CaO	MgO	TiO_2	SO_3	MnO	Na_2O	P_2O_3	LOI
Fly ash												
Zhang et.al	50.3	22.9	8.17	3.55	3.38	2	1.15	0.58	0.58			
Wongsa et.al	39.4	20.8	11.5	2.4	14.5	2.2	0.5	4.2		1.4	0.2	1.5
Mucsi et.al	45.85	16.8	12.05	1.83	13	2.9	0.48	3.76	0.18	0.5		
Azmi et.al	50	23.4	17.29	1.41	5.06		1.6	0.08	0.22			
luhar et.al	58.88	20.6	12.78			1.64		0.74		0.5		0.94
luhar et.al	58.88	20.6	12.78			1.64		0.74		0.5		0.94
Azrem et.al	50	23.4	16.92	1.41	5.06		1.6	0.08	0.22			
Aly et.al	39.79	11.2	1.2		34.4	7.61		0.46				
Yuwadee et.al	39.4	20.8	11.5	2.4	14.5	2.2	0.5	4.2		1.4	0.2	1.5
Rahendran et.al	61.75	24.6	6.47	0.55	3.45	1.53	0.91					
GGBS												
Zhang et.al	34.1	12.3	0.41	0.56	44.2	8.12	0.96	2.59	0.25			
Rahendran et.al	37.73	14.4	1.11		37.3	8.71		0.39	0.02			1.41
Kaolin												
Gandoman et.al	63.63	24.4	0.539	0.25	1.25	0.25		0.06			0.08	9.71
Metakaolin												
Gandoman et.al	69.3	26.9	1.471	0.71	1.47	0.31		0.06			0.09	0.55
OPC												
Yuwadee et.al	18.1	4.2	3	0.6	61.1	1.1	0.2	3.9		0.2	0.1	5.9

3. Properties of rubber-based geopolymer concrete

3.1 Mechanical properties

The mechanical properties of the rubber-based geopolymer mortar and concrete, such as compressive strength, split tensile strength, and flexural strength is the most vital features of the geopolymer mortar and concrete for its application.

3.1.1 Compressive strength

The superiority of the concrete determined by conducting a compressive strength test plays a dynamic role. Factors affecting incorporation of the crumb rubber in concrete are % off replacement for both fine aggregate and coarse aggregates, and both combine, the addition of steel fiber along with crumb rubber, various size of aggregate, NaOH molarity, SS/SH ratio, AAS/ Binder ratio, Elevated temperature. The Compressive strength test is conducted according to the certain standards by various codes BS EN:12390-3(2009)[4], ASTM C109/C109-16a (2016)[2], [14]–[16], MSZ EN 196-1:2005[17], IS:516-1959[18], ASTMC39[19]

Effect of the replacement of Fine Aggregate with the crumb rubber on *OPC & GPC:* The compressive strength of OPC and GPC for replacement of the fine aggregate by 0%, 10%, 20%, and 30% are studied [18]. Results show that the normal GPC shows similar strength than the OPC, which are 53.1 N/mm^2 and 54 N/mm^2 for the 28 days, 55.32 N/mm^{2,} and 56.23 N/mm^2 for 90 days and 56.9 N/mm^2 and 61.55 N/mm^2 for 365 days. For rubberised geopolymer concrete replacement of fine aggregate by 10% crumb rubber decreased the strength by 10-12% for curing period of 3,7, 28, 90,365 days, 20% replacement of fine aggregate by crumb rubber the strength of concrete decreased by 25-28 % for the curing period of 3,7, 28, 90 days and 30% replacement of fine aggregate by crumb rubber the strength of concrete decreased by 40-46 % for the curing period of 3,7, 28, 90 days.

Effect of replacement of Coarse aggregate and fine aggregates with crumb rubber*:* The compressive strength of rubberised GPC when replaced with both coarse aggregate and fine aggregates by 2%, 6%, 10%, 14% are studied[5]. It shows that the replacement of coarse and fine aggregates by 2% of crumb rubber leads to increment in strength by 6% whereas further increment 6%, 10%, 14% leads to reduce the strength by 20 %, 45% and 60%[5]. The use of crumb rubber by minimal amount may show better results compared to standard concrete. Fig 1 shows the effect of replacement of aggregates with crumb rubber[5]

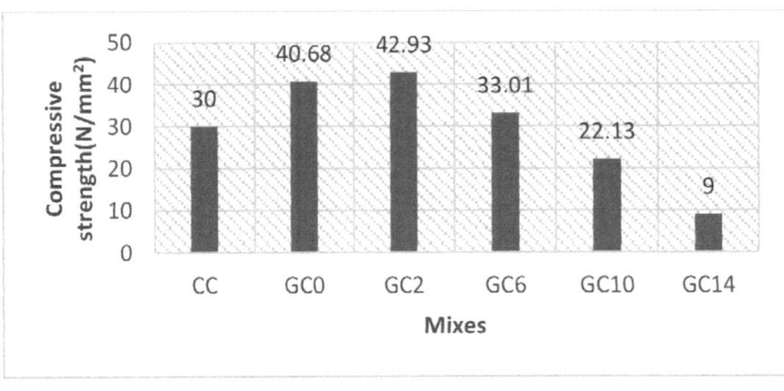

Figure 1. Effect of replacement of coarse and fine aggregates with crumbed rubber on compressive strength.

Effect of replacement of Fine aggregates with crumb rubber & steel fiber: The addition of steel fiber in GPC increases the compressive strength of concrete by a significant amount[4]. The study conducted on the rubberised geopolymer concrete replacement of fine aggregate by 5 %, 10%, and 15 % along with the addition of 0.5% and 1% steel fiber. The results suggest that the compressive strength is obtained maximum for 28 days in the 5% replacement of crumb rubber along with 0.5%

steel fiber in geopolymer concrete. It also shows that the loss of strength for 15% replacement of crumb rubber and 1% steel fiber for 28 days is 35%. But remaining mixes show the loss is very less[4]. Fig 2 shows that the compressive strength of rubberised geopolymer concrete with crumb rubber and steel fiber.

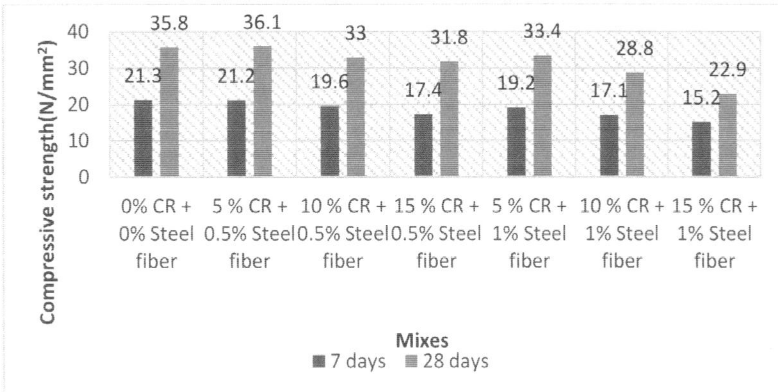

Figure 2. Effect of replacement of fine aggregate with crumb rubber along with steel fiber on compressive strength.

Effect of various sizes of crumb rubber: The compressive strength of rubberised geopolymer mortar for various sizes of aggregates conducted by Mucsi (2018)[17] explained here. It's evident from the Fig 3 that if the size of the aggregate decrease for the same proportion of replacement of rubber in GPC leads to increment in compressive strength of geopolymer mortar by 10-25 %. But the addition of crumb rubber by 5 % leads to the same strength for both size aggregates[17].

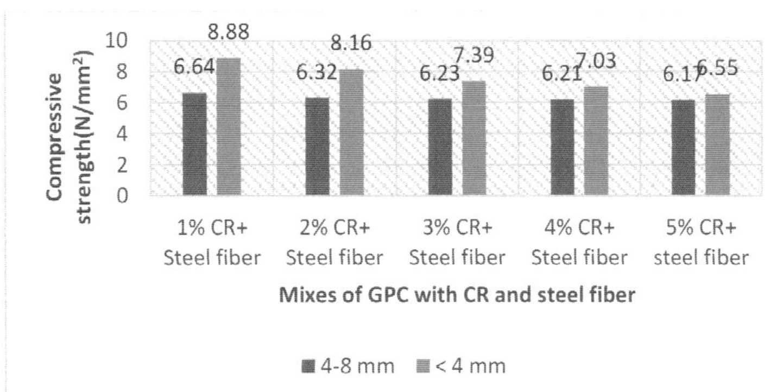

Figure 3. Effect of various size of aggregates on compressive strength.

Effect of elevated temperature on GPC: The compressive strength of rubberised GPC for elevated temperatures studied by Luhar (2018)[20] shows that the increase in temperature results in loss of strength in concrete by a significant amount. Fig 4 shows that the 5 % replacement of fine aggregate by crumb rubber leads to lower the compressive strength by 50% for 800^{0}C. But it shows that the 600^{0}C temperature shows the least compressive strength on geopolymer specimen.[20]

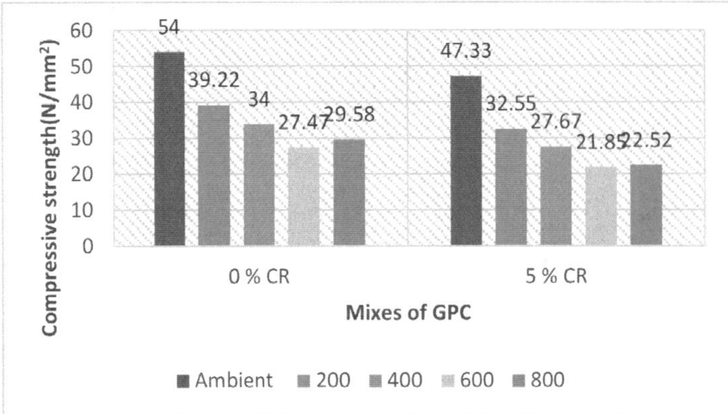

Figure 4. Effect of elevated temperature on compressive strength.

Effect of Sodium Silicate/Sodium Hydroxide (SS/SH) ratio: The effect of various SS/SH ratio on the compressive strength of rubberised GPC is studied[2]. The compressive strength of mortar declines with the increase in the SS/SH ratio. It shows little increase for early days curing period such as 3 days is 2.08 N/mm^2, 2.69 N/mm^2 and 2.65 N/mm^2 for 0.5, 1 and 1.5 SS/SH ratio and for 14 days is 2.53 N/mm^2, 2.69 N/mm^2, 2.86 N/mm^2 for 0.5, 1 and 1.5 SS/SH ratio. But for a curing period of 28, 90, 180 days, the decrement in compressive strength of mortar by 10-15 %[2]. Fig 5 shows the compressive strength of the GP mortar effect with the SS/SH ratio.

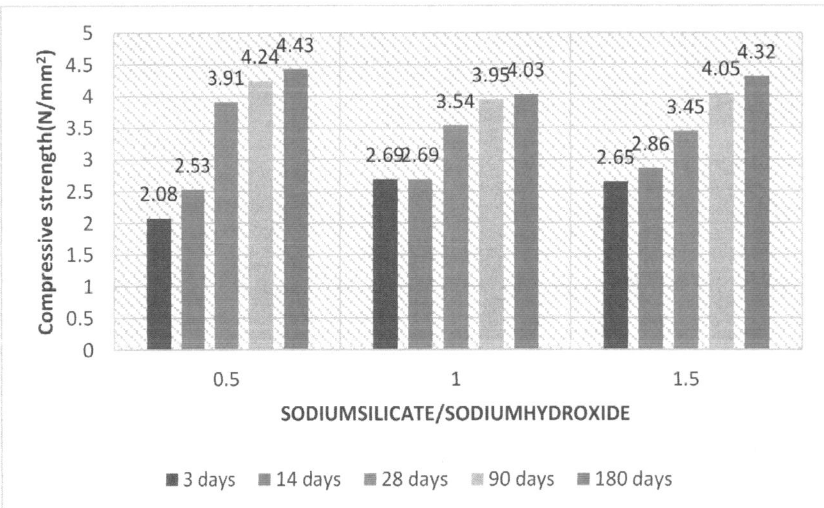

Figure 5. Effect of SS/SH ratio on compressive strength.

Effect of NaOH molarity: The effect of various molarity of NaOH on the compressive strength of rubberised GPC is studied[2]. The compressive strength of mortar increases with an increase in molarity of NaOH from 10 to 15, whereas further increment shows the decrement in strength for 20M. It shows the compressive strength of rubberized GPM for curing period of 3 days is 2.69 N/mm^2, 2.69 N/mm^2 and 2.44 N/mm^2 for 10, 15, 20 M of NaOH and for 14 days is 2.69 N/mm^2, 2.95 N/mm^2, 2.56 N/mm^2 for 10, 15, 20 M of NaOH. Compressive strength of rubber-based GPC increases 8% with molarity increment from 10-15 M. However, increment from 15 -20 M strength reduced to 5%[2]. Fig 6 shows the effect of various molarity of NaOH on the compressive strength of rubber-based geopolymer concrete.

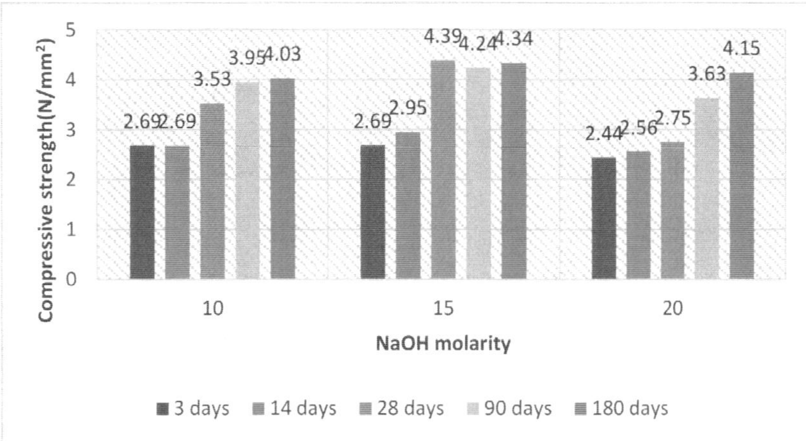

Figure 6. Effect of various molarity of NaOH on compressive strength.

3.1.2 Flexural strength

To determine the maximum stress of the concrete flexural strength test plays a vital role. Factors affecting the incorporation of crumb rubber in concrete are % off replacement for fine aggregate, the addition of steel fiber along with crumb rubber, various size of aggregate, NaOH molarity, SS/SH ratio, AAS/ Binder ratio, Elevated temperature. The flexural test is conducted according to the certain standards by various codes ASTMC1609[4], ASTM C78/C78-16 (2016)[2], [14]–[16], MSZ EN 196-1:2005[17], IS:516-1959[18], ASTMC78[19]

Effect of replacement of Fine Aggregate with crumb rubber on OPC & GPC: The addition of rubber in GPC increases the flexural strength of concrete both in OPC and GPC. The results show that the flexural strength of GPC is 6.45 $N/mm^{2,}$ which is greater than OPC is 5.35 N/mm^2. The addition of rubber shows the increment of flexural strength by 6% for 28 days and 90 days curing period and 8% for 365 days curing period. The flexural strength of rubberised GPC shows for 10%, 20%, and 30% replacement of fine aggregate with crumb rubber is 2%, 3.8%, and 6% when compared with nominal GPC.

Effect of replacement of Fine aggregates with crumb rubber & steel fiber: The addition of steel fiber in GPC increases the flexural strength of concrete by a significant amount. [4]. The study conducted on the rubberised geopolymer concrete replacement of fine aggregate by 5 %, 10%, and 15 % along with the addition of 0.5% and 1% steel fiber. The results suggest that the flexural strength is obtained maximum for 28 days in the 5% replacement of crumb rubber along with 1 % steel fiber in geopolymer concrete, which is 6.29 N/mm^2. It shows that the 5% replacement of crumb rubber shows an increment in flexural strength compared to nominal GPC, whereas further replacement results in the decrement of strength. But the addition of fiber to the rubber-based concrete shows better flexural strength[4]. Fig 7 shows the effect of replacement of fine aggregate with crumb rubber along with steel fiber on flexural strength.

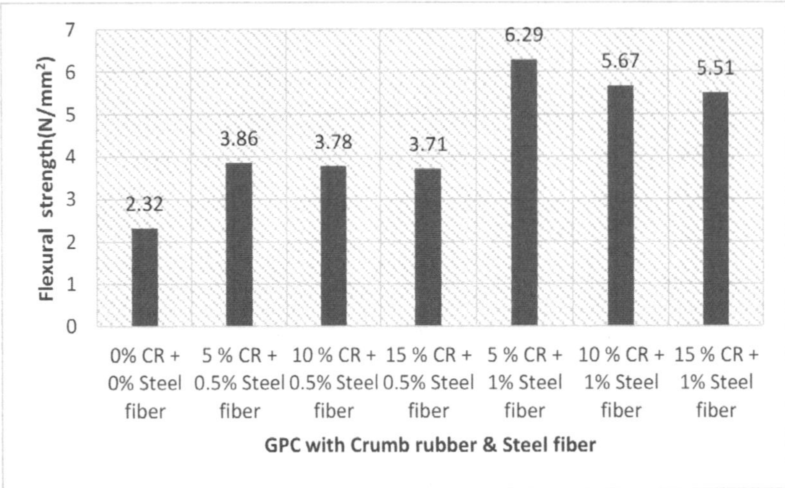

Figure 7. Effect of replacement of fine aggregate with crumb rubber along with steel fiber on flexural strength.

Effect of various sizes of crumb rubber: The flexural strength of rubberised geopolymer concrete for various sizes of crumb rubber is studied[17]. The results show that the increase in flexural strength with an increase in steel fiber until 3% steel fiber for 4-8 mm size of aggregate. Further increment of steel fiber for 4-8 mm leads to a decrease in flexural strength, whereas for < 4 mm size aggregates, the increment in steel fiber leads to increment in flexural strength. The addition of rubber reduced the flexural strength compared to steel fiber-based concrete. Flexural strength of rubber-based geopolymer increases with an increase in rubber content, whereas if the size of the aggregates decreases leads to further increment in strength. Fig 8 shows the compressive strength of rubberised geopolymer concrete with crumb rubber and steel fiber.[17]

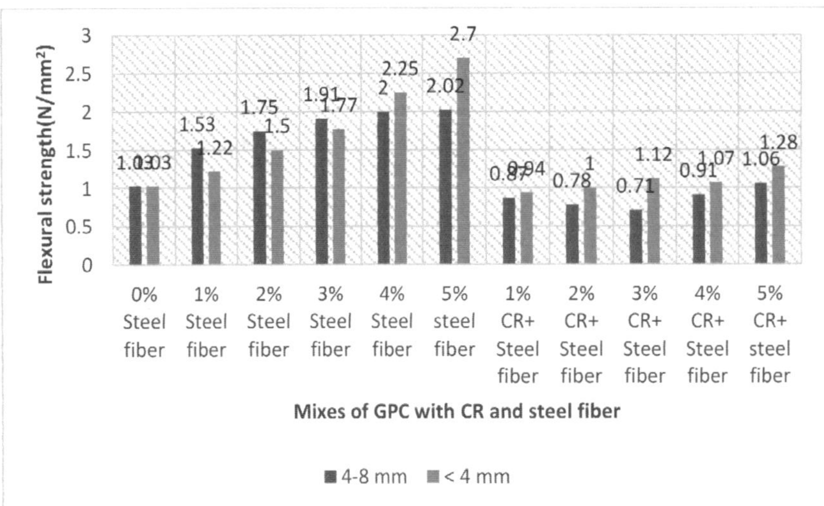

Figure 8. Effect of various size of aggregates on flexural strength.

Effect of various curing temperature on GPC: The flexural strength of rubberised GPC aimed at elevated temperatures are studied. It shows that the increase in temperature results in an increment of strength in concrete by a significant amount. Fig 9 shows that the for 100% spare of the fine aggregate by crumb rubber on 25^0C temperature shows Flexural strength of rubber-based geopolymer as 0.96 N/mm^2 whereas, for 60^0C and 90^0C, the flexural strength of mortar is 0.98 N/mm^2 and 1.89 N/mm^2 respectively. Various curing temperature displays that the surge in temperature primes to rise in flexural strength.[2]

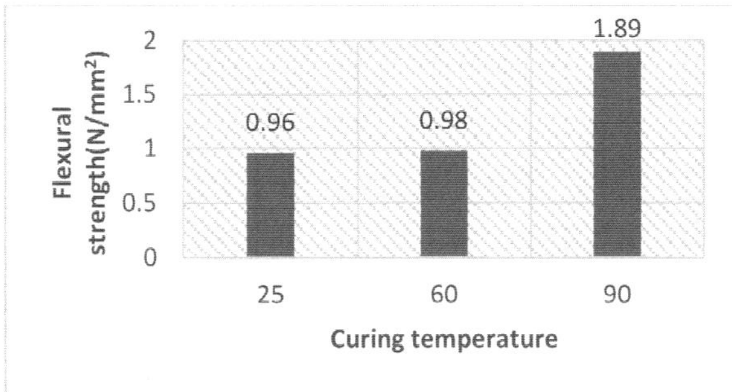

Figure 9. Effect of elevated curing temperature on flexural strength.

Effect of SS/SH: The Fig 10 displays that the surge in SS/SH leads to decline in flexural strength for SS/SH ratio of 1 which is 0.98 N/mm^2 when compared with SS/SH ratio of 0.5 which is 1.46 N/mm^2 and SS/SH of 1.5 leads to increase when compared with SS/SH ratio of 0.5 and 1 which is 1.29 N/mm^2 [2].

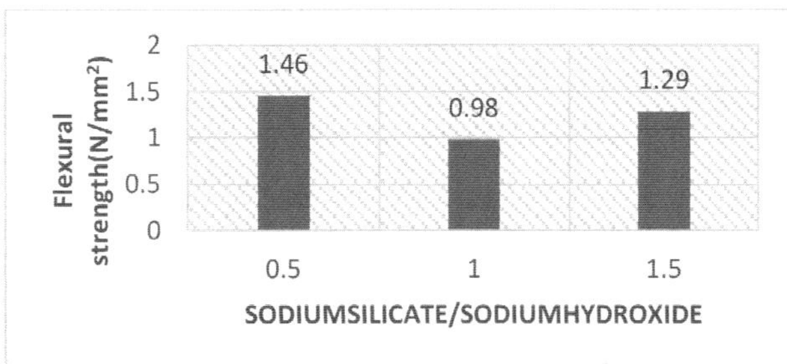

Figure 10. Effect of SS/SH on flexural strength.

Effect of NaOH molarity: The Fig 11 shows that the flexural strength of rubber-based geopolymer concrete is maximum 1.85 N/mm^2 when the molarity of NaOH is 15 whereas for 10 M and 20M is 0.98 N/mm^2 and 1.28 N/mm^2 correspondingly.[2]

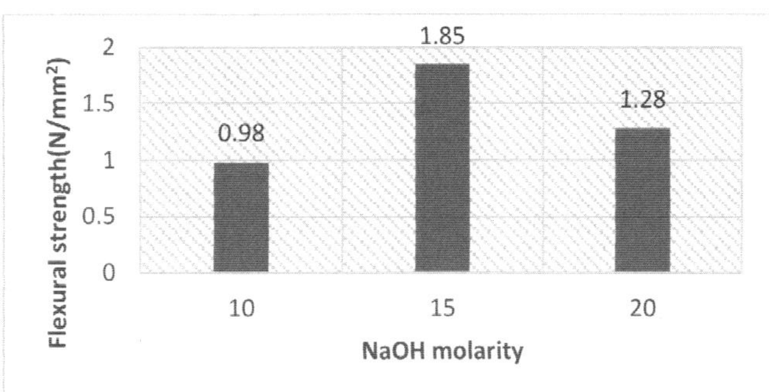

Figure 11. Effect of various molarity of NaOH on flexural strength.

IOP Conf. Series: Materials Science and Engineering **989** (2020) 012014 doi:10.1088/1757-899X/989/1/012014

3.1.3 Split tensile strength

The ductility of concrete can be attained by decisive the split tensile strength test. Factors affecting the incorporation of crumb rubber in concrete are % off replacement for both fine aggregate, size of aggregate.

Effect of replacement of Fine Aggregate with crumb rubber on OPC & GPC: The effect of replacement of fine aggregate by crumb rubber in proportions of 0%, 10%, 20%, 30%. The accumulation of crumbed rubber in OPC and GPC shows similar split tensile strength, but the GPC has split tensile strength advanced than OPC. The addition of rubber by 30% leads to obtained 5% increment in tensile strength compared to 0% replaced GPC[18]. This show that the tensile strength shows improvement by addition of rubber in GPC.

Effect of various sizes of crumb rubber: The scope of the various size crumb rubber affects the properties of rubberised GPC by means of a significant amount. Fig 12 shows that the tensile strength for 0% replacement shows that if the size of aggregates 1-4 mm, then the split tensile strength of rubberised GPC is 5.08, whereas, for size 2-4, the split tensile strength is 3.6 N/mm^2. Further addition of rubber by 10% shows the tensile strength of concrete as 5.13 for 1-4 mm size of aggregates whereas for size 2-4 mm is 2.355 N/mm^2. Results show if the size of the aggregate finer gives better split tensile strength[17].

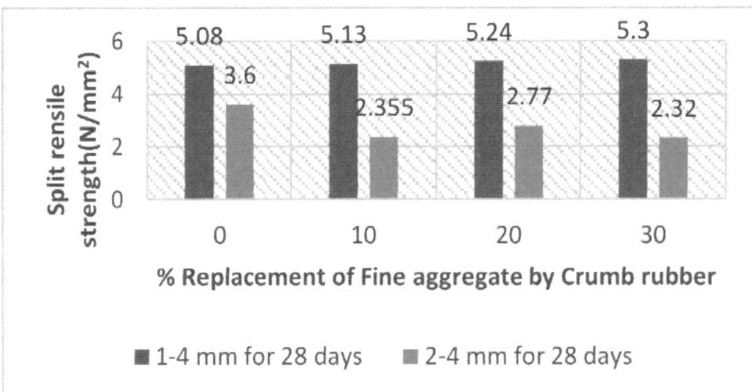

Figure 12. Effect of various size of aggregates on split tensile strength.

4. Conclusion

- Workability of the concrete increases with an surge in the rubber concrete but the steel fibre addition leads to decrement.
- Flow value decreases with an advance in rubber content in the geopolymer concrete.
- Compressive strength shows decreased with an surge in rubber concrete by whereas steel fibre addition leads to increment.
- The compressive strength of mortar reduced with the rise in crumbed rubber, but adding the steel fibre shows a noteworthy increment.
- Compressive strength of geopolymer concrete declined by 60% for 15% spare of the crumb rubber in concrete and 10% spare of the crumb rubber in mortar.
- 95 % of geopolymer concrete total compressive strength achieved for 7 days curing period, which shows the accumulation of rubber principals to accelerate the strength.
- An intensification in temperature shows similar properties for both GPC and rubberised geopolymer concrete.
- Thermal exposure primes to decrement in compressive strength up and around 600^{0}C, but then the further increment in temperature 800^{0}C shows an increment in compressive strength, but the hairline crack observed.

- Tensile strength for rubberised geopolymerized concrete increases with the rise in the rubber content.
- Flexural strength for rubberised geopolymer mortar and concrete surges with the rise in the rubber content.
- Flexural strength is observed extreme for 30 % standby of the Fine aggregate by the crumb rubber.
- Pre-treatment of the crumb rubber by NaOH leads to improve the compressive strength in geopolymer concrete.

5. References

[1] Leadbitter J 2002 PVC and sustainability *Prog. Polym. Sci* **27** 2197–2226

[2] Wongsa A, Sata V, Nematollahi B, Sanjayan J and Chindaprasirt P 2018 Mechanical and thermal properties of lightweight geopolymer mortar incorporating crumb rubber *Journal of Cleaner Production* **195** 1069-1080

[3] Thomas BS and Gupta RC 2015 Properties of high strength concrete containing scrap tire rubber *Journal of Cleaner Production* **113** 86-92

[4] Zhong H, Poon E W, Chen K and Zhang M 2019 Engineering properties of crumb rubber alkali-activated mortar reinforced with recycled steel fibres *Journal of Cleaner Production* **238** 117950

[5] Gandoman M and Kokabi M 2015 Sound barrier properties of sustainable waste rubber/geopolymer concrete *Iran Polym* **24** 105–112

[6] Malhotra P K and Mehta VM 2008 High-performance, High-Volume Fly ash concrete: Materials, Mixture, Proportioning, Properties, Construction Practice, and Case Histories *Suppl. Cem. Mater. Sustain. Dev. Inc. Ottawa*, **2** 3-9

[7] Hasanbeigi A, Price L and Lin E 2012 Emerging energy-efficiency and CO_2 emission-reduction technologies for cement and concrete production : A technical review *Renew. Sustain. Energy Rev.* **16 (8)** 6220–6238

[8] Praveen P U and Srinivasan K 2017 Self-compacting geopolymer concrete - A review *IOP Conf. Ser. Mater. Sci. Eng.* **263** 032024

[9] Duxson P, Provis J L and Si F 2008 Designing Precursors for Geopolymer Cements, *Journal of the american ceramic society* **91** 3864–3869

[10] Davidovits J and France S 2020 *Geopolymer Chemistry and Applications*, 5th edition (Saint-Quentin France: Institute of geopolymer) p 44

[11] Rose W I and Durant A J 2009 Fine ash content of explosive eruptions *J. Volcanol. Geotherm. Res.* **186 (1)** 32–39

[12] Tchakoute H K, Elimbi A, Yanne E and Djangang CN 2013 Utilisation of the volcanic ashes for the production of geopolymers cured at ambient temperature *Cem. Concr. Compos* **38** 75-81

[13] Rajendran M 2020 Performance of Crumb Rubber and Nano Fly Ash Based Ferro-Geopolymer Panels under Impact Load *KSCE Journal of civil engineering* **22** 1810-1820

[14] Azmi A A, Abdullah M M A B, Ghazali C M R, Sandhu AV and Hussain K 2016 Effect Of Crumb Rubber On Compressive Strength of Fly Ash Based Geopolymer Concrete *MATEC Web of Conferences* **78** 4–8

[15] Azmi A A, Abdullah M M A B, Ghazali C M R, Sandhu AV and Hussain K 2019 The Effect of Different Crumb Rubber Loading on the Properties of Fly Ash-Based Geopolymer Concrete *IOP Conf. Series: Materials Science and Engineering* **551** 012079

[16] Zaetang Y, Wongsa A, Chindaprasit P and Sata V 2019 Utilisation of crumb rubber as aggregate in high calcium fly ash geopolymer mortars *International Journal of GEOMATE* **17** (64) 158–165

[17] Mucsi G, Szenczi A and Nagy S 2018 Fibre-reinforced geopolymer from synergetic utilisation of fly ash and waste tire *J. Clean. Prod.* **178** 429-440

[18] Luhar S, Chaudhary S and Luhar I 2019 Development of rubberised geopolymer concrete : Strength and durability studies *Constr. Build. Mater.* **204** 740–753

[19] Muhammed A, El-feky M S, Kohail M and Nasr E A R 2019 Performance of geopolymer concrete containing recycled rubber *Constr. Build. Mater.* **207** 136–144

Mechanical properties of alkali-activated mortar in organic acid environment

Athira Ajay[1,2], Ramaswamy K P[1] and Anu V Thomas[1]

[1]Department of Civil Engineering, TKM College of Engineering, Kollam, Kerala, India.

[2]Email: athiajay96@gmail.com

Abstract. Ingression of acid into the concrete results in early degradation by microstructural alteration of phases leading to mass changes, weakening of mechanical properties and increase in porosity due to calcium leaching. This paper discusses the resistance of geopolymer composites (GP) and conventional concrete system (OPC) in 0.125M and 0.25M concentrations of acetic acid medium by investigating the changes in the mechanical properties of mortar. The results of the study reveal that the GP samples display increased resistance to acetic acid attack not only in terms of mass and strength loss but also with reference to the variation in ultrasonic pulse velocity and relative dynamic modulus of elasticity. The results also manifest that blends of ground granulated blast furnace slag (GGBS) and fly ash (class F) (FA) with red mud (RM) offer increased acid resistance due to the nature of the main binder gel produced in GP systems.

1. Introduction

Proliferation of agriculture and food industries like breeding, sugar and dairy industries produce large quantities of effluents that are major sources of organic acids. Concrete, being an extensively used construction material worldwide, often comes in contact with these untreated effluents and the organic acids in the effluents will eventually deteriorate the concrete structures. The interaction of these complex organic acids with the concrete structures leads to serious premature degradation of concrete matrix and thereby affects the durability of the structure.

Effluents such as liquid manure, silage effluents, winery wastewaters from the distillery and fermentation industries contain predominant organic acids like acetic, lactic, citric, butyric and tartaric acids, the pH of which is around 4.5 generally [1,2]. The silage acids namely acetic and lactic acids present in the silage effluents cause deterioration of conventional cast-in-place and precast concrete silos [3]. These acids react with the cement matrix and reduce the strength of the silo wall prompting the buckling of walls and eventually lead to failure. Majority of the previous research has focused on the degradation of concrete in the presence of inorganic acids. Only limited studies have reported the effect of organic acids on concrete structures. Regardless of organic acids being weak acids, its partial dissociative nature can destroy the alkaline concrete matrix. These acids instigate the corrosion of installed reinforcement demanding maintenance and repair, thereby disrupting the production facilities.

Concrete with Ordinary Portland Cement (OPC) has been reported to be less resistant to acidic media. Moreover, the use of OPC is discouraged nowadays as its production contributes alarmingly to carbon dioxide emissions and thus global warming. Contribution of CO_2 emissions worldwide by cement industry is approximated to be around 7% [4]. As we strive towards achieving sustainable development, the complete replacement of cement with alkali activated/ geopolymer composites can be considered as a viable alternative for cementitious materials exposed to aggressive acidic environments.

According to the Brazilian NBR 10004 standards, 66 million tons of red mud (RM) is produced annually across the world and it has been labeled as a hazardous pollutant [5, 6]. Therefore, the incorporation of such waste materials as an alternative binder material in alkali activated systems could lead to reduction of pollution level across the globe. Geopolymer composites (GP) system is an innovative technology that adopts the industrial solid wastes such as class F-fly ash (FA), ground granulated blast furnace slag (GGBS) and RM as raw material to form a 3D network of cementitious materials under alkaline conditions through polymerization [7]. Performance of alkali activated GGBS and FA against inorganic acids has been individually assessed by many authors. The organic acid resistance of these two binder types has not received much attention. This paper investigates the resistance of GP materials in acetic acid solution based on the study of degradation kinetics. The degradation kinetics was studied by monitoring the changes in the physico-mechanical properties of mortar specimens.

2. Mechanism of degradation by organic acids

Concrete being highly alkaline in nature has a pH around 12. In contact with acidic environment, the pH of the matrix decreases and the equilibrium of cement matrix get disturbed. The hydration products then undergo decomposition leading to severe degradation in the properties of the material. This basic mechanism of deterioration due to acid attack is acidolysis. In the case of organic acid attack, there is potential of two more mechanisms to happen simultaneously. Of the two other mechanisms, the first one is complexolysis, where the complexes are formed from the reaction between acid ions and metal ions extracted during the acidolysis phase. As a result, more of the solid phase is dissolved within the acid solution. The latter mechanism is the precipitation of expansive reaction products that is prominently significant when the salt product with a higher molar volume replaces the hydration products within the matrix and thus results in expansion and cracking [8, 9].

The rate of degradation of concrete system is influenced by the aggressiveness of the attacking acid which depends on numerous factors like the solubility of the salts, acid dissociation constant (pKa), mono/poly-acidity of the acids and the physical properties of the salts formed. Also, the alteration kinetics is influenced by these salts that may either precipitate or leach out depending on its solubility in the acid solution. While considering organic acids with soluble salts, the relative combativeness of the acids seems to be primarily influenced by the pKa of the acids. However, for organic acids with insoluble salts the relative aggressiveness is dominated by the poly-acidity factor [8, 9, 10]. The reaction of cement-based materials in acetic acid is comparable to that of strong acids like nitric acid or hydrochloric acid. Acetic acid, generally termed as a weak monocarboxylic acid, stimulates the dissolution of the calcium hydroxide present in concrete which gets leached away exposing fresh concrete surface to acid attack [10]. On reaction of acetic acid with C-S-H gel and aluminates, water soluble calcium and aluminum salts are formed, paving the way for increase in porosity of the matrix and reduction of mechanical strength. The rate of degradation by reaction with acetic acid proliferates at a rapid rate as no protective layer is formed on the external surface of specimens. This instates the need to investigate the effect of acetic acid further.

Moreover, as hydrates in conventional cementitious systems offer less resistance, the acid resistance potential of geopolymer (GP) composites as an alternative material is to be explored.

GP are prominently known to be the subset of alkali activated binder system which are amorphous in nature rather than crystalline [11, 12]. Geopolymerization is the sole reaction taking place within the geopolymer system unlike other alkali activated systems. This process is an exothermic reaction where reaction takes place between silica (Si)-alumina (Al) under alkaline condition which subsequently create three-dimensional polymeric chain of Si-O-Al-O bonds. The aluminium silicate bond influences the structural integrity of GP systems unlike the calcium silicate hydrate bond [13, 14]. Sodium aluminosilicate hydrate (N-A-S-H) (FA based system) or calcium aluminate hydrate silicate (C-A-S-H) (GGBS based system) or a combination sodium calcium aluminium silicate hydrate (N-C-A-S-H) gel are the main reaction products of the alkali-activated aluminosilicates [15]. On exposure to acid, an electrophilic attack of acid protons on Si-O-Al-O bonds is found. The sodium ions that act as charge compensation cations of the aluminosilicate framework, are leached out in exchange with hydronium ions from the solution. The atomic ratio of Si-to-Al is then found to increase, evidently indicating that the tetrahedral Al from the aluminosilicate framework is dissolved completely by breaking the Si-O-Al bond [13, 15, 16]. The acid resistance is found to increase with a decrease in CaO content, i.e., the resistance of C-S-H < C-A-S-H < N-A-S-H [17]. Acid attack itself is still an infant area of research. Moreover, the investigations on the effect of organic acids such as acetic acid resistance of geopolymer systems are rather limited and hence these aspects are explored in the present study.

3. Experimental program

3.1. Materials

The precursors used in the current investigation include FA (class F), GGBS and RM. Ordinary Portland Cement (OPC) 53 Grade is used in the control mix. RM, procured as wet sludge from HINDALCO, Belgaum, was oven dried at 100^0C for 24 hours. Later, the oven dried RM was subjected to 1000 revolutions in the Los-Angeles abrasion machine to attain the required fineness. Alkaline activators, sodium hydroxide (10M) and sodium silicate solution were adopted with Alkali silicate/Alkali hydroxide ratio as 2.5. Sulphonated Napthalene Formaldehyde (SNF) superplasticizer with 3% by weight of binders was used for the geopolymer mixes. Mortar ratio for mix design is maintained as 1:1.5 (binder: sand). Cylindrical moulds having 25 mm diameter and 50 mm height were used to prepare the specimens. The control mix was prepared using distilled water with water to cement ratio of 0.50. GP mix was prepared with Alkali/Binder ratio of 0.50. Manufactured sand (finer than 2 mm) was used in the mortar specimens as fine aggregates.

3.2. Mixing, casting, curing and testing procedures

The mixes adopted are OPC, FGP (100% FA), SGP (100% GGBS), FSGP (50% GGBS + 50%FA), RFGP (50% FA+ 50%RM) and RSGP (50% GGBS + 50%RM). The mixing was done using Hobart mixer. The plastic moulds were oiled and filling of mortar into the moulds were done manually with tamping. After casting, specimens were kept for 24 hours in atmospheric conditions for the initial set. For OPC mix, lime solution curing was done for 14 days to maintain the alkalinity and to ensure that no degradation reaction takes place with an alteration in the pH level. 14 days air (ambient) curing was done for SGP, FSGP and RSGP specimens. FGP and RFGP specimens were oven cured at 70^0C for 24 hours and then air cured for 14 days. No chemical admixture was used for the preparation of OPC mix investigated in the current study. To investigate the resistance of GP and OPC based systems to acetic acid attack, the mortar samples prepared were submerged completely in acetic acid solution of 0.125M and 0.25M concentrations for 8 weeks. The samples were monitored visually and their mass change, compressive strength, altered depth and ultrasonic pulse velocity variation were also recorded periodically. The acid solution was replaced once in two weeks to maintain the aggressiveness of the solution and to sustain the attack.

4. Results and discussions

4.1. Visual appearance

The extent of deterioration can be visually explored by taking digital photographs on a regular basis. Figure 1 showcases the photographs of specimens after 8 weeks of exposure in various concentrations of acetic acid solution. No formation or deposition of insoluble precipitate/products on the surface of the specimen was observed. On comparing the intensity of degradation with the concentrations of acids for all the binders, more damage was observed in the case of 0.25M acetic acid than 0.125M solution. This behaviour may be attributed to the fact that aggressiveness increases due to the increase in concentration. Again, on comparing the binders, extreme colour variation is found to be more predominant in OPC specimen. This may be due to the reaction of acetic acid with Portlandite and C-S-H gel to form water soluble calcium acetate monohydrate which leaches to the solution leaving amorphous silica gel on the specimen causing brown colour formation and thereby causing disintegration of the internal microstructure. Visual examination of RSGP and RFGP specimens is difficult because these are rich red in colour even before the onset of exposure. Notably, FGP and FSGP specimens did not depict extreme colour variations even after 8 weeks of exposure.

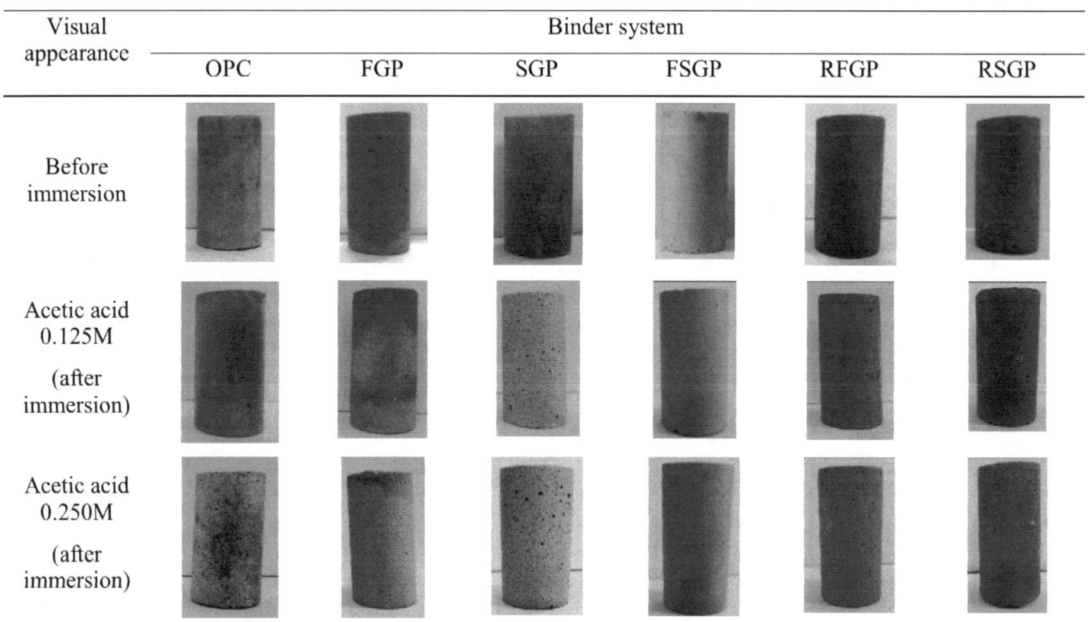

Figure 1. Visual observation of specimens immersed in various concentrations of acetic acid after 8 weeks of exposure.

4.2 Mass changes

On a weekly basis, mass changes (in saturated but surface dry condition) with respect to the initial saturated mass (noted before the acid exposure) was determined. The average of the mass changes of three specimens was calculated. The change in mass of the specimens with age of exposure in acid solution is shown in Figure 2. Typical graphs of the six mixes, i.e., OPC, FGP, SGP, FSGP, RFGP and RSGP on exposure to different concentration of acetic acid medium is represented in Figure 2.

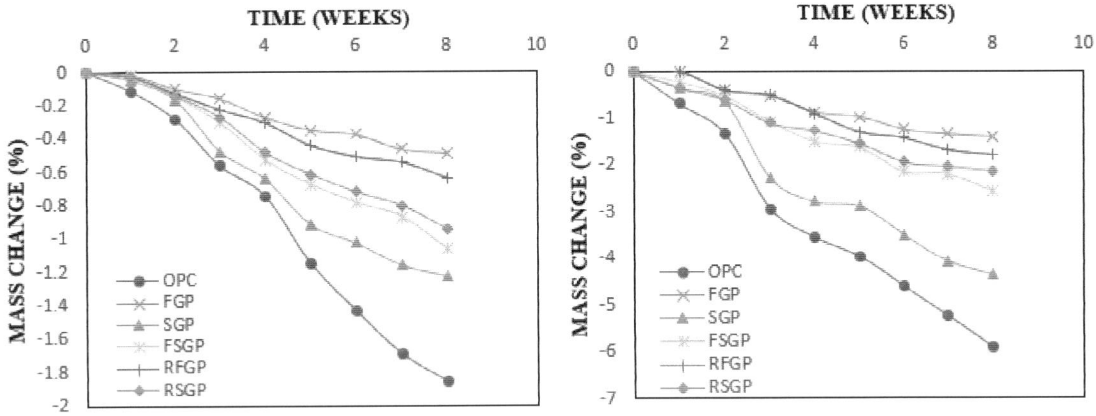

(a) 0.125M **(b) 0.25M**

Figure 2. Mass loss of mixes in various concentrations of acetic acid solution.

From Figure 2, it can be seen that there is a continuous decrease in the mass of all specimen in acetic acid medium irrespective of the medium concentrations. The continuous mass loss may be due to leaching of soluble salts formed as the reaction product. The mass loss proves that the intensity of degradation was high, the inner layers getting deteriorated by penetration of acid through the pores of the specimens. The decrease in mass is due to the formation of water-soluble calcium acetate which leaches out, leading to increase in the inherent porosity within the specimen and thus leading to mass loss. OPC specimens seems to have the highest mass loss in all acid medium with highest mass loss of about 5.90% in 0.25M and 1.85% in 0.125M acetic acid medium. On the contrary, FGP mixes have lowest mass loss in all acid medium, with a maximum of 1.41% and 0.48% mass loss in 0.25M and 0.125M medium respectively. This is due to presence of enhanced N-A-S-H gel and low calcium content within the FGP mix. Clearly RFGP mix showed improved resistance than RSGP and FSGP mix. The chemistry of GP system is different as there are no less stable hydrates such as Portlandite and CSH gel. Mass loss is found to increase in higher concentrations of acid medium. This is due to the increased concentration gradient that exists between the acid solution and the pore solution of the matrix causing higher rate of diffusion of acid ions into the matrix at higher concentrations of acid. Moreover, the variation of mass loss could be approximated to be linearly varying with the age of exposure.

4.3. Strength changes

The compressive strength was measured for all specimens on a weekly basis using a modified digital compression testing machine with a specially fabricated test rig. Initial compressive strength for OPC, FGP, SGP, FSGP, RFGP and RSGP are found to be 45.06, 20.50, 48.68, 26.70, 18.60 and 37.05 N/mm^2. SGP specimens are found to have higher strength than all mixes. The strength loss of mixes with respect to age of exposure in two acetic acid medium concentrations is presented in Figure 3.

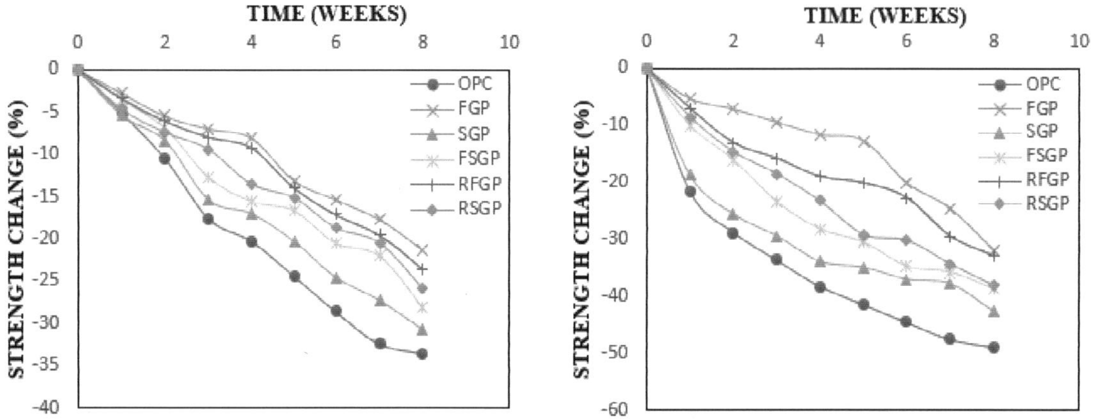

(a) 0.125M **(b) 0.25M**

Figure 3. Strength change of mixes in various concentration of acetic acid solution.

After 8 weeks of immersion, all mixes showed a significant loss in strength with time. However, OPC specimens showed the highest strength loss in all acid medium in spite of having good initial strength. Loss of strength for OPC mix varies from 33.60% in 0.125M medium to 48.86% in 0.25M medium. The calcium hydroxide being least stable in OPC system has been attacked by the acetic acid causing dissolution and leaching of calcium into solution. The dissolution of calcium hydroxide results in an increased porosity and contributes to a rapid decrease in compressive strength [18]. The FGP mix showed the lowest strength loss for both concentrations ranging from 21.3% in 0.125M acetic acid medium to 32% in 0.25M acetic acid solution. Although, the initial strength showcased by FGP is the lowest, its retainment of strength (residual strength) with regard to time and concentration seems to be more profound than all other mixes. These observations are in agreement with the mass loss where the OPC specimens also suffered larger mass losses when compared to GP mixes.

4.4. Altered depth

Altered depth is defined as the depth directly affected due to acid attack and is considered as a prime indicator of degradation kinetics. This can be estimated as the sum of depth of erosion and depth of reaction. The depth of reaction can be determined by spraying phenolphthalein solution to a freshly fractured surface of mortar specimens, and by measuring the depth of colorless layer in the specimen. Phenolphthalein changes from colorless to pink in the pH range of 8.2 to 10.0 [19]. The depth of erosion is deduced from the difference of diameter of specimens before and after exposure (at specific ages of exposure). The cross-section of specimens before and after acid attack is shown in Figure 4. The pink colour region after application of phenolphthalein solution shows the regions unaffected by acid exposure.

Visual appearance	Binder system					
	OPC	FGP	SGP	FSGP	RFGP	RSGP
Before immersion						
Acetic acid 0.125M (after immersion)						
Acetic acid 0.250M (after immersion)	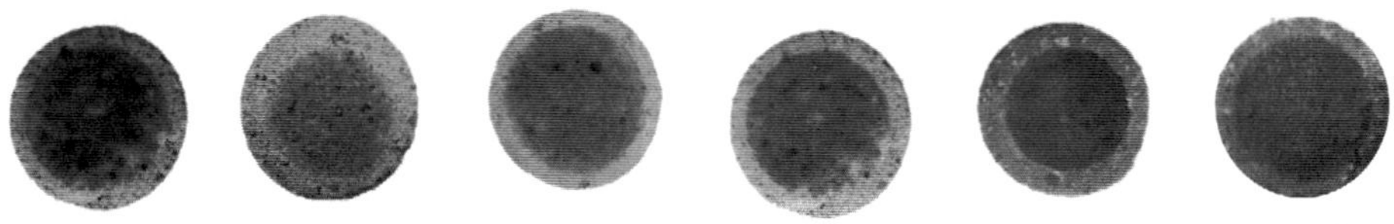					

Figure 4. Altered depth variation of mixes after 8 weeks of exposure in acetic acid medium.

From Figure 4, on keen observation, it can be concluded that the altered depth increased at a higher rate on exposure to 0.25M acid medium. For an elaborate understanding, a plot of altered depth of mixes in various concentrations with respect to the exposure age is plotted and shown in Figure 5. The depth of penetration is seen to be higher for FGP mix when compared to OPC and other GP mixes.

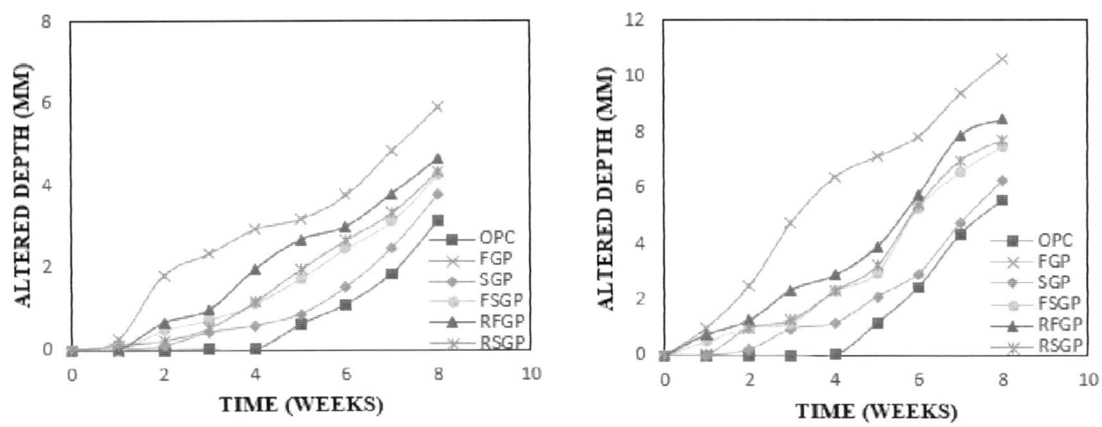

(a) 0.125M **(b) 0.25M**

Figure 5. Altered depth variation of mixes with age of exposure in different concentrations of acetic acid solution.

The FGP mix depicted the maximum depth of 10.62 mm in 0.25M and 5.93 mm in 0.125M acetic acid medium. On the contrary, the altered depth of OPC is determined to be the least, i.e., 5.57 mm in 0.25M and 3.13 mm in 0.125M acetic acid medium. Out of all GP mixes, the SGP mix showed the least altered depth in both concentrations of acetic acid solution. The lower altered depth in OPC may be ascribed to the additional calcium buffer available in OPC systems. Nevertheless, the relationship between the rate of advancement of the depth and time had a nonlinear nature [18]. It is concluded that the altered depths in geopolymer systems are higher despite the mass and strength losses being on the lower side when compared to OPC.

4.5. Ultrasonic Pulse Velocity (UPV) test

UPV test is a non-destructive test conducted to check the strength and microstructural quality of cementitious systems. This test enables us to determine the concrete uniformity, the presence of internal flaws and voids within the concrete specimens [19]. In the current study, the test was conducted once in two weeks with regard to the concentration of the acid medium. The difference in the UPV of specimens before and after exposure (8 weeks) in acetic acid solution is shown in Table 1. Among all mixes, the initial pulse velocity (measured just before exposure) was highest for SGP mix. But the initial pulse velocities of remaining GP mixes are lower compared to OPC mix. Higher velocities indicate good quality and continuity of the material, unlike lower velocities that indicates the presence of cracks or voids within the concrete. There is a significant decrease in pulse velocities of all mixes with exposure in both concentrations of acetic acid (0.125M and 0.25M). This indicates the progression of microstructural degradation within the concrete thereby increasing the voids or micro cracks within the structure. The outer zone is degraded probably owing to the formation of silica gel which has weak mechanical properties. The rate of decrease of pulse velocity increases for all mixes in higher concentration (0.25M) than in lower concentration (0.125M) of acetic acid.

Table 1. UPV of specimens after 8 weeks of exposure in acid medium.

Mix	Ultrasonic Pulse Velocity (km/sec)		
	Before immersion	*Acetic acid 0.125M*	*Acetic acid 0.25M*
OPC	4.12	3.59	3.07
FGP	3.34	2.98	2.76
SGP	4.39	3.73	3.35
FSGP	3.36	2.89	2.67
RFGP	3.20	2.84	2.64
RSGP	3.69	3.25	2.99

Based on the UPV values, the dynamic modulus of elasticity of specimens can be determined as per IS 13311 (Part 1): 1992 codal provisions. As the Poisson's ratio of the individual specimens is not known, the variation is expressed in terms of relative dynamic modulus of elasticity (E_{RD}) and is showcased in Figure 6. The relative dynamic modulus of elasticity is computed as the ratio of the dynamic modulus of elasticity of specimens at a particular age of exposure to the dynamic modulus of elasticity measured just before the exposure.

IOP Conf. Series: Materials Science and Engineering **989** (2020) 012015 doi:10.1088/1757-899X/989/1/012015

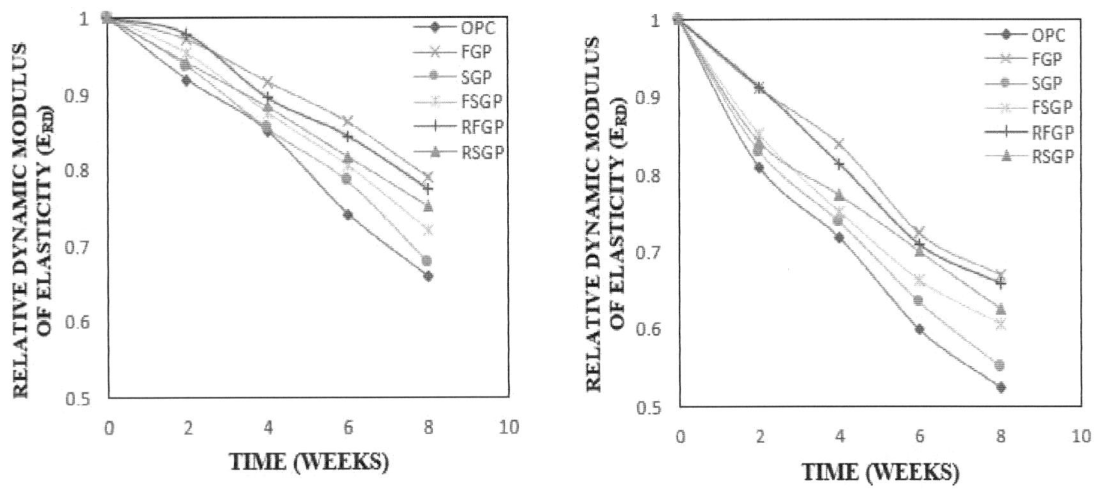

(a) 0.125M **(b) 0.25M**

Figure 6. Variation in relative dynamic modulus of elasticity of mixes in different concentrations of acetic acid.

The E_{RD} of OPC mix after 8 weeks of exposure was found to be 0.52 in 0.25M and 0.65 in 0.125M. The drop in the relative dynamic modulus of elasticity indicates the severity of degradation. The results indicate that the loss in dynamic modulus was highest in OPC mix. This correlates well with the strength and mass loss measurements. For FGP mix, the E_{RD} was 0.67 in 0.25M and 0.78 in 0.125M. Again, from the RM based GP mix, RFGP mix shows enhanced performance than RSGP mix. Thereby, with reference to the variation in UPV and E_{RD}, GP systems depicted superior performance in resisting acetic acid attack than OPC system in both concentrations of acetic acid.

5. Conclusions

A comprehensive experimental study was conducted to investigate the degradation kinetics and compare the resistance of GP and PC systems to silage effluent attack caused primarily by acetic acid for an exposure period of 8 weeks. Accelerated leaching test method was adopted for the study by selecting smaller sized specimens and frequent replenishment of acid solutions. There is a continuous decrease in mass, strength and dynamic modulus of elasticity due to the formation of water-soluble calcium acetate which leaches out, leading to increase in the inherent porosity within the specimen. OPC system has the highest mass loss (5.9%), strength loss (48.6%), and lowest relative dynamic modulus of elasticity (0.52) at higher concentration (0.25M) of acetic acid.

After 8 weeks of exposure to 0.25M and 0.125M acetic acid solution, FGP mix showed the least mass loss and strength loss. However, the altered depth was higher for FGP mortar samples in 0.125M and 0.25M acid medium unlike OPC which had the least altered depth. Therefore, evaluating degradation kinetics based on only one parameter i.e. either mass loss or altered depth is not suggested. New parameters are to be defined which captures both changes in the physio-mechanical properties and deterioration in microstructure. RFGP mix showed improved acetic acid resistance than RSGP and FSGP mix. Accordingly, RM material can be adopted as a binder or partial replacement for OPC as it promises both durability and sustainability for exposure to lower concentrations of acid solutions. Similarly, with regard to the variation in UPV and E_{RD}, GP systems delineate enhanced performance in resisting acetic acid attack than OPC system in both concentrations. Thereby, from the overall assessment, GP mortars have increased acetic acid resistance when compared with OPC samples.

The kinetics of degradation was more profound in the case of higher concentration of acids and the long-term effects can be observed by conducting an accelerated leaching test. The rate of degradation is found to be dependent on the concentration of acid, chemical composition of binders, and solubility and characteristics of the salts formed. The difference in behaviour between paste, mortar, and concrete has to be studied further, focusing on the role of interfacial transition zone in the durability of binder systems, in aggressive acidic environments.

6. References

[1] Bertron A, Alexander M G and De Belie N 2013 Performance of cement- based materials in aggressive aqueous environments *State-of-the-Art Report* RILEM TC 211 – PAE 131-173

[2] Larreur-Cayol S, Bertron A and Escadeillas G 2011 Degradation of cement-based materials by various organic acids in agroindustrial waste-waters *Cement and Concrete Research* **41** 882-892

[3] Dogangun A, Karaca Z, Durmus Aand Sezen H 2009 Cause of damage and failures in silo sStructures *Journal of Performance of Constructed Facilities* **23(2)** 65-71

[4] Pacheco-Torgal F, Abdollahnejad Z, Camoes A F, Jamshidi M and Ding Y 2012 Durability of alkali-activated binders: a clear advantage over Portland cement or an unproven issue? *Constr. Build. Mater.* **30** 400–405

[5] Arhina D D, Nuamaha R A, Tuffoura B A, Obadac D O and Awaso A Y 2017 Bauxite red mud-cement based composites: Characterisation for pavement applications *Case Studies in Construction Materials* **7** 45–55

[6] Deb P S, Nath P and Sarker P K 2014 The effects of ground granulated blast - furnace slag blending with fly ash and activator content on the workability and strength properties of geopolymer concrete cured at ambient temperature *Mater. Des.* **62** 32–39

[7] Li Y, Min X, Ke Y, Liu D and Tang C 2019 Preparation of red mud-based geopolymer materials from MSWI fly ash and red mud by mechanical activation *Waste Management* **83** 202–208

[8] Dyer T 2016 Influence of cement type on resistance to organic acids *Magazine of Concrete Research* **69(4)** 175-200

[9] Ramaswamy K P, Bertron A and Santhanam M 2017 Additional insights on the influencing factors and mechanism of degradation due to acid attack: Special case of acids forming soluble salts *Proceedings of International Conference on Advances in Construction Materials and Systems* (ICACMS), Chennai **4** 279–290

[10] Bertron A, Larreur-Cayol S and Escadeillas G 2009 Degradation of cementitious materials by some organic acids found in agro industrial effluents *Proceedings of the RILEM TC 211-PAE Final Conference* Toulouse France **1** 96-107

[11] Provis J L 2018 Alkali-activated materials *Cement and Concrete Research* **114** 40-48

[12] Provis J L and Deventer S J V 2009 Geopolymers *Structure, processing, properties and industrial applications* Woodhead Publishing Limited and CRC Press LLC

[13] Stephan D, Firdous R, Noël J and Djobo Y 2018 Natural pozzolan based geopolymers: A review on mechanical, micro-structural and durability characteristics *Construction and Building Materials* **190** 1251-1263

[14] Lodeiro G, Palomo A, Fernández-Jiménez A and Macphee D E 2011 Compatibility studies

between N-A-S-H and C-A-S-H gels Study in the ternary diagram Na2O–CaO–Al2O3–SiO2–H2O *Cement and Concrete Research* **41** 923–931

[15] Aiken T A, Sha W, Kwasny J and Soutsos M N 2017 Resistance of geopolymer and Portland cement-based systems to silage effluent attack *Cement and Concrete Research* **92** 52-65

[16] Gu L, Bennett T and Visintin P 2018 Evaluation of accelerated degradation test methods for cementitious composites subject to sulphuric acid attack; application to conventional and alkali activated concretes *Cement and Concrete Composites* **87** 187-204

[17] Gu L, Bennett T and Visintin P 2019 Sulphuric acid exposure of conventional concrete and alkali activated concrete: Assessment of test methodologies *Construction and Building Materials* **197** 681-692

[18] McPolin D O, Basheer P A M, and Long A E 2009 Carbonation and pH in mortars manufactured with supplementary cementitious materials *Journal of Materials in Civil Engineering* **21** 217–225

[19] Ghosh R, Sagar S P, Kumar A, Gupta S K and Kumar S 2018 Estimation of geopolymer concrete strength from ultrasonic pulse velocity (UPV) using high power pulser *Journal of Building Engineering* **16** 39–44

IOP Conf. Series: Materials Science and Engineering **989** (2020) 012016 doi:10.1088/1757-899X/989/1/012016

Mechanical properties of blended cement mortar systems exposed to ammonium chloride environment

Sheba Babu[1,2], Mohammed Sahad[1], Vijeesh K P[1], Syamjith C K[1], Mohammed Raneesh[1], Ramaswamy K P[1], Nazeer M[1] and Mohammed Thowsif[1]

[1] Department of Civil Engineering, TKM College of Engineering, Kollam-691005, India.

[2]Email: shebababu447@gmail.com

Abstract. There are number of areas where cement composites can be exposed to ammonium based compounds such as agricultural land, chemical fertilizer plants, waste waters, polluted and industrial environments, which can induce severe and premature deterioration of concrete systems. Based on the review of literature, the most common and deleterious ammonium based salts were found to be ammonium nitrate, ammonium chloride and ammonium sulphate. The cementitious systems with Ordinary Portland Cement (OPC) may be attacked easily by these ammonium based chemicals as the OPC hydrates such as Portlandite and C-S-H gel are considered to be unstable in these aggressive media. Thus, it is necessary to understand the alteration kinetics of cementitious systems and the effectiveness of Supplementary Cementitious Materials (SCMs) in improving the resistance when exposed to ammonium based environments for develpoing more durable and sustainable concrete systems. The aim of this experimental study was to investigate the durability aspects in terms of the changes in mechanical properties of cement mortar specimens made with OPC and different combinations of various Supplementary Cementitious Materials (SCMs) exposed to ammonium based salt solution prepared using chemical grade ammonium chloride (NH_4Cl). The mortar specimens were exposed to ammonium chloride solution in two different concentrations (1.25 M and 2.5 M) for a maximum exposure duration of two and half months. The SCMs used for the study include Class F Fly ash (FA), Ground Granulated Blast Furnace Slag (GGBS), and Silica fume (SF). The changes in the mechanical properties were evaluated on the basis of visual assessment, depth of penetration, changes in the compressive strength, changes in Ultrasonic Pulse Velocity (UPV) and relative dynamic modulus of elasticity. From all the test results, it can be concluded that the ternary blended mix with OPC, GGBS and Silica fume (OPC-GGBS-SF) has higher resistance to ammonium chloride attack. The results stress the need to include SCMs and the importance to tailor make the concrete for structures exposed to aggressive environment like ammonium chloride.

1. Introduction

Since cement concrete is the one of the most widely used construction material, it can come in contact with aggressive aqueous environments in many situations. Thus the concrete systems are more vulnerable to chemical attack by these aggressive environments. The understanding of the alteration kinetics and mechanisms of degradation of cementitious matrices by these aggressive media is an essential step in making concrete systems more durable which will then increase the service life of the concrete structures. The incorporation of Supplementary Cementitious Materials (SCMs) along with cement were found to be more effective in improving the durability of cement composites in aggressive chemical environments as it imparts a compact microstructure with low penetrability [2, 15, 16]. The replacement of cement with SCMs can also be considered as the most viable solution to make the cement concrete more sustainable.

The resistance of cementitious systems in ammonium environment and the effectiveness of SCMs in improving the durability can be considered as the one of the most important durability indices to be studied, as they can be exposed to ammonium based compounds in various areas such as chemical fertilizer plants, agricultural land, industrial environments, waste waters, ground waters and polluted environment [1, 3, 5, 10, 17]. Based on the review of literature, the most common and deleterious ammonium based compounds were found to be ammonium nitrate, ammonium chloride and ammonium sulphate [2, 4-17].

There are number of studies reported on the deterioration of cementitious matrices exposed to ammonium sulphate and ammonium nitrate solutions. The deterioration by ammonium sulphate ($(NH_4)_2SO_4$) is the one of the most aggressive corrosion, characterised by the intensive dissolution of cement hydrates, cracking and expansion due to the formation of gypsum and ettringite [7, 14]. It is caused by the combined action of acid hydrolysis, sulphate and ammonium corrosion with the formation of gypsum crystals on the surface of specimens [4, 6, 8, 12].

The mechanism of ammonium nitrate (NH_4NO_3)attack is characterised by the leaching, decalcification and the dissolution of hydration products [11]. The deterioration leads to the formation of soluble calcium nitrate, calcium nitro aluminate salt and ammonia emanation with rhombic calcite as surface deposits on the specimens [2, 12, 13].

But the studies reported on the deterioration of cementitious systems in ammonium chloride (NH4Cl) environment is very limited, characterised by the dissolution of portlandite and crystallization of thaumasite salt. Another intermediate phase is characterized by the formation of Friedel's salt. The final reaction products formed were found to be gypsum, calcite and vaterite [14]. The formation of calcium chloro aluminate hydrate and calcium hydroxychloride hydrate results in considerable increase in volume leading to expansion and disruption of hardened concrete [9]. However, the mechanism of ammonium chloride attack is not well understood and more micro-analytical characterization studies are to be performed to understand the mechanism in a better way.

Since there are very few works reported on the deterioration kinetics of cementitious matrices exposed to various concentrations of NH4Cl solution over different exposure duration, the objective of this experimental study was to investigate the effect of NH4Cl solution on the durability properties of cement mortar specimens with various combinations of SCMs over an exposure duration of two and half months on the basis of visual examinations, changes in compressive strength, degradation depth, Ultrasonic Pulse Velocity (UPV) and changes in the dynamic modulus of elasticity of mortar specimens.

2. Materials used

The various materials used in this experimental study include Ordinary Portland Cement (OPC) 53 Grade conforming to IS: 12269-1987, Manufactured Sand (M Sand) as fine aggregates conforming to IS 383, SCMs such as Class F Fly ash (FA), Ground Granulated Blast Furnace Slag (GGBS) and Silica fume (SF). The specific gravity of OPC, Fly ash, GGBS and Silica fume is 3.15, 2.20, 2.90 and 2.05 respectively. The ammonium based salt used in this experimental study was chemical grade ammonium chloride (NH_4Cl) in two different concentrations of 1.25 and 2.5 M (mol/L) respectively.

Six different mixes of cement mortar specimens made with OPC and different combination of SCMs (shown in Table 1) of 1:1.5 (binder: sand) mix proportion with water-binder ratio 0.50 were selected for the study. Mix 1 is kept as the reference or control mix. Mixes 2 to 4 are binary mixes whereas mixes 5 and 6 are ternary blended mixes. Fine aggregate (M Sand) which passed through 2.36 mm sieve and retained on 90 micron sieve were used for the study. Polycarboxylate Ether (PCE) admixture was used as superplasticizer with optimum dosage as 0.5% of weight of binder in order to improve the workability of the mortar mix. PCE was used only for Silica fume based mixes.

Table 1. Mortar mixes used for the study.

Mix	OPC (%)	FA (%)	GGBS (%)	SF (%)
Mix 1 (OPC)	100	-	-	-
Mix 2 (OPC+FA)	70	30	-	-
Mix 3 (OPC+GGBS)	50	-	50	-
Mix 4 (OPC+SF)	90	-	-	10
Mix 5 (OPC+FA+SF)	70	20	-	10
Mix 6 (OPC+GGBS+SF)	50	-	40	10

3. Methodology

The degradation kinetics of cement mortar specimens made with OPC and different combinations of SCMs exposed to two different concentrations 1.25 M and 2.5 M of ammonium chloride (NH_4Cl) solution were evaluated on the basis of changes in the mechanical properties of mortar specimens before and after exposure. The performance were evaluated periodically on the basis of visual assessment, depth of penetration, changes in the compressive strength and changes in the UPV and relative dynamic modulus of elasticity of specimens over an exposure duration of two and half months.

The raw materials required for the casting of mortar specimens were weighed accurately using precise weighing balance as per the mix proportion. All the ingredients were mixed well using a Hobart mixer for about 5 minutes, filled in the oiled plastic cylindrical moulds (diameter 25 mm and height 50 mm) and allowed to set for 24 hours. After 24 hours, the specimens were demoulded and kept for lime saturated curing for a duration of 14 days. For lime saturated curing, 3g/L of calcium hydroxide ($Ca(OH)_2$) solution was used. Cylindrical specimens having 25 mm diameter and 50 mm height were used for performing all the tests.

The initial appearance, compressive strength and UPV of the specimens (in saturated but surface dry condition) were noted before immersion in NH_4Cl solution to determine the changes in these parameters after the exposure. Then these specimens were exposed in two concentrations; 1.25 M and 2.5 M NH_4Cl solution. The NH_4Cl solution was replenished every 2 weeks in order to maintain the aggressiveness of the solution. The ratio of volume of ammonium solution to that of specimens was kept fixed at five for the evaluation of degradation kinetics for the entire exposure period. Thus, an accelerated immersion based leaching test is adopted in this experimental investigation in order to understand the degradation kinetics of various binders up on exposure to ammonium based solutions.

After each week of exposure, the specimens were then taken out of the solution and tested to determine the changes in the physical appearance, altered depth, compressive strength and UPV of the mortar specimens. The tests were carried out for a maximum exposure duration of 10 weeks for 1.25 M and 7 weeks (due to rapid deterioration) for 2.5 M NH_4Cl solution. The changes in the physical appearance of specimens were identified by taking digital photographs of specimens of various mixes. For measuring altered depth, 1% phenolphthalein solution was used as the indicator. Compressive strength test was carried out on the modified Marshall testing machine with testing rig. Ultrasonic Pulse Velocity (UPV) test were carried out as per the guidelines of IS 13311(Part I)-1992 using transducers having a frequency of 150 kHz. The mean value of the three specimens from each mix was taken for the each test result.

4. Results and discussions

4.1 Visual examinations

The cylindrical mortar specimens were visually examined to identify the changes in the physical appearance up on exposure in 1.25 M and 2.5 M NH_4Cl solution. The typical photographs of the specimens in (saturated and surface dry condition) taken on a weekly basis before and after exposure over the entire exposure duration is as shown in Table 2. It was observed that the specimens of all the six mixes exposed to both concentrations (1.25 M and 2.5 M) of NH_4Cl solution undergone only minimal changes in their physical appearance upon the exposure. The physical deterioration on the specimens was found to be minimal and they remained structurally intact. The variations in the physical appearance were found to be similar for all the mixes studied.

Table 2. Visual examination of mortar specimens in NH_4Cl solution (1.25 M and 2.5 M).

Before exposure	After exposure in NH_4Cl solution	
	1.25 M (10th week)	2.5 M (7th week)

4.2 Altered depth

In this test, the altered depth was determined by spraying 1% phenolphthalein indicator solution on the freshly cut and exposed surfaces of mortar specimens. The degradation depth measurement was started only after third week of exposure and continued for the entire exposure duration. A magenta coloured portion and a colourless portion were observed on the freshly cut surface. The altered depth is taken as the sum of depth of leaching and depth of reaction. The depth of leaching is taken as the half of the changes in diameter of specimens before and after exposure and the depth of reaction is taken as the half of the depth of colourless portion obtained on the specimens after exposure in ammonium chloride solution. It was found that the degradation depth increased with the increase in the concentration (the altered depth higher for 2.5 M as compared to 1.25 M) of NH_4Cl solution. Table 3 shows the photographs of the freshly cut surface of specimens after spraying the phenolphthalein solution. The variations in the altered depth of specimens of all the six mixes obtained after exposure in 1.25 M (for 10 weeks) and 2.5 M (for 7 weeks) NH_4Cl solution is as shown in Figure 1 and 2 respectively.

Table 3. Images showing altered depth in NH_4Cl solution (1.25 M and 2.5 M).

Mortar mix	After exposure in NH_4Cl solution	
	1.25 M (10th week)	2.5 M (7th week)
OPC	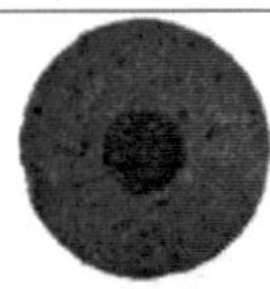	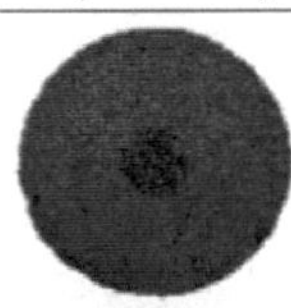
OPC+FLYASH		
OPC+GGBS	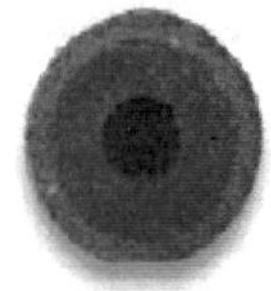	
OPC+SF		
OPC+FLYASH+SF	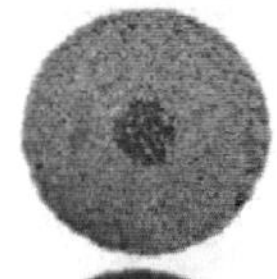	
OPC+GGBS+SF	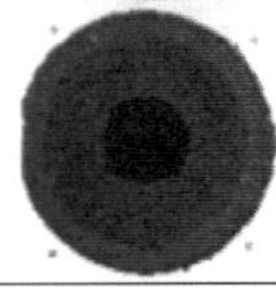	

The degradation depth was found to be higher for fly ash based mixes (Mix 2 and Mix 5) and lower for control mix OPC (Mix 1) in both the concentrations. Among the two mixes incorporated with flyash, the altered depth was found to be higher for binary blended mixes with OPC and Flyash as compared to ternary mix with OPC, Flyash and SF. For binary blended mix OPC-FLYASH (Mix 2), the altered depth in 1.25 M and 2.5 M was found to be 11.6 mm and 12.83 mm respectively. But for the mix OPC-FLYASH-SF (Mix 5), the degradation depth was found to be 10.02 mm and 12.80 mm respectively in 1.25 M and 2.5 M concentrations. The altered depth for control mix OPC was found to be only 7.90 mm and 11.06 mm respectively in 1.25 M and 2.5 M concentrations of NH_4Cl solution. After the control mix OPC, the altered depth was found to be higher for GGBS based mixes (Mix 3 and Mix 6). For binary mix with OPC and GGBS, the altered depth was found to be 8.31 mm and 11.94 mm respectively and for ternary mix with OPC, GGBS and SF, the altered depth was found to be 8.30 mm and 11.58 mm respectively in 1.25 M and 2.5 M concentrations of NH_4Cl solution. From this result, the altered depth of Fly ash based mix (both binary and ternary) was found to be almost same in both 1.25 M and 2.5 M NH_4Cl solution. Similarly, the altered depth of GGBS based mix (both binary and ternary) was also found to be same in 1.25 M and 2.5 M NH_4Cl solution. Based on the degradation depth, it can be ascertained that the performance of control mix OPC was better compared to the other mixes. But among the blended mixes with SCMs, the performance of OPC-GGBS-SF based mix was found to be better.

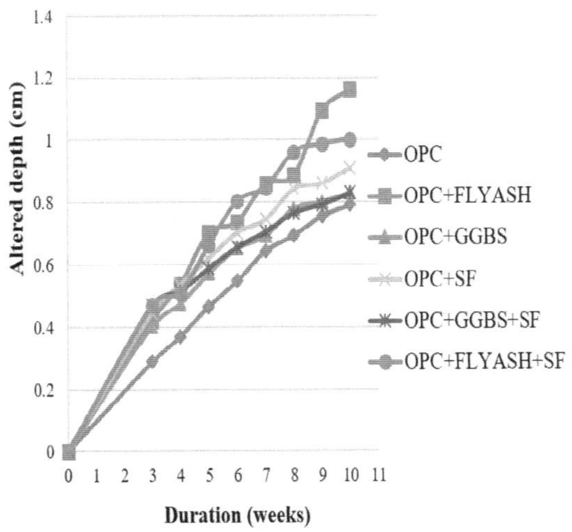

Figure 1. Altered depth in 1.25 M NH_4Cl solution.

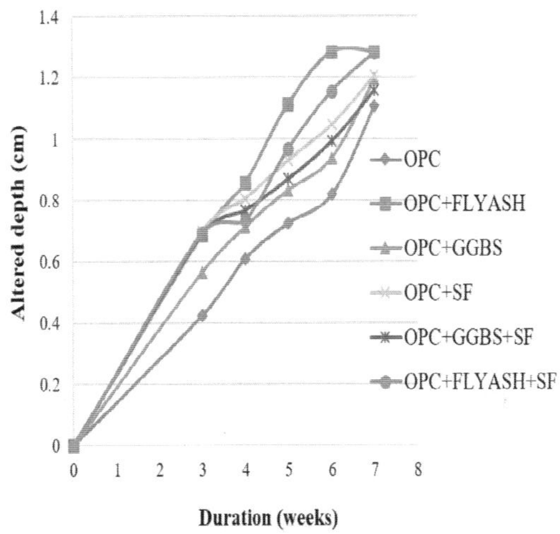

Figure 2. Altered depth in 2.5 M NH_4Cl solution.

4.3 Compressive strength test

The compressive strength variation of cylindrical mortar specimens of all the six mixes of mortar specimens when exposed to 1.25 M (maximum duration 10 weeks) and 2.5 M (maximum duration 7 weeks) NH_4Cl solution is presented in Figure 3 and 4 respectively. It was found that the initial compressive strength (after 14 days of curing) of control mix OPC was found to be higher when compared to other binary and ternary blended mixes. Among the binary blended mixes, the initial compressive strength was found to be higher for mix with OPC and SF as compared to GGBS and FLYASH based mixes. Among ternary blended binder mixes, the high initial compressive strength was obtained with Mix 5 (OPC-FLYASH-SF) when compared with Mix 6 (OPC-GGBS-SF). It was found that specimens of all the mixes have a undergone significant reduction in strength when exposed to both 1.25 M and 2.5 M NH_4Cl solution. After exposure in NH_4Cl, all the specimens have undergone a reduction in compressive strength. The strength loss was found to be higher for the control mix OPC and lower for GGBS based mixes. The strength loss for OPC exposed in 1.25 M and 2.5 M NH_4Cl solution was found to be about 77.58% and 79.24% respectively. It was found that the compressive strength of both the binary (OPC-GGBS) and ternary (OPC-GGBS-SF) blended mixes with GGBS was reduced by about 70% in both concentrations at the end of the exposure duration. Based on the strength loss, it can be ascertained that the performance of OPC-GGBS-SF blended mix was better compared to the other mixes.

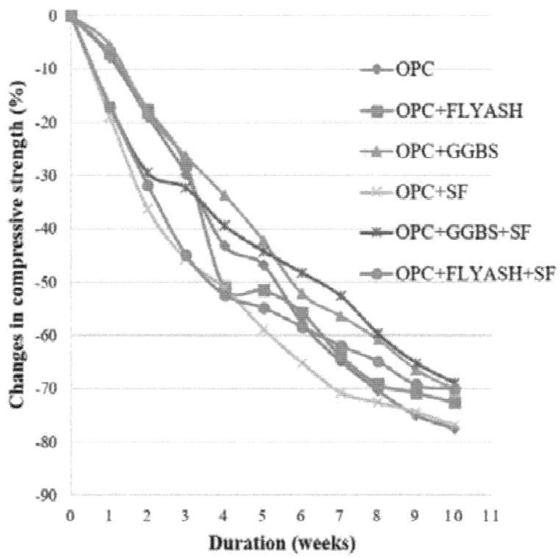

Figure 3. Changes in compressive strength in 1.25 M NH_4Cl solution.

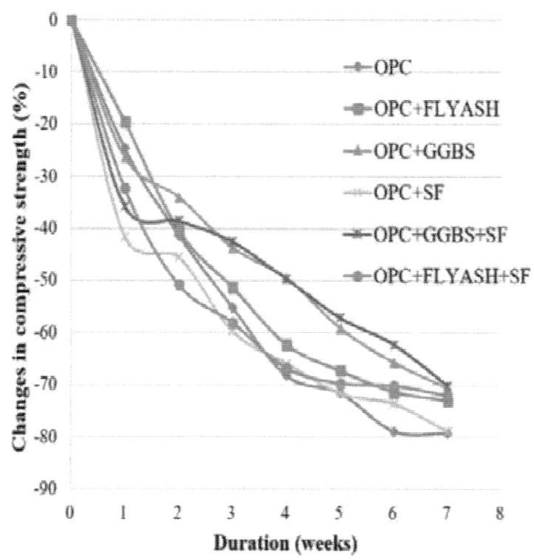

Figure 4. Changes in compressive strength in 2.5 M NH_4Cl solution.

4.4 Ultrasonic Pulse Velocity (UPV) test

The UPV test, being a non-destructive test helps in assessing the strength and quality of specimens by measuring the changes in the pulse velocity. The changes in the UPV of specimens of various mixes before and after exposure in ammonium chloride solution are as shown in Table 4. Before exposure, the velocities of specimens of all the six mixes were found to be above 3.5 km/sec, which indicates the good quality specimens as per IS 13311(Part I)-1992. Among the various mixes, the initial pulse velocity was found to be high for the control mix OPC. The pulse velocities of binary blended mixes was high as compared to ternary mixes. The higher velocities indicates good quality and homogeneity of the material while lower velocities indicates the presence of micro cracks or voids or deterioration in microstructure. After exposure in the ammonium chloride solution, a significant reduction in the pulse velocities were observed on all the specimens of various mixes in both concentrations (1.25 M and 2.5 M) which gives information about the microstructural deterioration. The reduction in pulse velocity was found to be higher for the control mix OPC and lower for GGBS based mixes (both binary and ternary). Among the GGBS mixes, the velocity reduction was found to be lower for ternary blended mix OPC, GGBS and SF as compared to binary mix with GGBS and SF.

From these changes in the pulse velocity, the changes in the dynamic modulus of elasticity of specimens can be determined as per the codal provisions. Since the poisson's ratio of the specimens were not known, the changes in the relative dynamic modulus of elasticity of specimens were analysed and shown in Figure 5 and 6 respectively. Relative dynamic modulus of elasticity was calculated as the ratio of dynamic modulus of elasticity at a particular age of exposure to the dynamic modulus of elasticity just before the exposure. Here also, the variation was found to be similar as in the case of changes in pulse velocity. The relative dynamic modulus of elasticity of control mix OPC was found to be 0.730 and 0.674 respectively in 1.25 M and 2.5 M NH_4Cl solution. For the ternary blended mix OPC-GGBS-SF, the relative dynamic elastic modulus was about 0.853 and 0.756 respectively in 1.25 M and 2.5 M concentrations and this higher value indicates better resistance. But for the binary mix OPC-GGBS, it was about 0.776 and 0.746 respectively in 1.25 M and 2.5 M NH_4Cl solution. Thus based on the changes in the UPV and relative dynamic modulus of elasticity, the performance of ternary blended mix with GGBS and SF as SCM (OPC-GGBS-SF) was found to be better as compared to other mixes.

Table 4. Changes in the UPV of specimens in NH_4Cl solution (1.25 M and 2.5 M).

Mortar Mix	Ultrasonic Pulse Velocity (km/sec)		
	Before exposure	After exposure in NH_4Cl solution 1.25 M (10th week)	2.5 M (7th week)
OPC	3.89	3.43	3.38
OPC+FLYASH	3.77	3.38	3.29
OPC+GGBS	3.75	3.48	3.43
OPC+SF	3.74	3.43	3.36
OPC+FLYASH+SF	3.73	3.32	3.28
OPC+GGBS+SF	3.73	3.49	3.46

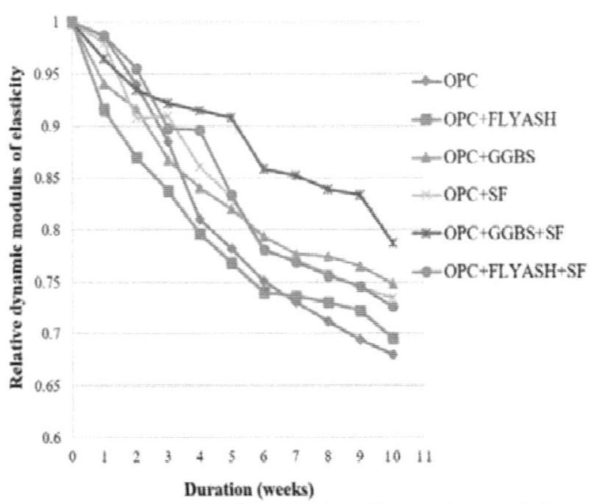

Figure 5. Changes in relative dynamic modulus of elasticity in 1.25 M NH_4Cl solution.

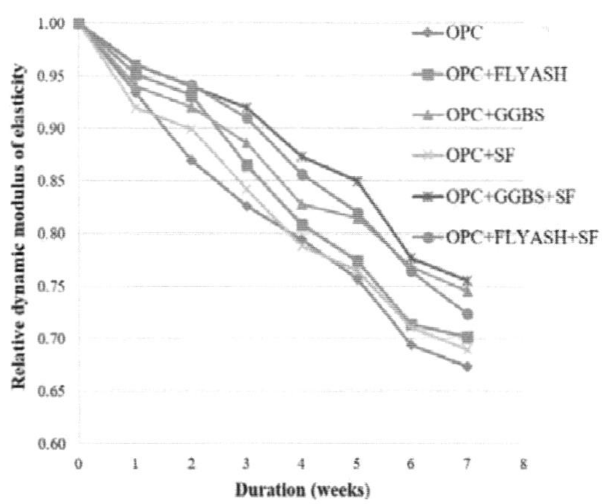

Figure 6. Changes in relative dynamic modulus of elasticity in 2.5 M NH_4Cl solution.

5. Conclusions

Concrete systems exposed to the chemical fertilizer plants, agricultural land, sewage plants and many other industrial applications are more prone to chemical attacks by ammonium based compounds. Thus the resistance of cementitious systems in ammonium environment has to be comprehensively investigated as the exposure to these aggressive environments can induce severe and premature deterioration of cementitious systems.

In this experimental programme, NH_4Cl solution in two different concentrations (1.25 M and 2.5 M) was used to study the effect of ammonium based salt solution on the mechanical properties of OPC cement mortar specimens made with different combinations (both binary and ternary) of various SCMs. The degradation kinetics was evaluated in terms of visual examinations, changes in the compressive strength, degradation depth and changes in the dynamic modulus of elasticity of mortar specimens before and after the exposure.

From the test results, it was found that specimens of all the mixes have undergone a significant reduction in their compressive strength when exposed to 1.25 M and 2.5 M NH_4Cl solution. Based on the compressive strength loss, it can be ascertained that the performance of OPC-GGBS-SF blended mix was better compared to the other mixes. The reduction in compressive strength may be ascribed to the leaching phenomenon and decalcification of hydrates such as Portlandite and CSH gel.

The measurement of degradation depth helps in identifying the degraded zone from the sound zone, as the magenta coloured portion indicates unattacked sound portion (pH above 9) and colourless portion indicate deteriorated region (pH below 9). In terms of the depth of degradation, it can be ascertained that the performance of control mix OPC was better compared to the other mixes. But among the blended mixes with SCMs, the performance of OPC-GGBS-SF based mix was found to be better.

It was found that specimens of all the mixes have also undergone a significant reduction in their pulse velocities and relative dynamic modulus when exposed to 1.25 M and 2.5 M NH_4Cl solution. Based on the changes in the UPV and relative dynamic modulus, the performance of ternary blended mix with GGBS and SF as SCM (OPC-GGBS-SF) was found to be better as compared to the other mixes. The changes in UPV gives information about the variations in the quality and microstructural deterioration that the specimens have undergone upon exposure to NH_4Cl solution.

Based on all the test results, it can be concluded that the ternary blended mix with OPC, GGBS and Silica fume (OPC-GGBS-SF) possess better mechanical properties after the exposure, and has higher resistance to ammonium chloride attack. The results stress the need to include SCMs in concrete mixture formulations and the importance to tailor make the concrete for structures exposed to aggressive environment like ammonium chloride. The changes in mechanical properties shall be further validated by conducting a detailed microanalytical characterization study.

Acknowledgement

The authors thank the TEQIP II for the financial assistance received through seed money funding in 2019-20.

6. References

[1] Alexander M, Bertron A and De Belie N 2013 Performance of cement-based materials in aggressive aqueous environments *RILEM State-of-the-Art Reports* Springer **10** 1- 469

[2] Amin M and Bassuoni M T 2017 Performance of concrete with blended binders in ammonium sulphate solution *Journal of Sustainable Cement based Materials* **7** 15-37

[3] Arafa M, Alqedra M and Shubari T 2015 Quantifying effect of ammonium nitrate attack on mechanical and physical properties of cement mortars *Journal of Scientific Research and Reports* **7** 137-145

[4] Bassuoni M T and Nehdi M L 2012 Resistance of self consolidating concrete to ammonium sulphate attack *Materials and Structures* **45** 977-994

[5] Fiertak M and Stryszewska T 2013 Resistance of three component binders in highly chemically corrosive environments *11th International Conference on Modern Building Materials* Procedia Engineering **57** 278-286

[6] Jauberthie R and Rendell F 2003 Physicochemical study of the alteration surface of concrete exposed to ammonium salts *Cement and Concrete Research* **33** 85-91

[7] Lea F M 1965 The action of ammonium salts on concrete *Magazine of Concrete Research* **17 (52)** 115-116

[8] Mbessa M and Pera J 2001 Durability of high strength concrete in ammonium sulfate solution *Cement and Concrete Research* **31** 1227-1231

[9] Ohama Y, Kawakami M and Fukuzawa K 1997 *Polymers in concrete* Proceedings of the Second East Asia Symposium on Polymers in concrete

[10] Pepenar I 2009 Degradation of reinforced concrete Structures in ammonium nitrate environments *Concrete in aggressive aqueous environments, Performance, Testing and Modeling* 442-448

[11] Rehman A, Qudoos A, Jakhrani S H, Kim H G and Ryou J S 2019 Influence of nano-silica on the leaching attack upon photocatalytic cement mortars *International Journal of Concrete Structures and Materials* **13(1)** 11-20

[12] Rendell F and Jauberthie R 1999 The deterioration of mortar in sulphate environments *Construction and Building Materials* **13(6)** 321–327

[13] Schneider U and Chen S W 2009 The leaching effect of concrete immersed in ammonium nitrate solution *Proceedings of the RILEM TC 211-PAE Final Conference* **1** 54-64

[14] Slupik B S, Podworny J and Staszuk M 2018 Corrosion of cement pastes made of CEM I and CEM III/A caused by a saturated water solution of ammonium chloride after 4 and 25 days of aggressive immersion *Construction and Building Materials* **170** 279-289

[15] Wachira J M 2019 Effects of chlorides on corrosion of simulated reinforced blended cement mortars *International Journal of Corrosion* **2019** 1-7

[16] Wagan F H, Hisbani N Q and Wagan G H 2019 Effect of strength of concrete by partial replacement of ordinary Portland cement with silica fume *American Journal of Mechanics and Applications* **7** 10-14

[17] Wong L L, Asrah H, Rahman M E and Mannan M A 2013 Effects of aggressive ammonium nitrate on durability properties of concrete using sandstone and granite aggregate *International Journal of Civil and Environmental Engineering* **7** 49-53

IOP Conf. Series: Materials Science and Engineering **989** (2020) 012017 doi:10.1088/1757-899X/989/1/012017

Seismic performance of exoskeleton structures in irregular buildings

Vivek Vijayan[1,2] and Sajeeb R[1]

[1] Department of Civil Engineering, TKM College of Engineering, Kollam-691005, India

[2]Email: vivekvvijayan@gmail.com

Abstract. Exoskeleton Structure is a diagrid like structure used for retrofitting existing structures. This paper evaluates the seismic performance of exoskeleton structures on buildings with plan irregularity. Linear dynamic analysis using Response Spectrum method is carried out. L shaped buildings frames with A/L ratios of 0.15, 0.30 and 0.50 are considered in the present study. The results show that exoskeleton structure is capable of reducing response parameters such as shear force, floor displacement and forces in columns.

1. Introduction

In the current scenario of stringent seismic requirements imbibed in various codes, retrofitting of existing buildings to required performance is quite challenging to practicing structural engineers. Conventional retrofitting techniques including addition of shear walls, addition of infill walls and addition of braces are quite cumbersome both from construction and functional view point. These conventional techniques affect the functional aspect of the structures and in most cases involve strengthening of existing foundation. These conventional techniques can make the structure temporarily out of use during strengthening which is not feasible, at least for some structures. All these issues could be overcome by external retrofitting strategies. Majority of conventional retrofitting make use of strengthening concept. While techniques like base isolation make use of reducing seismic demand concept. In external retrofitting strategies also, seismic demand reduction is mainly adopted.

Exoskeleton structures are external self-supporting diagrid like structural system suitably connected to primary structure in order to improve the seismic performance of primary structure. Exoskeleton structure acts as sacrificial element and absorbs forces and rigid links are used to connect exoskeleton structures with the building to be protected.

The structural control using exoskeleton structures largely depends on the structural properties of rigid links and their connection with the main building. The connection between the existing and external structure could be rigid or flexible. A higher stiffness of the link is required to alter the time period of higher modes[1]. Acceleration of floors were found to be increased with inclusion of exoskeleton structures [2]. There must be a trade-off between displacement control and acceleration control. In addition to displacement/acceleration control, exoskeleton structures were found to be effective in reducing base shear too. A considerable part of the base shear induced due to seismic actions was found to be shared by the exoskeleton structure, reducing the base shear of the building frame to be controlled [3]. As of now only studies with rigid connections are available in case of exoskeleton structures.

IOP Conf. Series: Materials Science and Engineering **989** (2020) 012017 doi:10.1088/1757-899X/989/1/012017

Exoskeleton structures can be modelled as a coupled primary-secondary oscillator system as shown in Figure 1. M1, K1, C1 are mass, stiffness, damping coefficient of primary oscillator and M2, K2, C2 are mass, stiffness, damping coefficient of secondary oscillator. K is the stiffness of the connecting link.

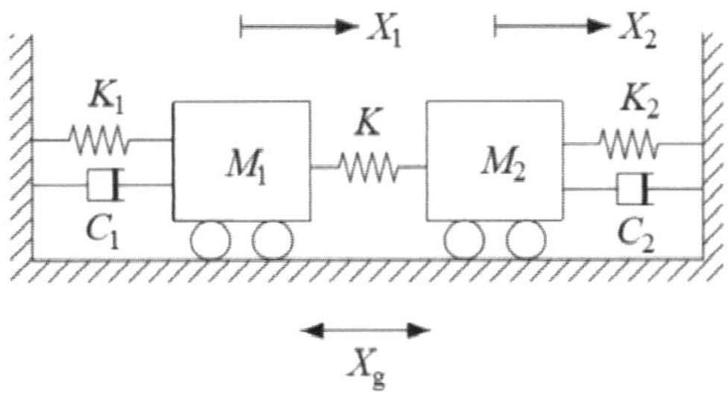

Figure 1. Structural model (Source: [3]).

No studies have so far been reported on exoskeleton structures connected to irregular structures. The present paper discusses the efficiency of exoskeleton structures on plan irregular structures.

2. Details of building configuration

In the present study, typical RCC frames of regular and irregular configurations are considered. The regular frame considered is having a square plan. Irregular frames considered are L- shaped frames (in plan). i.e., irregularity is introduced by a re-entrant corner as illustrated in IS 1893 (Part 1):2016 (Figure 2). The overall plan dimensions (total width in either directions) of the irregular frames are kept same as those of the regular frame.

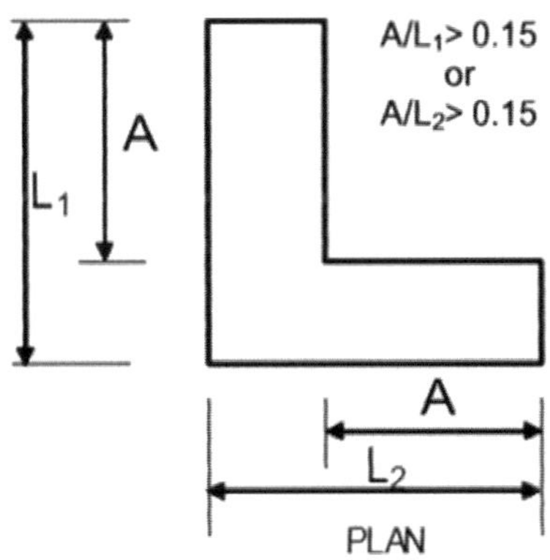

Figure 2. Irregularity due to re-entrant corner (Source: [4]).

The regular frame considered in the study has a square plan (L/B = 1) having dimensions of 24 m × 24 m, with 4 bays of 6 m span along both directions. The height of the building considered is 12 m, with 4 stories of 3 m height. Exoskeleton structure is provided outside the periphery of the building at a distance of 100 cm. The exoskeleton structure is connected to the building at various floor levels by links.

The plan and elevation of the regular building, considered in the study, with exoskeleton structure is shown in Figure 3.

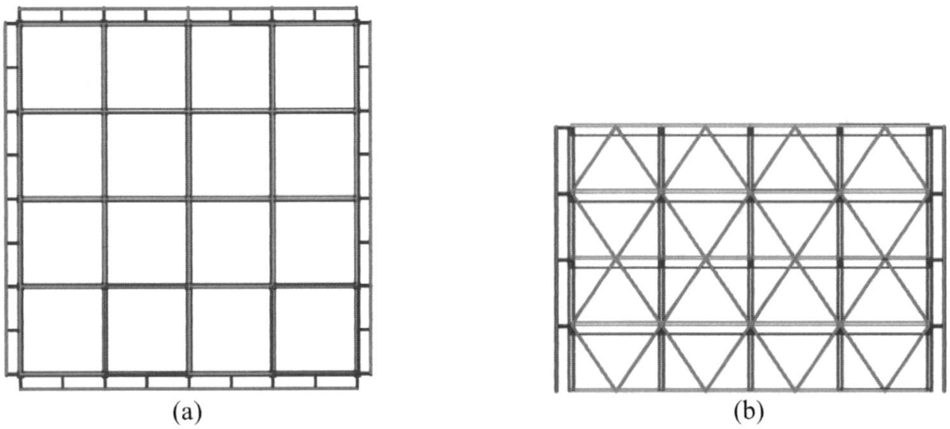

Figure 3. Configuration of regular building frame (L/B=1) connected with Exoskeleton Structure (a) Plan (b) Elevation.

As one may observe (Figure 2), the A/L ratio defines the degree of irregularity. As per IS 1893, the frame is considered irregular when A/L exceeds 0.15. Frames with different values of A/L ratios of 0.15, 0.30 and 0.50 are considered. The plans of these frames are shown in Figure 4.

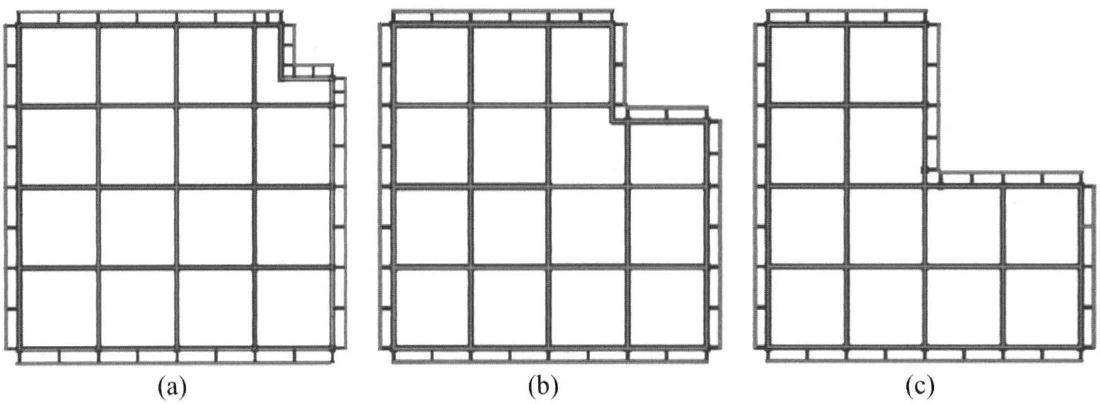

Figure 4. Plan of irregular frames with different A/L ratios (a) A/L=0.15 (b) A/L=0.3(c) A/L=0.5.

The number of stories (4) and storey height (3 m) of all frames are kept the same. The sizes of beams, columns and slabs of the frame are: Beam size - 400 mm × 300 mm, Column size- 400 mm × 400 mm, Slab thickness- 120 mm. Grade of concrete and rebar are M25and Fe 415 respectively.

Exoskeleton structures are made of Fe250 grade steel. Columns are made of I sections (ISLB 200), diagonal beams in exoskeleton structures is hollow circular beams (114.3 mm × 5mm) at inclination of 45 degrees. Location of exoskeleton structures is 100 cm from outside perimeter of building. The building frame (bare frame) is connected to the exoskeleton structures using links. The link is a rod of 10 cm diameter and 100 cm length. The links are modelled as axially rigid links by releasing moments. The Young's Modulus (E) of the link is assumed to be 2 × 1013N/mm^2. A high value of E is chosen to increase the axial stiffness of the link to make it as rigid as possible.

3. Seismic analysis

Linear dynamic analysis using Response Spectrum method was carried out in ETABS for frames of all four configurations – regular frame and L-shaped frames with A/L=0.15, 0.30 and 0.50. The frame without exoskeleton is referred as bare frame in the following sections.

Total eight models – four bare frames and four frames with exoskeleton structures – were analysed. The response spectrum given in IS code for Zone III was considered in the analysis. The importance factor (I) adopted is 1.5 and response reduction factor (R) of 3 is adopted. Response Spectrum analysis involves calculation of responses in each mode of vibration and combining them using suitable modal combination rules. In the present analysis, Square Root of Sum of Squares (SRSS) method as suggested in IS 1893 (Part 1):2016 is adopted for modal combination.

4. Results and discussions

Responses such as base shear and floor displacements of both the bare frame and exoskeleton with rigid link are considered for comparing the seismic performance of frames. Irregularity induces torsion and hence torsional responses are also considered for investigating the control effectiveness of exoskeleton structures. The maximum values of bending moment, twisting moment and shear force in the corner columns are considered for the study.

To indicate the control effectiveness, a control parameter (Cr) is defined as given below:

$$C_r = \frac{r_{uncontrolled} - r_{controlled}}{r_{uncontrolled}} \quad (1)$$

Here r is the response parameters considered in the present study such as base shear, storey displacement and forces in columns. Also the uncontrolled response refers to the response of the frame without any control measures (bare frame) and controlled response refers to the response of frame strengthened with exoskeleton structure.

4.1. Floor displacements

The maximum displacements of various floors for various configurations of frames are shown in figure 5.

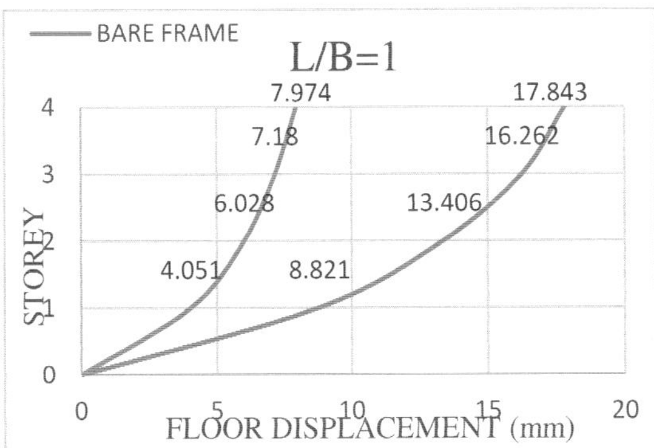

(a)regular frame(L/B=1)

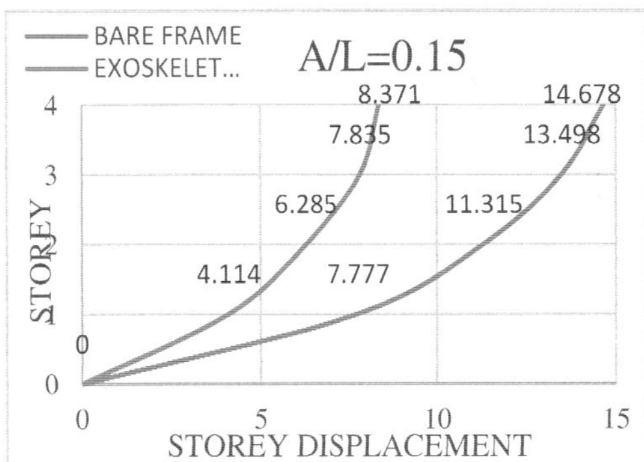

(b)L-frame with A/L=0.15

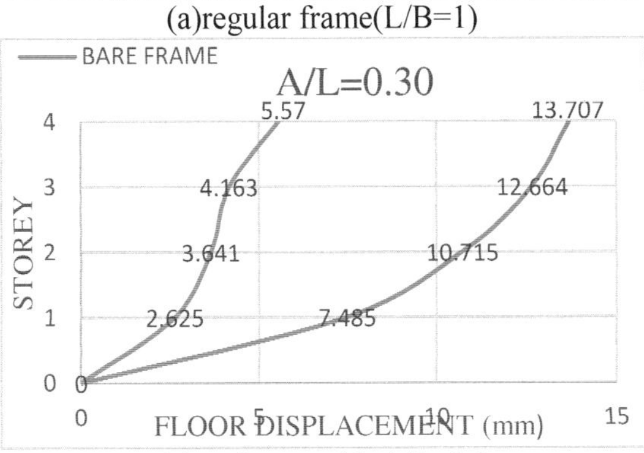

(c) L-frame with A/L=0.3

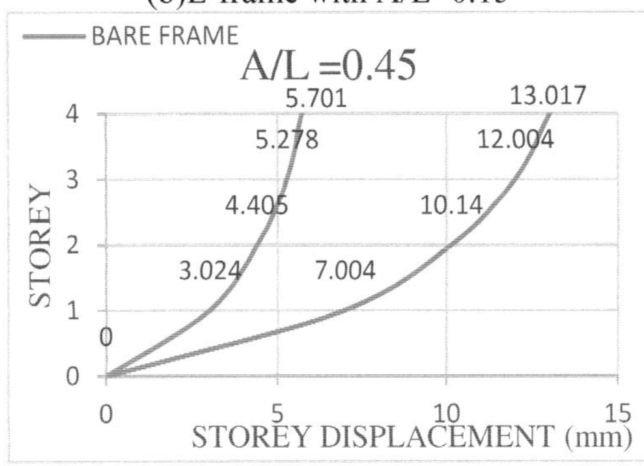

(d) L-frame with A/L=0.5

Figure 5. Variation of floor displacements different types of frames.

IOP Conf. Series: Materials Science and Engineering **989** (2020) 012017 doi:10.1088/1757-899X/989/1/012017

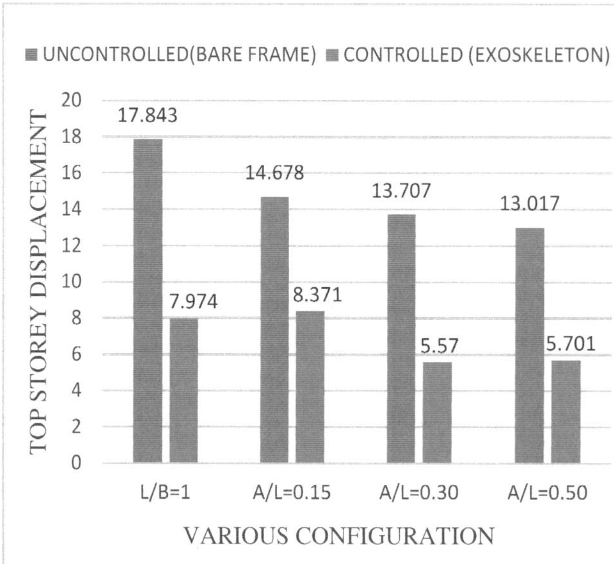

Figure 6. Displacement control of exoskeleton structures for frames of different configurations.

From the above graphs it is evident that displacement of the storey is effective controlled by using exoskeleton structures. The control effectiveness with respect to top floor displacement (C_{TFD}) - i.e. percentage reduction in top floor displacement of exoskeleton frame compared to the bare frame – for frames with L/B=1, A/L=0.15, A/L=0.30 and A/L=0.50 are respectively 55.31%, 42.96%, 59.36% and 56.20%.

4.2. Base shear

The maximum base shear induced in frames of different configurations are provided in Table 1.

Table 1. Maximum base shear induced in frames.

Frame Configuration	Base shear (bare frame) (kN)	Base shear (exoskeleton) (kN)	Control Effectiveness (C_{BSH})
L/B=1	590.39	196.47	66.72 %
A/L=0.15	651.34	355.79	45.37%
A/L=0.30	645.14	252.14	60.87%
A/L=0.50	518.08	180.73	65.11%

From the above table it is observable that the exoskeleton structure is effective in controlling the base shear for frames of all configurations considered in the present study.

4.3. Forces in corner columns

Maximum values of BM, TM and SF in the columns located at all corners are taken and the maximum of each are noted. The maximum values of each of these responses in corner columns are tabulated in Table 2. The control effectiveness is also shown.

Table 2. Maximum response of corner columns.

Details		Maximum response		
		BM (kNm)	SF (kN)	TM (kNm)
Regular B/L = 1	Bare frame	54.00	20.79	0
	Exoskeleton	25.54	9.85	2.45
	Control effectiveness	52.7 %	52.6 %	--
Irregular A/L = 0.15	Bare frame	70.63	52.68	4.33
	Exoskeleton	27.69	19.62	2.85
	Control effectiveness	60.79%	62.75%	34.11%
Irregular A/L = 0.3	Bare frame	79.10	30.45	3.08
	Exoskeleton	28.32	10.89	0.61
	Control effectiveness	64.19%	64.23%	80.19%
Irregular A/L = 0.50	Bare frame	74.74	30.50	1.65
	Exoskeleton	21.61	8.82	2.18
	Control effectiveness	71.08%	71.08%	-32.1 %

The above tables show that there is considerable reduction in the maximum values of bending moment and shear forces induced in the corner columns in frames of all configurations. However, exoskeleton structure is not found to be effective for controlling torsional moments for frames with L/B=1 and A/L=0.50. For the other two frames, the control effectiveness with respect to torsional moment, is found to be reducing with increase in A/L ratio.

5. Conclusion

Exoskeleton structure is a suitable retrofitting strategy especially in situations where space constraints and functional constraints are present. Performance of exoskeleton structures in the seismic response control of regular and plan irregular frames are considered in the present study. Of the several ways of connecting the exoskeleton with the bare frame, rigid links with hinged connection is adopted in the present study. The study suggests that the exoskeleton structure is capable of reducing floor displacements and base shear in plan irregular as well as regular structures. Control effectiveness is found to be excellent. However, exoskeleton structure is not found to be effective for controlling the torsional moments induced in the corner columns of plan irregular frames. Only a linear dynamic analysis is presently chosen. More insights into the seismic behavior of exoskeleton structures may be obtained from nonlinear dynamic analysis.

6. References

[1] Westermo B D 1989 The dynamics of interstructural connection to prevent pounding *Earthquake Engineering & Structural Dynamics* **18(5)** 687-699
[2] Reggio A, Restuccia L and Ferro G A 2018 Feasibility and effectiveness of exoskeleton structures for seismic protection *Procedia Structural Integrity* **1(9)** 303-10
[3] Reggio A, Restuccia L, Martelli L and Ferro G A 2019 Seismic performance of exoskeleton structures *Engineering Structures* **1(198)** 109459
[4] IS1893 B I S 2016 Indian Standard criteria for earthquake resistant design of structures (part 1): General provisions and buildings (sixth revision) Bureau of Indian Standards New Delhi

IOP Conf. Series: Materials Science and Engineering **989** (2020) 012018 doi:10.1088/1757-899X/989/1/012018

Durability studies on blended cement paste exposed to ammonium chloride environment

Sheba Babu[1,2], Mohammed Sahad[1], Vijeesh K P[1], Syamjith C K[1], Mohammed Raneesh[1], Ramaswamy K P[1], Nazeer M[1] and Mohammed Thowsif[1]

[1]Department of Civil Engineering,TKM College of Engineering, Kollam-691005, India.

[2]Email: shebababu447@gmail.com

Abstract. The exposure of concrete systems in ammonium based environments can be considered as an aggressive medium which can induce severe and premature deterioration, thus adversely affecting the durability of concrete structures. This paper aims to understand the degradation kinetics of cementitious systems made with Ordinary Portland Cement (OPC) and various combinations of Supplementary Cementitious Materials (SCMs) exposed to chemical grade ammonium chloride (NH_4Cl) solution. For this study, the cement paste specimens were exposed to ammonium chloride solution in two different concentrations (1.25 M and 2.5 M) for an exposure duration of two and half months. The kinetics of degradation were evaluated on the basis of changes in mass, visual assessment, changes in thickness and changes in the pH of ammonium chloride solution. Based on the test results, the ternary blended mix with OPC, Ground Granulated Blast Furnace Slag (GGBS) and Silica fume (SF), that is, OPC(50%) - GGBS(40%) - SF(10%) is found to have higher resistance to ammonium chloride attack. The results stress the need to include SCMs and the importance to tailor make the cementitious system for structures exposed to aggressive environment like ammonium chloride.

1. Introduction

Since concrete is the most widely construction material, it can be subjected to various chemical attacks when exposed to aggressive environment conditions. So it is essential to understand the durability aspects of cement based materials in various aggressive environments, in order to find the most promising exploratory alternative for cement, since the durability issues related to cement based materials are more severe. The addition of Supplementary Cementitious Materials (SCMs) along with cement were found to be more effective in improving the performance of cement composites in aggressive chemical environments as it imparts a compact microstructure with low penetrability [2, 15]. The partial replacement of cement with SCMs can also be considered as the most viable solution to make the cement concrete more sustainable.

The resistance of cementitious systems in ammonium environment can be considered as one of the most important durability indices to be studied, as they can be exposed to ammonium based compounds in various areas such as chemical fertilizer plants, agricultural land, industrial environments, waste waters, ground waters and polluted environment [1, 3, 5, 10, 16]. The most aggressive and deleterious ammonium based compounds were found to be ammonium nitrate, ammonium sulphate and ammonium chloride [2, 4-14, 16].

A number of studies have been carried out on the deterioration of cementitious matrices exposed to ammonium sulphate and ammonium nitrate solutions. The deterioration by ammonium sulphate ($(NH_4)_2SO_4$) is characterised by the intensive dissolution of cement hydrates and expansion of cement systems due to the formation of gypsum and ettringite [14]. It is caused by the combined action of acid hydrolysis, sulphate and ammonium corrosion, with needle like insoluble gypsum crystal formation on the surface of specimens [4, 6-8, 12].

The mechanism of deterioration by ammonium nitrate (NH_4NO_3) is characterised by the leaching of calcium ions resulting in the dissolution of hydration products [11]. The final reaction products due to ammonium nitrate attack were soluble calcium nitrate, calcium nitro aluminate salt and ammonia gas emanation [2, 13]. On exposure to ammonium nitrate environment, the surface deposits on the specimens were observed to be rhombic calcite, which is formed due to the chemical reaction of calcium ions with dissolved carbon dioxide (CO_2) [12].

Ammonium chloride (NH_4Cl) attack is characterised by the dissolution of Portlandite, followed by the crystallization of thaumasite salt, Friedel's salt and ettringite. The final reaction products were found to be gypsum, calcite and vaterite [14]. From the experimental study conducted by Ohama et al. [9], the ammonium chloride attack is characterised by the formation of salts like hydrated calcium chloride, calcium hydroxychloride hydrate and calcium chloro aluminate hydrate. However, the mechanism of degradation of cement based materials in ammonium chloride environment is not explored much and more kinetics and micro-analytical characterization studies are to be performed for the better understanding of the deterioration.

Since there are only limited works reported on the degradation kinetics of cementitious matrices and the effectiveness of SCMs in improving the resistance when exposed to various concentrations of NH_4Cl solution over different exposure duration, the objective of this experimental study is to evaluate the effect of NH_4Cl solution on the durability properties of cement paste specimens with various combinations of SCMs over an exposure duration of two and half months.

2. Experimental programme

This experimental study was conducted on cement paste specimens made with Ordinary Portland Cement (OPC) and various combinations of SCMs exposed to two different concentrations (1.25 M and 2.5 M) of ammonium chloride (chemical grade) solution in order to understand the nature of the deterioration of cement based materials in ammonium environments. The extent of deterioration were evaluated in terms of visual examinations, changes in mass, changes in thickness and changes in the pH of the NH_4Cl solution before and after exposure. The various materials used for the preparation of paste specimens were as follows:

- Type of cement used is 53 grade Ordinary Portland Cement (OPC) conforming to IS: 12269-1987.
- SCMs used include Class F Fly ash (FA), Ground Granulated Blast Furnace Slag (GGBS) and Silica fume (SF).

The specific gravity of OPC, Fly ash, GGBS and Silica fume is 3.15, 2.20, 2.90 and 2.05 respectively. The water to binder ratio chosen for all the mixes was 0.50. The six different mixes with various combinations of SCMs adopted for this study is shown in Table 1. Poly Carboxylate Ether (PCE) was used as the superplasticizer, with an optimum dosage of 0.5% by weight of cement (used only for Silica fume based mixes).

Table 1. Paste mixes used for the study.

Mix	OPC (%)	FA (%)	GGBS (%)	SF (%)
Mix 1 (OPC)	100	-	-	-
Mix 2 (OPC+FA)	70	30	-	-
Mix 3 (OPC+GGBS)	50	-	50	-
Mix 4 (OPC+SF)	90	-	-	10
Mix 5 (OPC+FA+SF)	70	20	-	10
Mix 6 (OPC+GGBS+SF)	50	-	40	10

The raw materials required for the casting of paste specimens were weighed accurately using a precise weighing balance as per the mix proportion. All the ingredients were mixed well using a Hobart mixer for about 5 minutes, filled in the oiled plastic cylindrical moulds (diameter 25 mm and height 50 mm) and allowed to set for 24 hours. After that, the specimens were demoulded and kept for lime saturated curing for a duration of 14 days. For lime saturated curing, 3g/L of calcium hydroxide ($Ca(OH)_2$) solution was used. The cylindrical paste specimens having 25 mm diameter and 50 mm height were used for performing all the tests.

The initial appearance, mass and thickness (in saturated but surface dry condition) and pH of the solution were noted before immersion in NH_4Cl solution to determine the changes in these parameters after the exposure. Then these specimens were exposed to two different concentrations - 1.25 M and 2.5 M NH_4Cl solution. The NH_4Cl solution was replenished once in every 2 weeks in order to maintain the aggressiveness of the solution. The ratio of volume of ammonium solution to that of specimens was kept fixed at five for the evaluation of degradation kinetics for the entire exposure period. Thus, an accelerated immersion based leaching test is adopted in this experimental investigation in order to understand the degradation kinetics of various binders upon exposure to ammonium based solution.

The pH of NH_4Cl solution (initial and weekly readings) was measured using a digital pH meter of 0.01 accuracy. The specimens were then taken out of the solution after each week of exposure and tested to determine the changes in the physical appearance, mass, thickness and pH. The tests were carried out for a maximum exposure duration of 10 weeks for 1.25 M and 7 weeks for 2.5 M NH_4Cl solution. Mass changes of specimens (in saturated but surface dry condition) were measured using a precise weighing balance of 1 μg accuracy. The changes in the physical appearance of specimens were identified by taking photographs of specimens on a weekly basis. The thickness changes were measured using a precise digital vernier caliper of 0.01 mm least count. The mean value of the three specimens from each mix was taken for the each test result.

3. Results and Discussions

3.1 Changes in mass

The changes in mass of the cement paste specimens of various mixes exposed to 1.25 M (for 10 weeks) and 2.5 M (for 7 weeks) NH_4Cl solution is shown in Figure 1 and 2 respectively. From the graphs, it is clear that the all the specimens have undergone a significant reduction in their mass when exposed to both concentrations of NH_4Cl solution as compared to their initial mass. The reduction in mass could be attributed to the leaching of alkali ions and the progressive decalcification of cement hydrates. It was found that the percentage reduction in mass of specimens increased with the increase in the concentration (the mass loss found to be higher for 2.5 M as compared to 1.25 M) of NH_4Cl solution.

The mass loss was found to be less for GGBS based mixes (Mix 3 and Mix 6) and higher for the control mix OPC (Mix 1) in both concentrations of NH_4Cl solution. Among the Mix 3 (OPC-GGBS) and Mix 6 (OPC-GGBS-SF), the mass loss was found to be lower for ternary blended mix (Mix 6) as compared to the binary mix (Mix 3). The mass loss for Mix 6 in 1.25 M and 2.5 M was only 9.83% and 10.08% respectively, and that of Mix 3 was found to be 11.16% and 11.85% respectively. The mass loss for control mix OPC (Mix 1) was found to be about 17.67% in 1.25 M and 20.28% in 2.5 M NH_4Cl solution. The mass loss for binary blended OPC-FLYASH based mix (Mix 2) was found to be 12.38% and 12.96% respectively and that of ternary blended mix was found to be 13.09% and 13.78% respectively in 1.25 and 2.5 M of NH_4Cl solution. After OPC, the mass loss was found to be higher for binary mix with OPC and SF (Mix 4) as a binder which was about 15.59% and 17.11% respectively in these two concentrations of NH_4Cl solution. Based on the mass loss, it can be ascertained that the performance of OPC-GGBS-SF blended mix was better compared to the other mixes.

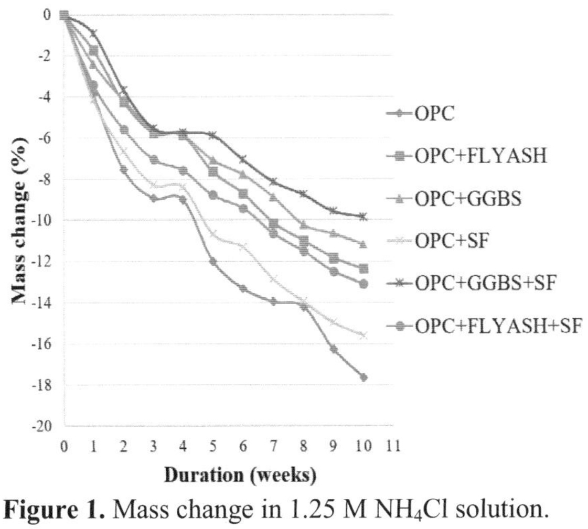

Figure 1. Mass change in 1.25 M NH_4Cl solution.

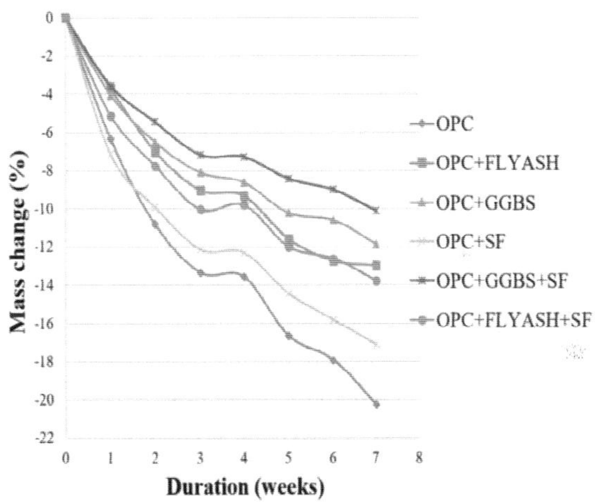

Figure 2. Mass change in 2.5 M NH_4Cl solution.

3.2 Changes in diameter

Figure 3 and 4 shows the changes in thickness of cylindrical paste specimens of various mixes exposed to 1.25 M (for 10 weeks) and 2.5 M (for 7 weeks) NH_4Cl solution. From the graphs, it is clear that the specimens of all the six mixes have undergone a significant reduction in their thickness when exposed to both concentrations of the salt solution. The reduction in thickness could also be attributed to the leaching and the progressive decalcification of hydrates and the decalcification shrinkage resulting thereby. It was found that the percentage reduction in the thickness of specimens increased with the increase in the concentration. The thickness loss was found to be higher for 2.5 M as compared to 1.25 M of NH_4Cl solution.

Among the various mixes, the thickness loss was found to be higher for the control mix OPC (Mix 1) and lower for GGBS based mixes (Mix 3 and Mix 6) in both concentrations of NH_4Cl solution which is similar to the variation in mass loss. The thickness loss for control mix OPC (Mix 1) was found to be about 2.80 % in 1.25 M and 3.06 % in 2.5 M NH_4Cl solution. Among the Mix 3 (OPC-GGBS) and Mix 6 (OPC-GGBS-SF), the thickness loss was found to be lower for ternary blended mix (Mix 6) as compared to the binary mix (Mix 3) with GGBS as the SCM. The thickness loss for Mix 6 in 1.25 M and 2.5 M was only 2.41 % and 2.46 % respectively, and that of Mix 3 was found to be 2.49 % and 2.57 % respectively. Thus in terms of thickness loss, the ternary blended mix OPC-GGBS-SF was found to be more resistant in ammonium chloride environment as compared to the other mixes.

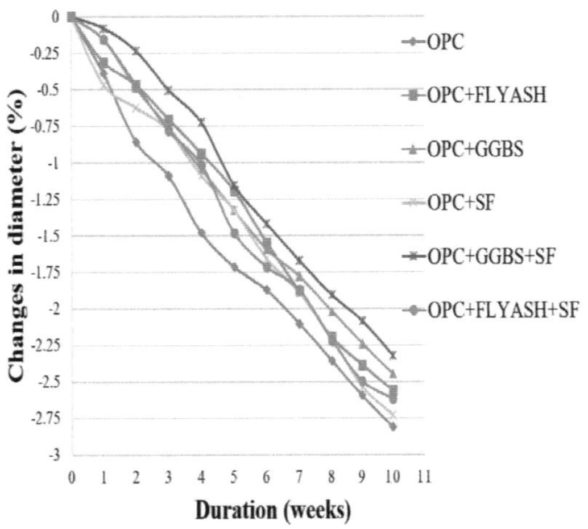

Figure 3. Changes in diameter in 1.25 M NH_4Cl solution.

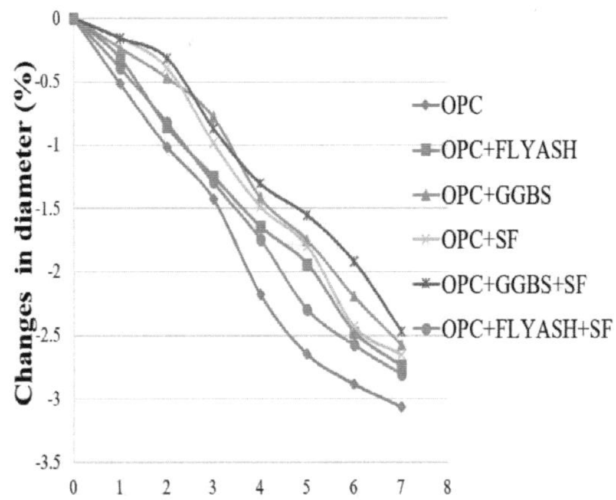

Figure 4. Changes in diameter in 2.5 M NH_4Cl solution.

3.3 Visual examinations

The physical appearances of the specimens before and after exposure were as shown in Table 2. The cylindrical paste specimens were visually examined to identify the changes in the physical appearance up on exposure in 1.25 M and 2.5 M NH_4Cl solution. The photographs of the specimens were taken over the entire exposure duration. It was observed that the specimens of all the six mixes exposed to both concentrations (1.25 M and 2.5 M) of NH_4Cl solution undergone minimal changes in their physical appearance up on exposure. The physical deterioration on the specimens was found to be minimal and they remained structurally intact. But it was observed that all the specimens have undergone significant reduction in their mass and thickness in both concentrations of NH_4Cl solution which gives information about the microstructural deterioration occurred on the specimens.

Table 2. Visual examination of paste specimens in NH_4Cl solution (1.25 M and 2.5 M).

Paste mix	Before exposure	After exposure in NH_4Cl solution	
		1.25 M (10th week)	2.5 M (7th week)
OPC	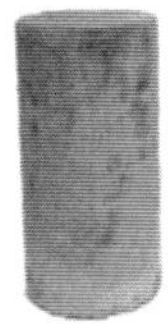		
OPC+FLYASH			
OPC+GGBS			
OPC+SF			
OPC+FLYASH+SF			
OPC+GGBS+SF			

3.4 Changes in pH

The general trend in the variation of pH of 1.25 M and 2.5 M NH_4Cl (up to 6 weeks duration) solution resulted from the immersion of different mixes of paste specimens is shown in Figure 5. It was found that the initial pH of the solution depends on the concentration of NH_4Cl solution, pH value being higher for 1.25 M as compared to 2.5 M NH_4Cl solution. Since NH_4Cl solution is acidic, the initial pH of 1.25 M and 2.5 M was found to be 5.61 and 5.38 respectively. The initial pH value of 2.5 M was found to be high as compared to 1.25 M, which can be explained by the fact that, higher the molarity of salt solution, the concentration of hydrogen ions will be higher, resulting in lower pH and vice versa. After a week, a rapid rise in the pH of the ammonium chloride solution was observed which may be due to the diffusion of alkali and hydroxyl ions from the paste specimens into the solution. The rapid increase in the pH gives information about the higher rate of leaching and decalcification occurred due to the immersion of specimens. The pH rise was found to be similar for all the mixes studied or in other words, the variation in pH was minimal between the mixes.

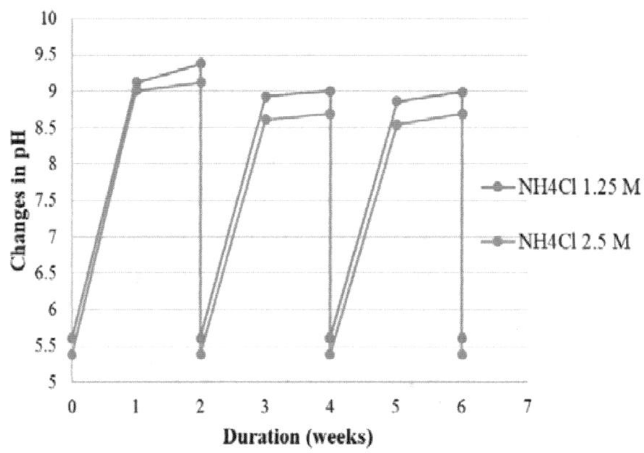

Figure 5. Variation of pH of NH_4Cl solution (both 1.25 M and 2.5 M).

4. Conclusions

The rapid degradation of concrete structures in aggressive environments is a major threat to the durability of structures. The exposure of concrete systems in ammonium based environment can be considered as one such aggressive media which can induce severe and premature deterioration. There are a number of areas where the cementitious systems can be exposed to ammonium based salts such as chemical fertilizer plants, agricultural land, sewage plants and many other industrial applications. Thus the resistance of cementitious systems in ammonium environment can be considered as the one of the most important durability indices to be studied.

Since limited studies were reported on the effectiveness of SCMs in imparting durability in the aggressive ammonium chloride environment and its degradation kinetics, an experimental study (accelerated test) was conducted to analyze the changes in the physical and mechanical properties of the cement paste specimens made with OPC and different combinations of SCMs exposed to 1.25 M and 2.5 M ammonium chloride solution. The degradation kinetics were evaluated in terms of visual examinations, changes in the pH of NH_4Cl solution, changes in mass and changes in thickness of paste specimens before and after exposure.

From the test results, it was found that there is appreciable early mass and thickness loss when specimens were exposed to 1.25 M and 2.5 M NH_4Cl solution. Based on the mass and thickness loss, it can be ascertained that the performance of OPC-GGBS-SF blended mix was better compared to the other mixes. The reduction in mass and thickness may be ascribed to the leaching phenomenon and decalcification of hydrates such as Portlandite and CSH gel. It is found that there was a considerable increase in the pH when the paste specimens were immersed in both 1.25 M and 2.5 M NH_4Cl solution, which might be due to the migration of alkali ions from the specimens into the solution. The rapid increase in the pH at the early ages of exposure indicates that the rate of leaching and decalcification is significantly higher during the initial weeks of immersion.

From all the test results, it can be concluded that the ternary mix with OPC, Ground Granulated Blast Furnace Slag (GGBS) and Silica fume (SF), that is, OPC(50%) - GGBS(40%) - SF(10%) performed better in both concentrations of ammonium chloride as compared to other mixes. A detailed microstructural characterization study is essential to validate the changes in these physico-mechanical properties. These results indicate the effectiveness of SCMs in making the concrete systems more durable in aggressive and leaching environments such as ammonium chloride.

Acknowledgement
The authors thank the TEQIP II for the financial assistance received through seed money funding for this work in 2019-20.

5. References

[1] Alexander M, Bertron A and De Belie N 2013 Performance of cement-based materials in aggressive aqueous environments *RILEM State-of-the-Art Reports* Springer **10** 1- 469

[2] Amin M and Bassuoni M T 2017 Performance of concrete with blended binders in ammonium sulphate solution *Journal of Sustainable Cement based Materials* **7** 15-37

[3] Arafa M, Alqedra M and Shubari T 2015 Quantifying effect of ammonium nitrate attack on mechanical and physical properties of cement mortars *Journal of Scientific Research and Reports* **7** 137-145

[4] Bassuoni M T and Nehdi M L 2012 Resistance of self consolidating concrete to ammonium sulphate attack *Materials and Structures* **45** 977-994

[5] Fiertak M and Stryszewska T 2013 Resistance of three component binders in highly chemically corrosive environments *11th International Conference on Modern Building Materials Procedia Engineering* **57** 278-286

[6] Jauberthie R and Rendell F 2003 Physicochemical study of the alteration surface of concrete exposed to ammonium salts *Cement and Concrete Research* **33** 85-91

[7] Lea F M 1965 The action of ammonium salts on concrete *Magazine of Concrete Research* **17 (52)** 115-116

[8] Mbessa M and Pera J 2001 Durability of high strength concrete in ammonium sulfate solution *Cement and Concrete Research* **31** 1227-1231

[9] Ohama Y, Kawakami M and Fukuzawa K 1997 *Polymers in concrete* Proceedings of the Second East Asia Symposium on Polymers in concrete

IOP Conf. Series: Materials Science and Engineering **989** (2020) 012018 doi:10.1088/1757-899X/989/1/012018

[10] Pepenar I 2009 Degradation of reinforced concrete structures in ammonium nitrate environments *Concrete in aggressive aqueous environments, Performance, Testing and Modeling* 442-448

[11] Rehman A, Qudoos A, Jakhrani S H, Kim H G and Ryou J S 2019 Influence of nano-silica on the leaching attack upon photocatalytic cement mortars *International Journal of Concrete Structures and Materials* **13(1)** 11-20

[12] Rendell F and Jauberthie R 1999 The deterioration of mortar in sulphate environments *Construction and Building Materials* **13(6)** 321–327

[13] Schneider U and Chen S W 2009 The leaching effect of concrete immersed in ammonium nitrate solution *Proceedings of the RILEM TC 211-PAE Final Conference* **1** 54-64

[14] Slupik B S, Podworny J and Staszuk M 2018 Corrosion of cement pastes made of CEM I and CEM III/A caused by a saturated water solution of ammonium chloride after 4 and 25 days of aggressive immersion *Construction and Building Materials* **170** 279-289

[15] Thomas M D A, Hooton R D, Scott A and Zibara H 2012 The effects of Supplementary Cementitious Materials on chloride binding in hardened cement paste *Cement and Concrete Research* **42** 1-7

[16] Wong L L, Asrah H, Rahman M E and Mannan M A 2013 Effects of aggressive ammonium nitrate on durability properties of concrete using sandstone and granite aggregate *International Journal of Civil and Environmental Engineering* **7** 49-53

Performance assessment of hexagrid structural system

Sooriya P[1,2] and Bushra M A[1]

[1] Department of Civil Engineering, TKM College of Engineering, Kollam, Kerala, India.

[2]Email: sooriyapurushothaman48@gmail.com

Abstract. The progression of innovation and advancement of economy of the world have brought new period of tall structures. Different types of structural systems pose challenges which are being faced by the structural engineers and architects in terms of innovation in structural system. Hexagrid is one among them. This system is adopted in tall buildings due to their structural efficiency and architectural potentials. Hexagrid encloses intersection of diagonal and horizontal or vertical members. This system is very similar to that of diagrid system and is configured by locating hexagons along the periphery of the building. This study aims to compare between hexagrid structural systems with different patterns and to find the optimal pattern which resists the lateral loads in hexagrid structural system. The use of hexagrids in a skyscraper is a relatively new idea. As such, more is yet to be explored in such a system.

1. Introduction

1.1. General

In recent years, fundamental economic growth of tall buildings is due to the quick development of populace, significant expense of land, improvement in the perspective on city and limitation in the horizontal growth due to less space. Today, it is difficult to envision a significant city without tall structures. Recent design trends pose new challenges in providing a structural framework with requirements of strength, stiffness and stability against the loads that are being experienced by the structures over the life cycle.

A significant purpose of this arrangement is to present a framework that is capable of minimizing lateral loads. The need for opposing lateral loads is more significant contrasted with withstanding gravity for elevated structures. These days, the structural systems ought to be so that it responds to the architecturally aesthetic prerequisites as well as it is structurally efficient. A framework which opposes both gravity and lateral loads with the same structural component can be considered as a productive structural framework.

The most effective framework for tall structures has been the tube-type structural systems. The diagrid system with tubular behaviour was found to be structurally efficient and architecturally sound. The diagonals provided around the perimeter of the building acts as a façade, which helps in providing the aesthetic view of the structure. To enhance the effectiveness of the tube type structures, another framework known as "Hexagrid" was introduced and was first proposed in literature published in 2012 [1]. Hexagrid system is the new emphasis of the exoskeleton tubular framework. The framework is comprised of a system of multi-storey tall hex-angulated truss system [2]. It highlights modules composed of hexagon-shaped structural elements on the periphery of the structure to determine both lateral and gravitational forces to a similar fashion as that of diagrids. Hexagrid structures are rising as a new trend for elevated structures in the modern period of architecture.

In contradistinction to diagrid systems, the amount of research conducted on hexagrid systems is very few. The hexagonal pattern was first introduced in Sino-steel International Plaza located in China. A couple of other structures in which hexagonal patterns implemented are as shown in the Figure 1.

(a) (b) (c)

Figure 1. (a) Sino-steel International Plaza, China **(b)** Al Bahr Towers, Abu Dhabi **(c)** Nano Towers, Dubai.

1.2. Hexagrid system

This system bears a resemblance to diagrid system, in which the hexagons are situated along the outermost periphery of the building which bears the purpose of enhancing the structural productiveness and aesthetic appearance. Therefore, this system resembles a tube. Hexagrids encompass intersection of diagonal and horizontal or vertical members to form six node hexagonal grids which are obliged by mega columns located on the corner of the structure. They transfer both lateral as well as gravity loads. This type of innovative system eradicates the necessity of external vertical columns in elevated structures that offer a great scenic view from the inner side of the structure. The columns present at the corner can also be eliminated to achieve architectural demand. Also, the load paths are uninterrupted and continuous [3]. Hexagrids helps in 10-15% of reduction in steel. The diagonal assumes a significant role in the behaviour of hexagonal framework as it determines the stress distribution resisting internal forces. Hexagrids are more compelling in minimization of shear deformation as they carry shear by axial action of diagonal members, unlike conventional tall buildings which carry shear by the bending of vertical columns and horizontal spandrels. The hexagrid structural system provides both bending as well as shear rigidity. But, it does not require high shear rigidity cores as shear is being carried by hexagon elements that are located on the exterior periphery of the structure. Subsequently, hexagrid framework can be reviewed as an ideal framework for elevated buildings. Hexagrid framework shall take the form in two patterns are shown in Figure 2 [7].

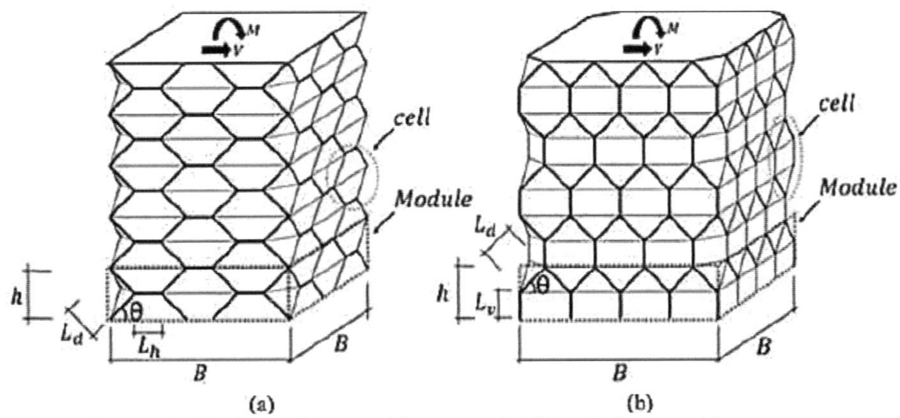

Figure 2. Horizontal hexagrid pattern **(b)** Vertical hexagrid pattern.

2. Building configuration

2.1. Building description and modelling

A square building of plan area 1296 m^2 and size 36m x 36m is considered. The building consists of 36 stories with uniform storey height of 3.6 m. The inclined columns are provided at a spacing of 6 m along the perimeter. The support conditions of interior columns are fixed and that of exterior hexagrid is hinged. The floor slab is subjected to a live load of 2.5 kN/m^2. The computations for the design earthquake load is based on zone factor of 0.16, importance factor 1, medium soil and response reduction factor 5 [4]. The structure is modelled and analyzed in ETABS 2016. The plan of the structure is as shown below in the Figure 3.

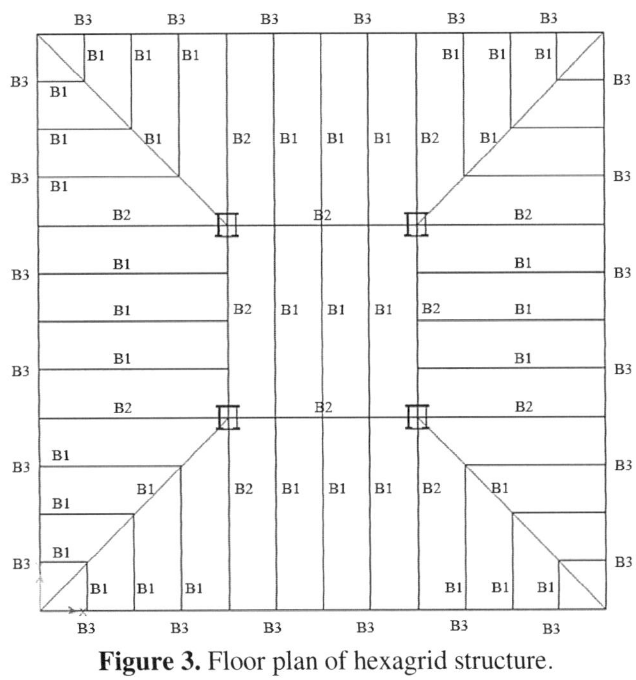

Figure 3. Floor plan of hexagrid structure.

Table 1. Size of typical members of 36 storey hexagrid structure.

Storey	Diagonal Columns (same for all storey)	Interior Columns	Beams (same for all storey)
From 30th – 36th storey	825 mm Pipe sections with 50mm thickness	1500 mm x 1500 mm	B1 and B3 = ISMB550 to ISWB600 B2 = ISWB600 with
From 12th – 29th storey		1650 mm x 1650 mm	top and bottom cover
From 1st – 11th storey		1800 mm x 1800 mm	plate of 220 x 50 mm

Figure 4 shows the details of internal column which opposes the gravity loads [5]. The member sizes can be lowered by the consideration of higher strength of structural steel.

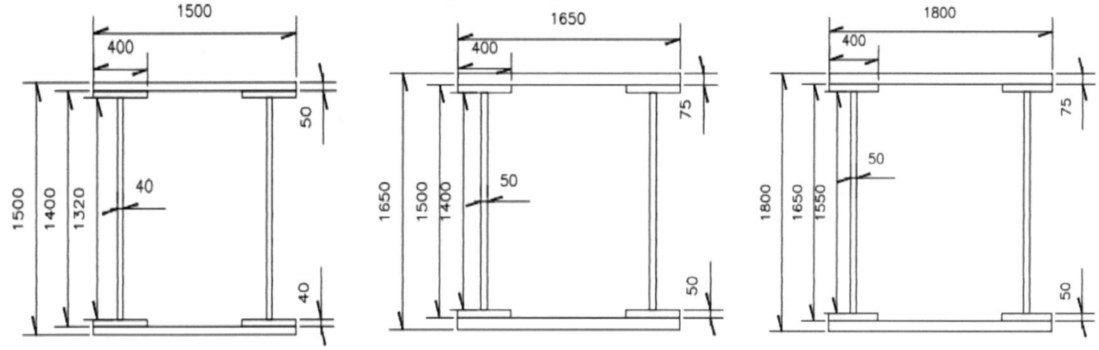

Figure 4. Details of internal column of 36 storey hexagrid structure.

2.2. Building patterns

The patterns considered for the study are listed below in figures 5 and 6. The models are divided based on the patterns, storey height and hexagrid density. A total of 12 patterns are observed and analyzed.

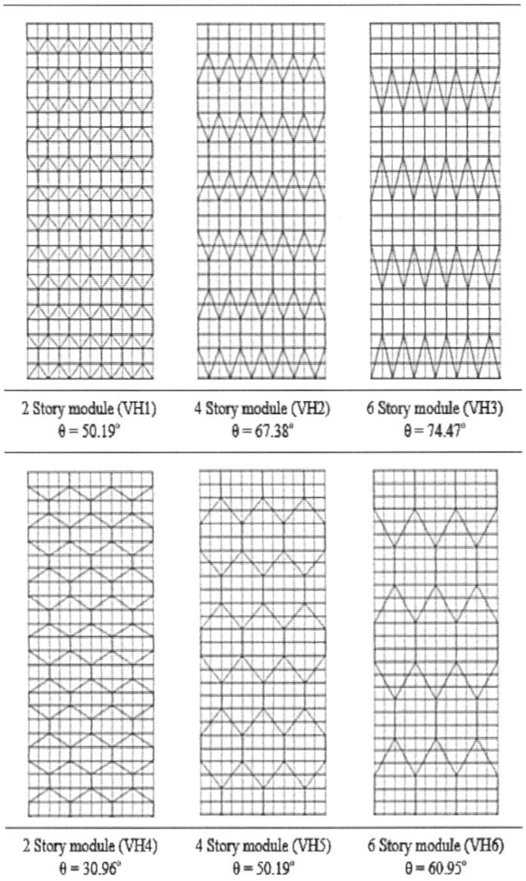

Figure 5. Vertical Patterns of Hexagrid structure.

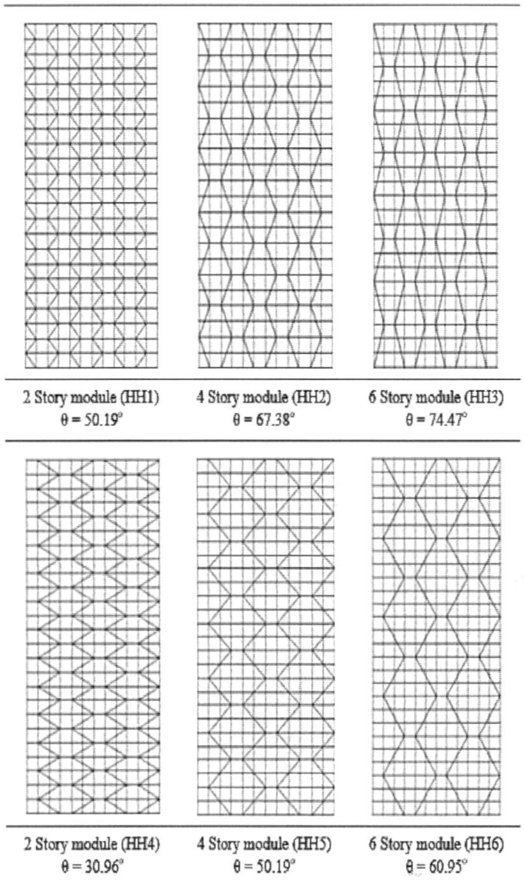

Figure 6. Horizontal Patterns of Hexagrid structure.

3. Results and discussion

The response spectrum analysis results of all hexagrid patterns are compared and presented in this section.

3.1. Time period

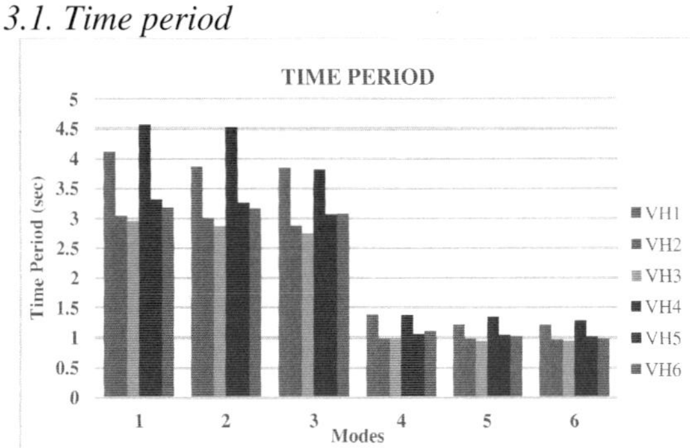

Figure 7. (a) Time period of vertical hexagrid buildings.

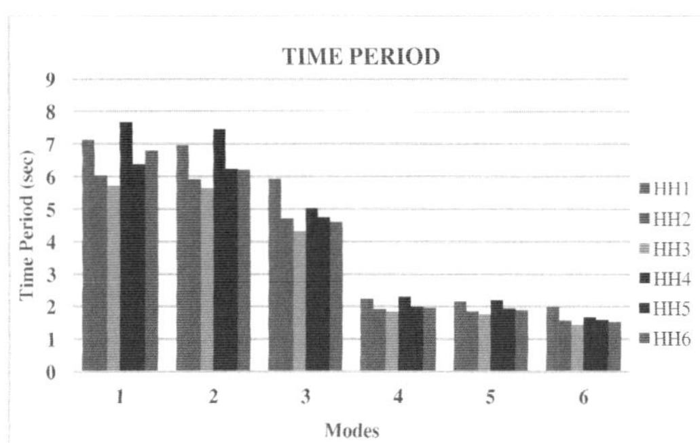

Figure 7. (b) Time period of horizontal hexagrid buildings.

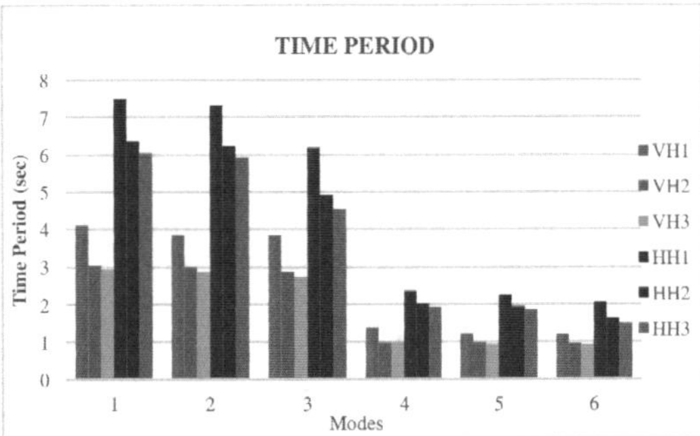

Figure 7. (c) Time period of vertical and horizontal hexagrid buildings.

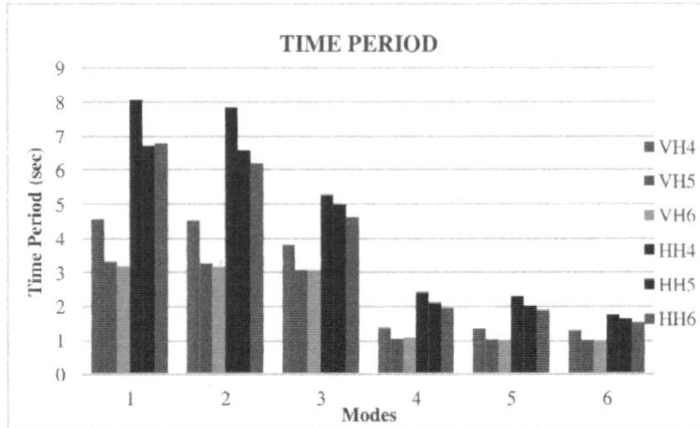

Figure 7. (d) Time period of vertical and horizontal hexagrid buildings with angles decreased.

In vertical hexagrid, as the module size increases, time period decreases as shown in Figure 7(a). In horizontal hexagrids, as the module size increases, time period decreases as shown in Figure 7(b). From figures 7(c) and 7(d) it is found that vertical hexagrids give lesser time period when compared to horizontal hexagrids.

3.2. Storey displacement

The displacement decreases in vertical hexagrid when both the module size and angle increases as shown in Figure 8(a). The same is noticed in horizontal hexagrid as shown in Figure 8(b).

Figure 8. (a) Storey Displacement of vertical hexagrid buildings.

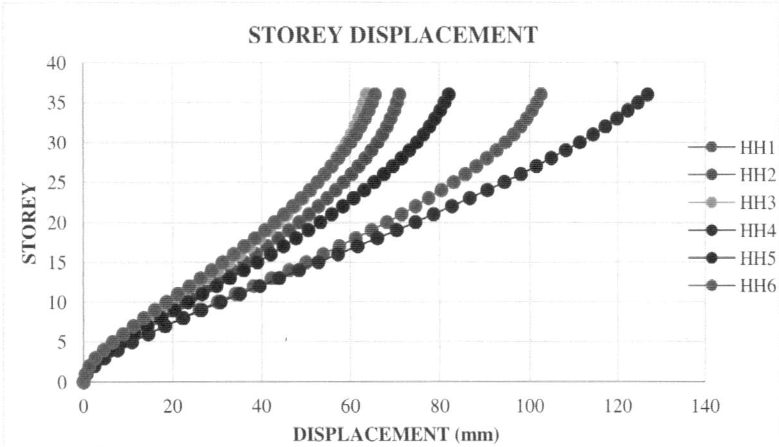

Figure 8. (b) Storey Displacement of horizontal hexagrid buildings.

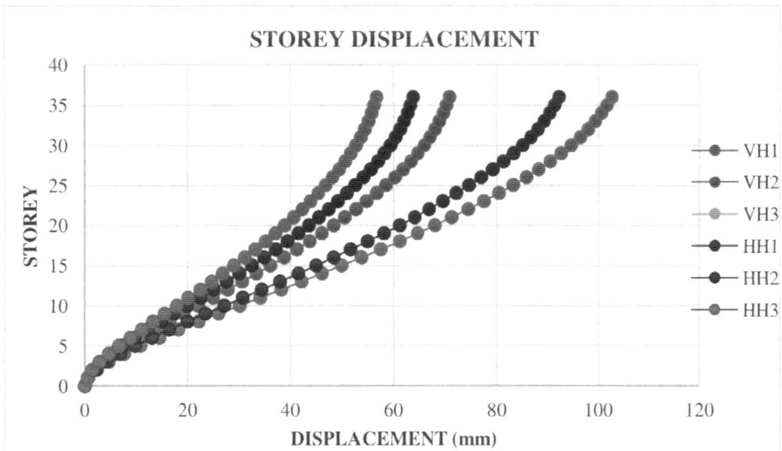

Figure 8. (c) Storey Displacement of vertical and horizontal hexagrid buildings.

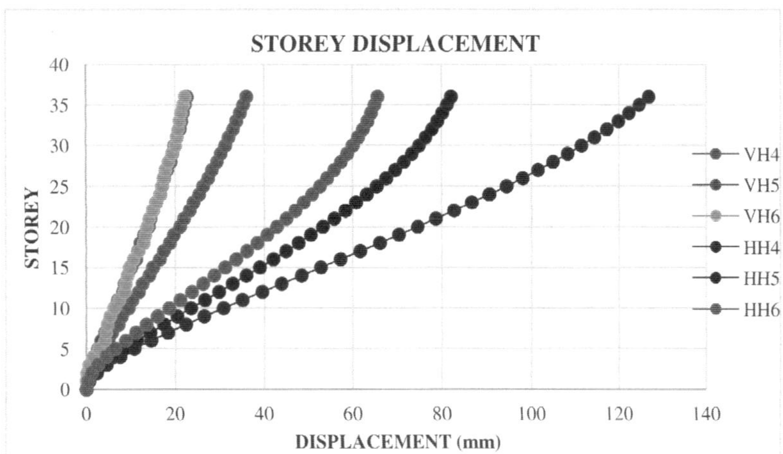

Figure 8. (d) Storey Displacement of vertical and horizontal hexagrid buildings with angles decreased.

In both the figures 8(c) and 8(d), it is noted that vertical hexagrids displaced lesser than that of the horizontal hexagrids. It is observed that the displacement increases in both vertical as well as horizontal hexagrids when the angle decreases as shown in Figure 8(d).

It is concluded that the vertical hexagrids of larger module size are preferred over horizontal hexagrids.

3.3. Storey drift

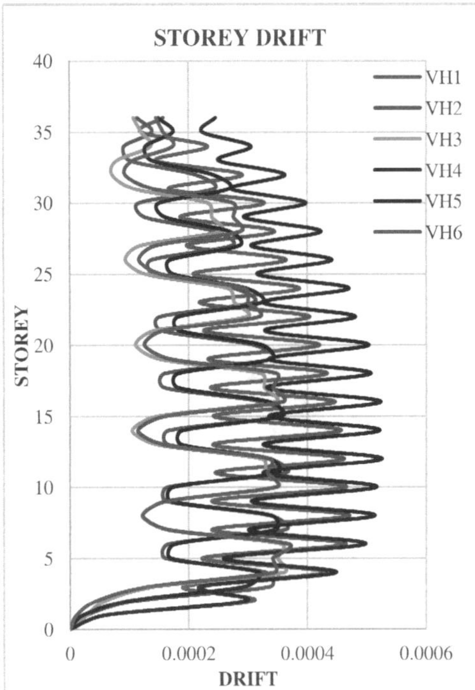

Figure 9. (a) Storey Drift of Vertical hexagrid buildings.

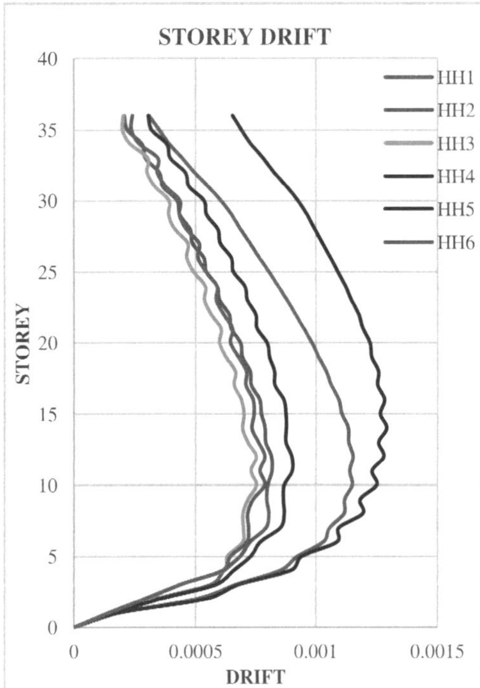

Figure 9. (b) Storey Drift of Horizontal hexagrid buildings.

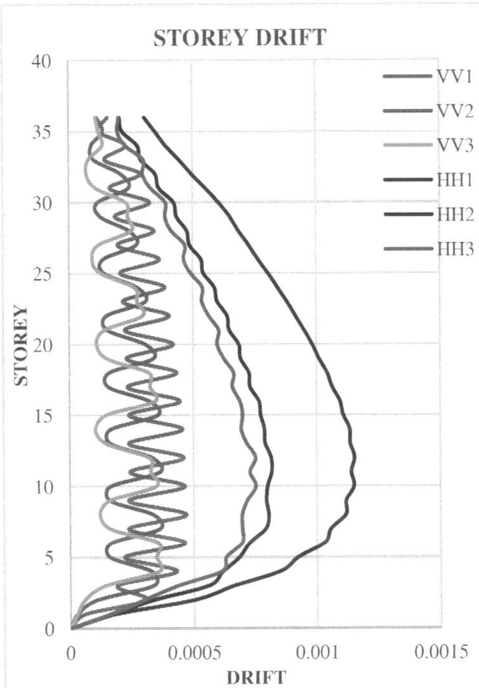

Figure 9. (c) Storey Drift of vertical and horizontal hexagrid buildings.

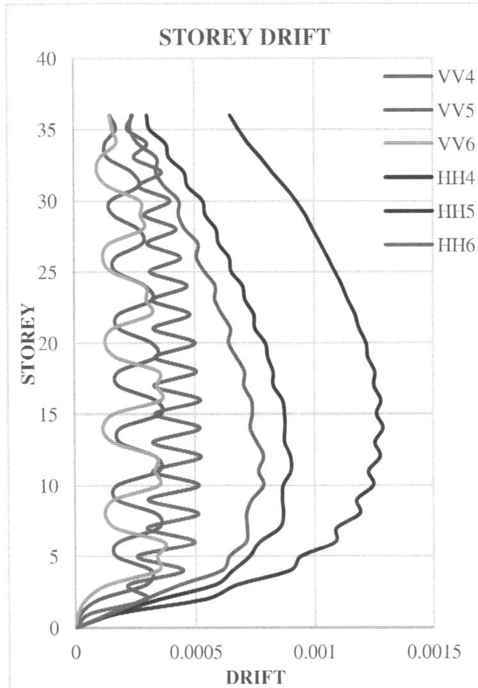

Figure 9. (d) Storey Drift of vertical & horizontal hexagrid buildings with angles decreased.

In vertical hexagrid, with an increase in module size, there is a decrease in the drift as shown in Figure 9(a). Also, the drift decreases with an increase in module size and angle in horizontal hexagrid as shown in Figure 9(b). From both figures 9(c) and 9(d), the vertical hexagrids are found to have lesser storey drift when compared to horizontal hexagrids. It is observed in Figure 9(d) that as the angle decreases, the storey drift is found to increase in both vertical and horizontal hexagrid structures.

Therefore, it is concluded to prefer vertical hexagrids with larger module size over horizontal hexagrids.

3.4. Load distribution

As seen in Figure 10(a), the gravity load is evenly distributed between the internal core and perimeter columns in horizontal hexagrid. The vertical hexagrid turned out to take a higher gravity load compared to horizontal hexagrid. In Figure 10(b), as the angle is decreased, it is seen that the gravity load is distributed equally between the interior core and perimeter columns in vertical hexagrid. While in horizontal hexagrid, the distribution of gravity load is more towards the interior columns.

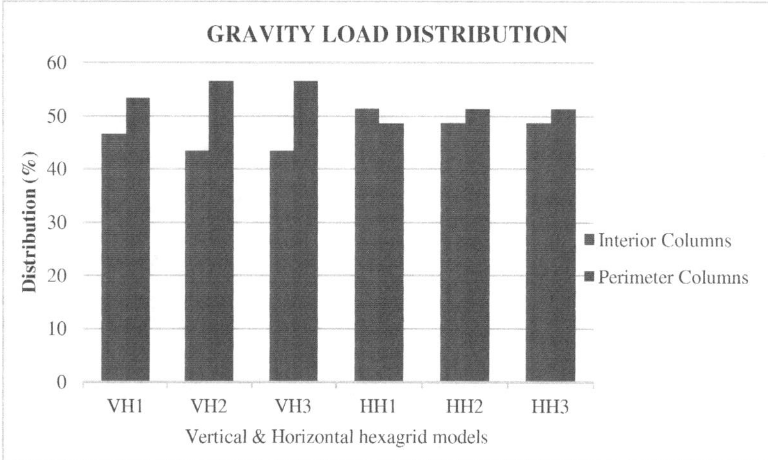

Figure 10. (a) Gravity load distribution in both vertical & horizontal hexagrid buildings.

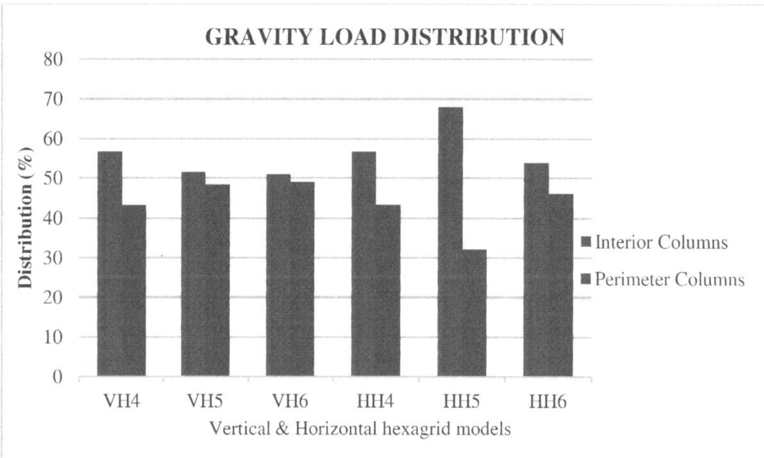

Figure 10. (b) Gravity load distribution in both vertical & horizontal hexagrid buildings with angles decreased.

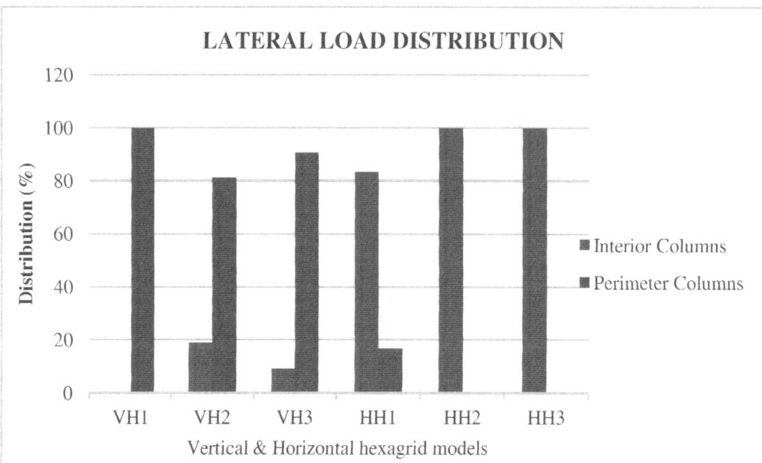

Figure 10. (c) Lateral load distribution in both vertical & horizontal hexagrid buildings.

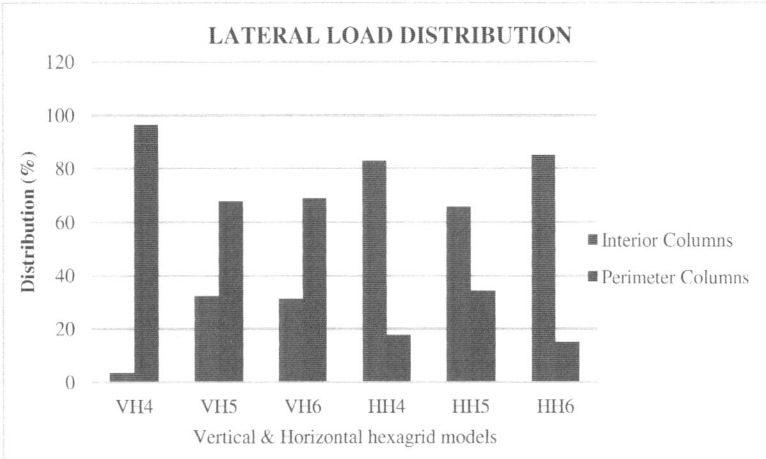

Figure 10. (d) Lateral load distribution in both vertical & horizontal hexagrid buildings with angles decreased.

As seen in Figure 10(c), the majority of the lateral loads are opposed by the perimeter diagonals in vertical hexagrid. But the lateral load is found to be resisted by the interior columns in horizontal hexagrid. In Figure 10(d), as the angle decreases, it is noted that the majority of the lateral loads are taken by the perimeter diagonals in vertical hexagrid. Whereas in horizontal hexagrid, some of the lateral load is found to be distributed to the perimeter diagonals.

The conclusion derived from the response spectrum analysis by observing all the above results is that vertical hexagrid building of larger module size is preferred over horizontal hexagrids.

4. Conclusions

The purpose of the work is to contrast between hexagrid frameworks with different patterns and to obtain the optimal pattern. Four parameters are considered for the comparative study. The following results obtained from the study are:

- The displacement values decrease as module size increases in both vertical and horizontal hexagrids. It is observed that as the angle decreases, displacement values increases in both vertical and horizontal hexagrids.
- The drift values decrease in both vertical and horizontal hexagrids as module size increases. As the angle decreases, it is observed that the drift values increases in both vertical and horizontal hexagrids.
- There is an equitable distribution of gravity load between the internal core and perimeter columns in horizontal hexagrid. The vertical hexagrids turned out to take a higher gravity load compared to horizontal hexagrids.
- Most of the lateral load is found to be resisted by the perimeter columns in vertical hexagrid. But the lateral load is found to be resisted by the interior columns in horizontal hexagrids. As the angle decreases, some of the lateral load is found to be distributed to the perimeter diagonals in the horizontal hexagrid structure.

Vertical hexagrids of larger module size gives better performance compared to horizontal hexagrid structure.

IOP Conf. Series: Materials Science and Engineering **989** (2020) 012019 doi:10.1088/1757-899X/989/1/012019

5. References

[1] Mashhadiali N and Kheyroddin A 2012 Proposing the hexagrid system as a new structural system for tall buildings *The Structural Design of Tall and Special Buildings* **22** 1310-1329

[2] Ebin D S and Prakash D A 2016 Study on Structural Behaviour of Hexagrid Structural Systems in Multi Storey Buildings *International Journal of Chemical Sciences* **14** 330-340

[3] Nejad P A 2016 New Innovated Structural System for Tall Buildings *International Journal of High-Rise Buildings* **5(4)** 251-262

[4] IS 1893 Part-I 2016 Criteria for Earthquake Resistant Design of Structures

[5] Jani K and Patel P V 2013 Analysis and Design of diagrid structural system for high rise steel buildings *Procedia Engineering* **51** 92-100

[6] Lee H U and Kim Y C 2017 Preliminary design of tall building structures with a hexagrid system *Procedia Engineering* **171** 1085-1091

[7] Vagadiya R S, Jivani D K, Dhamsaniya R G, and Sanghani M V 2017 A study on seismic behaviour of hexagrid type structural system *International Journal of Advance Research and Innovative Ideas in Education* **3(2)** 4412- 4422

[8] Mathews M R, George J M, Issac B M and Davis D 2016 A Study on the seismic performance of hexagrid system with different patterns *Applied Mechanics and Materials* **857** 30-35

[9] Thomas M and Mathai A 2018 Seismic analysis of hexagrid structure with various patterns *International Research Journal of Engineering and Technology* **5(4)** 4147-4149

IOP Conf. Series: Materials Science and Engineering **989** (2020) 012020 doi:10.1088/1757-899X/989/1/012020

Analytical study on seismic response of RC and steel structures with fluid viscous dampers

Subasini Y[1,2] and Sivakumar C G[1]

[1]Department of Civil Engineering, PSG College of Technology, Avinashi Road, Peelamedu, Coimbatore, India

[2]Email: yogarajasubasini@gmail.com

Abstract. This paper presents a comparative study of the performance of Fluid Viscous Damper (FVD) in mitigating the seismic response of both Reinforced Concrete (RC) and Steel structures. The aim of the study is to understand the behavior of the seismic resistant structures and to study the advantages of FVD over other dampers. The study is done to determine the optimum positioning of FVD in order to reduce the storey displacements and storey drifts. G+3, G+5, G+7 RC and steel structures were modeled in SAP2000 software and were subjected to non linear time history analysis under three different earthquake ground motions such as Imperial Valley – 02 (1940), Loma Prieta (1989) and Northridge-01 (1994). The dampers were placed in four different configurations such as alternate (AT), corner (CR), diagonal (DL) and middle (ML). The seismic response parameters such as the storey displacements and the storey drifts were determined for the structures. It is found that the presence of FVD reduces the storey displacement and storey drift values to about 40-70% from the uncontrolled structure. It is also understood that the optimum positioning of the dampers is greatly influenced by the aspect ratio of the building.

1. Introduction

Earthquake is one of the natural phenomenon that challenge most of the engineers of the construction industry. Thus designing the structures to resist seismic forces is of vital importance. Protection of tall and large structures against seismic disturbances is a major topic of research for many years. Generally, buildings possess low inherent damping capacity and thus they are easily susceptible to earthquake forces.There are many seismic design strategies and devices for the structures such as moment resisting frames, construction of diaphragms, shear wall, braced frames etc. Dampers provide passive energy dissipation and they reduce the seismic energy by friction or by deformation. The damping devices take up the seismic energy to perform its function rather than transmitting it to the structure. Isolating the base is also a best method of passive energy dissipation. But providing dampers is more advantageous compared to base isolation because base isolation cannot be used for tall buildings since the building might overturn. When existing buildings are retrofitted, the process becomes easier with dampers than with the base isolators. This is more advantageous only when the application is external and the residents are not disturbed.

Murat Diclelia, et al., [4] studied the behaviour of the fluid viscous dampers by comparing the performance of steel frames with bracings provides in chevron configuration. Authors studied the effect of the damping ratio and the velocity exponent on the seismic performance of the structure. When FVDs were installed, it improved the performance by maintaining the elastic behaviour. The dampers with larger damping ratios and smaller velocity exponents were more effective in seismic mitigation. Moreover the FVDs with damping ratios more than 50% did not significantly improve the seismic performance of the CBFs.

Mehdi Banazadeh, et al., [1] investigated the seismic performance of steel moment resisting frames installed with both linear and non linear FVDs with the same damping ratio. Incremental dynamic analysis was carried out to determine the collapse probability, under far field records. It was noted that the use of the damper improves the performance and reduces the collapse probability when compared to that of the conventional steel frames. It was also understood that the frames with linear FVDs performed better than the steel frames with non linear FVDs with an equal damping ratio and thus the collapse probability was lower.

Ras A., et al., [10] conducted a 3D numerical investigation on a twelve storey steel moment resisting frame by performing non linear time history analysis. The frame was provided with fluid viscous dampers in diagonal configuration because it possesses linear relationship between force and velocity. From the results, it was observed that the maximum displacements of the damped frame structure decreased to about 32% when compared to the braced frame and also the maximum acceleration was also reduced to about 37%, which in turn resulted in reduction of the base shear force values.

The main objective of this study is to develop knowledge on seismic resistant structures by modelling and analysing RC and steel structures against seismic loads using the software SAP2000 and interpret the results, to find the optimum positioning of the energy dissipating dampers in the structures to reduce the storey displacements and inter storey drifts by carrying out various trials and to understand the behavior of FVD and the advantages of the damper over other types of dampers. In this study, G+3, G+5, G+7 RC and steel structures were compared and the efficiency of FVD in reducing the storey displacements and storey drifts was also determined. The buildings were analysed for three different earthquake ground motions and the optimum positioning of the damper was found using SAP2000 software.

2. Fluid viscous damper

2.1. Selection of the damper

Seismic dampers are used in buildings and other structures to mitigate the vibrations induced by earthquakes. It is a mechanical device that dissipates the energy of the seismic waves affecting the structure. The dampers are classified into various types. They are friction dampers, yielding dampers, viscous dampers, tuned mass dampers and magnetic dampers. Viscous damper was selected for the study because it is passive device that does not have any sensors or actuators and does not consume any power, instead, it relies on the damping properties of materials used in its design.

2.2. Principle of working

The basic principle with which FVD works is the dissipation of energy because of the fluid flowing through the orifices. The detailed view of FVD is depicted in Figure 1. In FVD, the fluid flows from one chamber to the other chamber through orifice when the piston moves from left to right or from right to left. There will be dissipation in energy due to the loss in head that occurs when the fluid circulates from a larger area i.e. cylinder to a smaller area i.e. the orifice and then again from the smaller area to a larger area. The damping force of FVD is proportional to the pressure difference across the piston head.

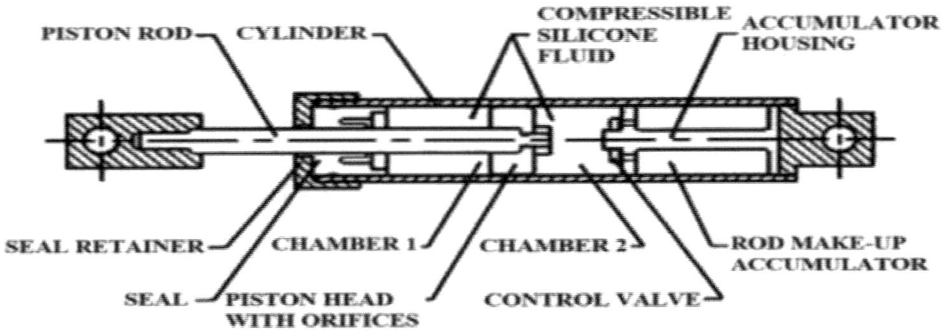

Figure 1. Fluid viscous damper [10].

F is directly proportional to V^α, where α is a constant parameter called as the velocity exponent and this constant is influenced by the shape of the orifice that changes the characteristics of the fluid along with the speed of the fluid. Generally the value of α is taken from 0.3 to 1.0, for seismic protection. When the value of α is taken as one, the device becomes a linear FVD as force varies linearly with velocity. When the value of α is taken less than one, the force has non-linear relationship with velocity and hence called as non-linear FVD. Non-linear damper with the value of α greater than one is generally not used for practical applications.

3. Methodology

3.1. Selection of ground motion data

ATC- 63 guidelines categorised the strong ground motion parameters and as per the guidelines, three of the most commonly used earthquake records in vibration control are selected for the study to understand the performance of fluid viscous dampers.The details of the ground motion are tabulated in Table 1.

Table 1. Details of the ground motion.

RSN Number	Earthquake name	Station name	Magnitude	PSA (g)
6	Imperial valley – 02 (1940)	El Centro Array #9	6.95	0.68
752	Loma prieta (1989)	Capitola	6.93	1.14
953	Northridge-07 (1994)	Beverly Hills – 14145 Mulhol	6.69	1.82

Source : http://ngawest2.berkeley.edu./

3.2. Damper configuration

Out of the many configurations like chevron, upper toggle, lower toggle bracing systems, the diagonal bracing system was chosen since it minimizes the usage of materials and it reduces the cost of the structure.

3.3. Method of analysis

Non linear time history analysis is a step by step procedure to determine the dynamic response of the structure that varies with time for the applied seismic loading of the respective earthquake. Direct integration method is used under which the value of γ is taken as 0.5 and β is taken as 0.25.

IOP Conf. Series: Materials Science and Engineering **989** (2020) 012020 doi:10.1088/1757-899X/989/1/012020

4. Design procedure of FVD

Generally, the existing design formulae and procedure for FVD are provided by FEMA 273 and many other research papers. New design formulae were derived by Hwang et al., [6] considering the relative vertical deformations as they are comparable with the horizontal relative deformations of the end locations of the damper. The stiffness of the supporting bar were determined from the paper by Lu et al., [8]

- Initially modal analysis was done without the installation of the dampers and the fundamental time period (T) and the horizontal deformation (ϕ_{hi}) for the first mode were noted down.
- The expected damping ratio (ρ) for the structure and the damping exponent (α) for the damper were assumed as 20% and 0.5 respectively.
- For the expected damping ratio, the displacement at the roof level (A) was determined by conducting non linear time history analysis for all the considered earthquake ground motions.
- Generally the value of A is taken as the maximum value among the considered ground motions.
- Then for the first vibration mode, the horizontal (ϕ_{hj}) and vertical displacements (ϕ_{vj}) between the end positions of the damper to be fixed, were determined.
- The mass at the roof level was determined and noted as m_i.
- Then the damping coefficient 'C' of the damper can be determined from the formula:

$$\rho = \frac{\{T^{(2-\alpha)}\}\eta j\ Cj\ \lambda j\ (fh\ \emptyset hj - fv\ \emptyset vj)^{1+\alpha}}{2\pi^{(3-\alpha)} A^{(1-\alpha)} \Sigma\ mi\ (\emptyset\ hj)^2} \quad (1)$$

where

η_j – number of identical dampers with the same damping coefficient in each storey.
Cj – damping coefficient for the damper j.
α - damping exponent taken as 0.5.
ρ - expected damping ratio for the structure taken as 20%.
ϕ_{hj} - horizontal displacement between the ends of the jth damper
ϕ_{vj} - relative vertical displacement between the ends of the jth damper
f_h - horizontal magnification factor for diagonal damper – cos θ
f_v - vertical magnification factor for diagonal damper – sin θ

$$\lambda_j - 2^{2+\alpha} \frac{\Gamma^2(1+\alpha/2)}{\Gamma(2+\alpha)} \quad (2)$$

Lu et al., [8] derived the following formula to determine the stiffness of the supporting bar with 95% energy efficiency as the index,

$$\frac{K^2}{K^2 + (Ce\ \omega)^2} \geq 95\% \quad (3)$$

where C_e is the equivalent linear damping coefficient that is calculated using the formula,

$$C_e = \frac{2\ C\alpha u^{\alpha-1} \omega^{\alpha-1}}{\alpha+1} \quad (4)$$

K is the design stiffness of the supporting bar
ω is the vibration frequency.

5. Modelling and analysis of RC and steel structures

G+3, G+5, G+7 storied RC and steel structures were modelled for the same plan. The plan of the structure considered is shown in Figure 2.

Figure 2. Structural plan of RC and steel structures.

The structures are considered to be located at Imphal (Manipur) which is in seismic zone 5 with medium stiff soil condition. The details of both RC and Steel Structures are tabulated in Table 2. The support conditions were considered as fixed at the base. The live load on the slab was taken as 3 kN/m^2. All the load combinations were defined as per IS 1893-2016 [11]. The bay length in X direction is taken as 7.35 m and the bay length in Y direction is taken as 6.6 m.

Table 2. Details of RC and steel structures.

	RC Structure	Steel Structure
Columns	0.4 m x 0.4 m	ISMB 250
Beams	0.3 m x 0.4 m	ISMB 300
Thickness of slab	150 mm	150 mm RC slab
Storey height	3 m	3 m
Grade	M30	Fe 345
Importance factor (I)	1	1
Soil type	II (medium or stiff soils)	II (medium or stiff soils)
Response Reduction factor (R)	5(RC building with SMRF)	5(Steel building with SMRF)
Zone factor	0.36 (Zone V)	0.36 (Zone V)

The dampers are positioned in four different configurations to determine the optimum positioning of the dampers. They are depicted in Figure 3.

IOP Conf. Series: Materials Science and Engineering **989** (2020) 012020 doi:10.1088/1757-899X/989/1/012020

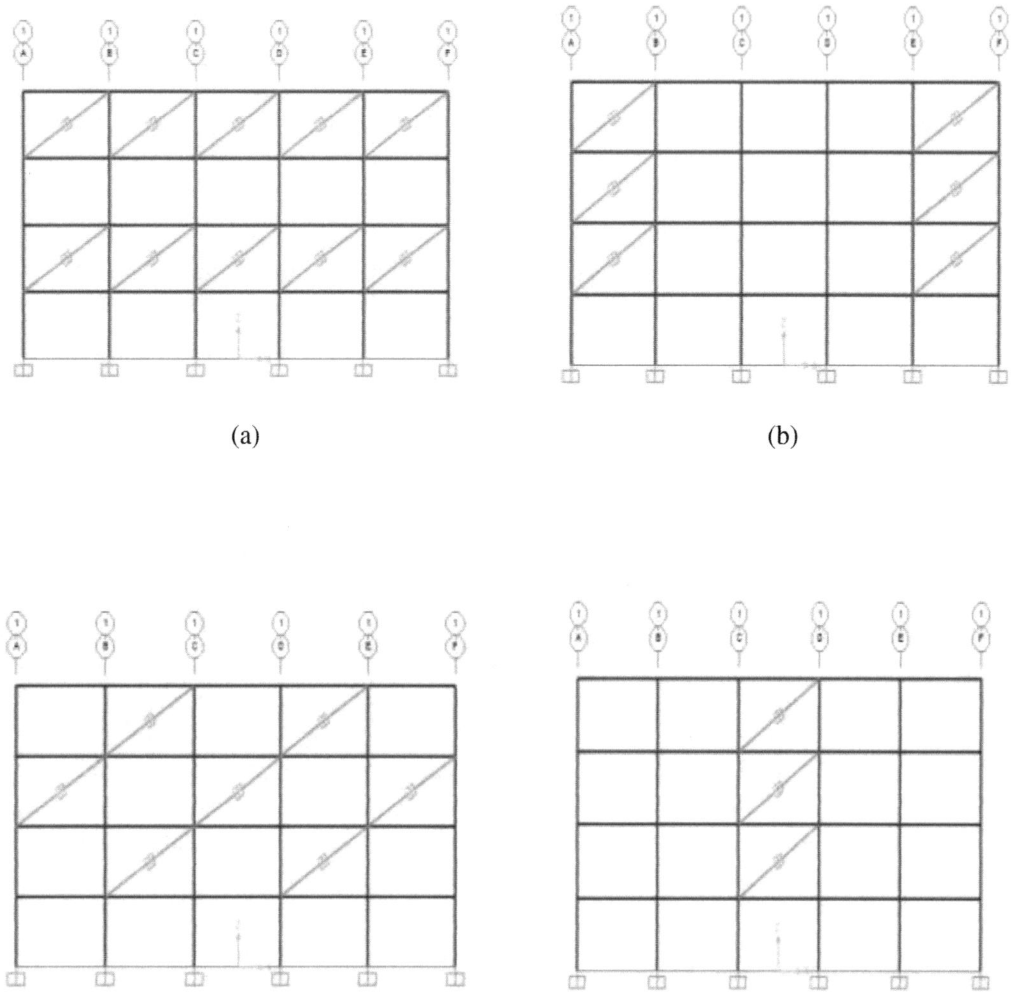

Figure 3. Damper positioning: (a) Alternate arrangement [AL], (b) Corner arrangement [CR], (c) Diagonal arrangement [DG] and (d) Middle arrangement [ML].

The buildings without the dampers were analysed for all the ground motions and the uncontrolled response was determined. The dampers were designed for all the buildings separately and the dampers were positioned and analysed. The responses such as storey displacements and storey drifts were determined for all the cases.

6. Results and discussion

6.1. Storey displacements

From the test results, it was found that the presence of FVD reduces the storey displacements. The results of the RC and steel buildings are depicted in Figure 4. These results were obtained by taking the average of the results of all the three earthquake ground motions.

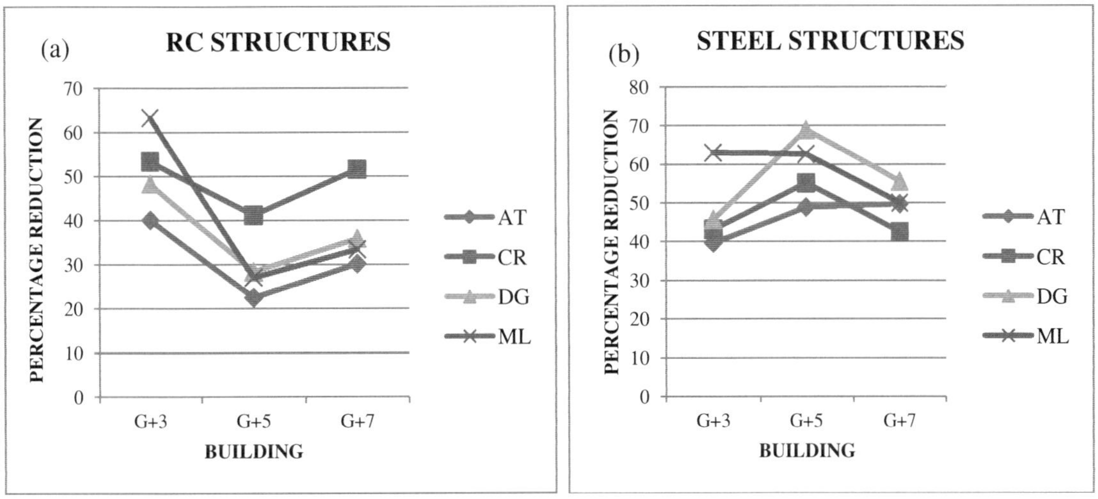

Figure 4. Percentage reduction of storey displacements in: (a) RC structures and (b) Steel structures.

It is found that the middle arrangement reduces the storey displacement to about 63.28% compared to the other arrangements for the G+3 RC structure. For G+5 and G+7 structures, the corner arrangement reduces the storey displacement to about 41.19% and 51.53% compared to the other arrangements respectively. It can be clearly understood that the optimum arrangement of the dampers varies with the storey height.

For the G+3 steel structure, it is found that the middle arrangement reduces the storey displacement to about 62.98% compared to the other arrangements. For G+5 and G+7 structures, the diagonal arrangement reduces the storey displacement to about 68.94% and 55.56% compared to the other arrangements respectively. FVD reduces the storey displacements of the structures to about 40-70% as per the results obtained. It is found that the aspect ratio of the building affects the optimum positioning of the dampers.

6.2. Storey drift

Storey drift is usually called as the inter- storey drift and it is defined as the lateral displacement of one level relative to the other. From the test results, it is also found that FVD reduces the storey drift to a greater extent. The percentage reduction of the storey drift values for the RC and steel structures are depicted in Figure 5. These results were obtained by taking the average of the results of all the three earthquake ground motions.

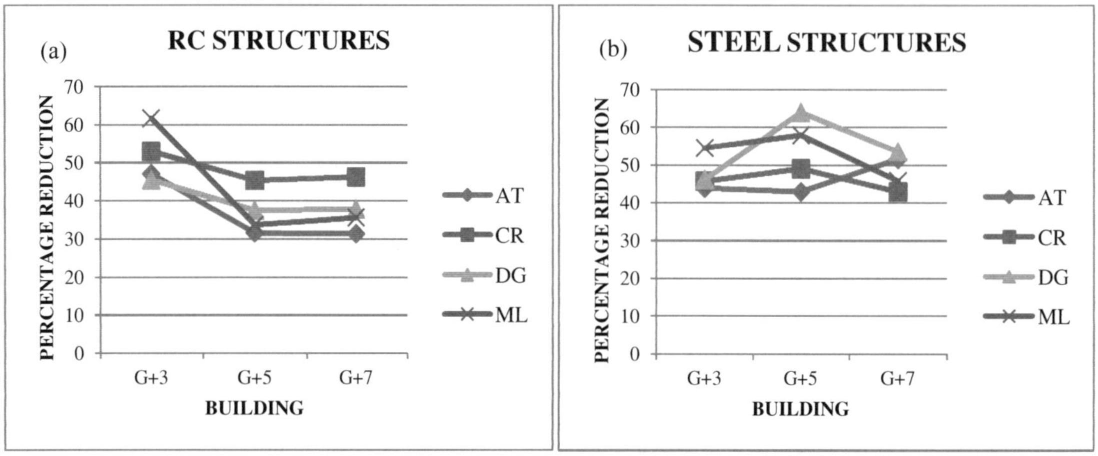

Figure 5. Percentage reduction of storey drifts in: (a) RC structures and (b) Steel structures.

It is found that the middle arrangement reduces the storey drift values to about 61.64% compared to the other arrangements for the G+3 RC structure. For G+5 and G+7 structures, the corner arrangement reduces the storey drift values to about 45.29% and 46.26% compared to the other arrangements respectively. From the determined results, it can be understood that the optimum arrangement to reduce storey drift values for a building is the one that reduces the storey displacements for the same building. Both are dependent to each other.

For the G+3 steel structure, it is found that the middle arrangement reduces the storey drift values to about 54.54% compared to the other arrangements. For G+5 and G+7 structures, the diagonal arrangement reduces the storey drift values to about 63.99% and 53.60% compared to the other arrangements respectively. FVD reduces the storey drift values of the structures to about 40-65% as per the results obtained.

7. Conclusion

From the analysis conducted, the following conclusions can be made.

- FVD is a passive device that does not have any sensors or actuators and does not consume any power. Instead, it relies on the damping properties of materials used in its design and hence it is more advantageous.
- The presence of FVD reduces the storey displacement and storey drift values for both RC and steel structures to about 40-70%.
- It was understood that the optimum positioning of the dampers depends on the aspect ratio of the building. Since the plan is kept the same, it directly depends on the height of the building.
- For the G+3 RC structure whose aspect ratio was 0.54, the optimum damper placement was found to be the middle placement with a reduction percentage of 63.28% for storey displacement and 61.64% for storey drift respectively.
- For the G+3 steel structure whose aspect ratio was 0.54, the optimum damper placement was found to be the middle placement with a reduction percentage of 62.98% for storey displacement and 54.54% for storey drift respectively.
- For the G+5 RC structure whose aspect ratio was 0.82, the optimum damper placement was found to be the corner placement with a reduction percentage of 41.19% for storey displacement and 45.29% for storey drift respectively.
- For the G+5 steel structure whose aspect ratio was 0.82, the optimum damper placement was found to be the diagonal placement with a reduction percentage of 68.94% for storey displacement and 63.99% for storey drift respectively.

IOP Conf. Series: Materials Science and Engineering **989** (2020) 012020 doi:10.1088/1757-899X/989/1/012020

- For the G+7 RC structure whose aspect ratio was 1.09, the optimum damper placement was found to be the corner placement with a reduction percentage of 51.53% for storey displacement and 46.26% for storey drift respectively.
- For the G+7 steel structure whose aspect ratio was 1.09, the optimum damper placement was found to be the diagonal placement with a reduction percentage of 55.56% for storey displacement and 53.6% for storey drift respectively.

Acknowledgement
The authors wish to record their sincere and heartfelt thanks to the Management, Principal in Charge, Head of the Department, Former Head of the Department, Department of Civil Engineering, PSG College of Technology, Coimbatore.

8. References

[1] Banazadeh Mehdi and Ghanbari Ali 2017 Seismic performance assessment of steel moment-resisting frames equipped with linear and nonlinear fluid viscous dampers with the same damping ratio *Journal of Constructional Steel Research* **136** 215–228
[2] Chukka Naga Dheeraj Kumar Reddy and Muthumani K 2019 Comparison of X-shaped metallic dampers with fluid viscous dampers and influence of their placement on seismic response of the building *Asian Journal of Civil Engineering* **20** 869–882
[3] Chopra Anil K 2012 Earthquake Dynamics of Base-Isolated Buildings Dynamics of structures: Theory and Applications to Earthquake Engineering *Fourth Edition Pearson Prentice Hall, United States of America*
[4] Diclelia Murat and Mehta Anshu 2007 Seismic performance of chevron braced steel frames with and without viscous fluid dampers as a function of ground motion and damper characteristics *Journal of Constructional Steel Research* **63** 1102–1115
[5] Esra Mete Guneyisi and Gulay Altay 2008 Seismic fragility assessment of effectiveness of viscous dampers in R/C buildings under scenario earthquakes *Structural Safety* **30** 461–480
[6] Hwang Jenn-Shin, Huang Yin-Nan, Shy-Lian Yi and Song-Yen Ho 2008 Design Formulations for Supplemental Viscous Dampers to Building Structures *J. Struct. Eng.* **134(1)** 22-31
[7] Liu Yanhui, Wu Jinbiao and Dona Marco 2018 Effectiveness of fluid viscous dampers for improved seismic performance of inter-storey isolated buildings *Engineering structures* **169** 276-292
[8] Lu Yun-xiang, Ca Yuan-qi, Qu Qing-fei and Zhan Qian-hua 2012 Study on the Effect of Supporting Stiffness on Energy Dissipation Efficiency of Viscous Dampers *Applied Mechanics and Materials* **105-107** 96-101
[9] Moradpoura S and Dehestanib M 2019 Optimal DDBD procedure for designing steel structures with nonlinear fluid viscous dampers *Structures* **22** 154–174
[10] Ras A and Boumechra N 2016 Seismic energy dissipation study of linear fluid viscous dampers in steel structure design *Alexandria Engineering Journal* **55** 2821–2832
[11] IS 1893 (Part 1) 2016 Criteria for Earthquake Resistant Design of Structures - General Provisions and Buildings.

IOP Conf. Series: Materials Science and Engineering **989** (2020) 012021 doi:10.1088/1757-899X/989/1/012021

Seismic performance of eccentrically braced frames

Rinu G David[1,2] and Sarif N[1]

[1]Department of Civil Engineering, TKM College of Engineering, Kollam, Kerala, India.
[2]Email: rinugdavid@gmail.com

Abstract. The main objective of this research is to assess the seismic performance of Eccentrically Braced Frames of different configurations. Modelled Eccentrically Braced Frames subjected to both linear and nonlinear analysis in SAP 2000. The linear analysis gives an insight to mode shapes and mass participation ratios. The nonlinear analysis includes the pushover analysis, which provides information about the collapse mechanisms and performance points. This study also extended to compare the performance of Eccentrically braced frames with the performance of Special Moment Resisting Frames that helps to understand the structural efficiency of both systems. This study concludes that all selected configurations of Eccentrically braced frames undergo small roof displacement that is well below the target drift whereas SMRF frames will show large displacement. 2D braced EBFs shows a better post yielding behaviour and is more ductile compared to other frame systems. Shear links in the EBF frames increase the stiffness of the frame which in turn results in high base shear demand. Post yielding behaviour of EBF frames can be improved by proper detailing of beam-column joints and link connections which confirms that performance of EBF frames is superior to SMRF frames in seismic areas.

1. Introduction

In the present era, the occurrence of earthquakes is quite often compared to ancient times. This increases the challenge in the construction industry to design the seismic-resistant structures which should reduce the human casualty, without much compromising the performance of the building. This resulted in the development of different types of lateral force resisting frames. In high seismic hazard areas, steel structures face the problem of Lateral stability. Eccentrically Braced Frame (EBF) is a combination of concentrically braced frame and moment-resisting frame which brings high elastic stiffness and high inelastic responses to the system [1, 2]. EBFs are one of the systems which resist the lateral load usually adopted in high seismic areas. Eccentrically braced systems have a 'structural fuse' which is the link element that dissipates energy through yielding. AISC Seismic Provisions for Structural Steel Buildings (1997) changed the link-to-column connections design requirements of EBF after the Northridge Earthquake. Currently, the AISC Seismic Provisions suggest the designer to either use a connection after the qualification testing or use a short shear link with reinforcement which exclude the inelastic action in the link-to-column connection. However, data available for the cyclic loading performance of EBF link-to-column connections, including connections with reinforcement are limited.

The present research work mainly focusing to understand the more realistic the behaviour of the EBF structures by the nonlinear static pushover analysis. This study also compares the performance of different Eccentrically Braced Frames with Special Moment Resisting Frames and thereby assess the efficacy of the EBF as a lateral force resisting structure. This study is limited to single bay nine-storey bare frame having horizontal shear yielding links.

2. Special moment resisting frames (SMRF) and Eccentrically braced frames (EBF)

In high seismic areas, Steel Moment Resisting Frames is preferred because of its high capacity of ductility, speed and ease of their construction. Lateral loads acting on this structural system is resisted by structural frame members and rigid joints. The lateral stiffness and strength of the whole structure depends on the strength of the frame and bending rigidity. Comparing to braced systems, moment frames requires members of larger size to keep the deflection in the lateral direction within the code limits, which may increase the cost of construction. The inherent flexibility of moment resisting frames may also induce non-structural damages under the earthquake excitation. After the occurrence of 1994Northridge earthquake, it is seen that the performance of the special moment resisting frames is not up to the satisfactory level, because the high ductility property of the system is challenged by the brittle failure of beam-column connections [3]. Minimum inelastic rotation capacity of 0.03 radian is required for Beam-to-column connections in SMRFs as per AISC. Proper placing of semi-rigid connections along with the rigid connection could improve the performance of moment frames.

The unique characteristic of an EBF is that one end of the brace is joined to another brace or a column through a beam section called a link. Brace force is transmitted either through the shear and bending action of the link of an eccentric brace. Lateral stiffness of an EBF frame varies by the ratio of the length of the link to that of the beam. As the length of the link decreases, it makes the frame stiffer, thereby approaching the stiffness equivalent to a Concentrically Braced Frame. Long links make the frame more flexible thereby reduces the stiffness equivalent to that of a moment frame. The foremost function of the link is to provide a weak section in the frame so that its plastic deformation capacity can be utilised and thereby dissipates more energy due to earthquakes. Ductility and energy dissipation capacity of EBFs is explained, by evaluating the behaviour of EBF frames in cyclic load [4]. Well-designed EBF prevents buckling of braces and also link can withstand large deformations without loss in strength, and make stable hysteretic loops analogous to SMRF.

Short links yields in shear whereas the long links will form plastic moment hinges at the ends. An intermediate link shows significant amounts of both shear and moment yielding occur. Several shear link specimens are failed by fracturing the link web before achieving the required inelastic rotation levels [5] and finds that AISC 2002 loading protocol is inadequate. The revised loading protocol AISC 2002 for shear links achieves the inelastic rotations and thereby avoids the premature failure. Increasing the link beam length leads to a reduction in the stiffness of the structure whereas the ductility increases [6]. Premature failure of specimens is observed in the experimental investigation of link-to-column connections in eccentrically braced frames [7]. The inelastic behaviour of six Eccentrically Braced Frames under Non-Linear Range is studied in 2014 [8]. The V-braced frame is more economical whereas Eccentric Inverted-V frame shows a better performance. Stiffness, ultimate capacity, and inelastic member deformation requirements of an active beam links in Eccentrically Braced Frames were compared [9]. The behaviour of long links in eccentrically braced frames is analysed in 1989 [10]. Long links attached to columns is not usually preferred in EBFs. The present study aims to assess the performance of different configurations of Eccentrically Braced Frames in seismic areas. This study also aims to compare the seismic performance of Eccentrically Braced Frames and Special Moment Resisting Frames.

3. Methodology and Numerical modelling of frames

3.1 General

From the past researches, a suitable methodology is adopted to analyse the performance of the frames in the seismic environment as shown in Figure 1. The numerical modelling is done with the help of SAP 2000.The effective time period, roof displacement, performance points, lateral stiffness, post yielding behaviour, the formation of the hinge mechanism is discussed from the results of the analysed model. Link damage is stated in terms of post-yield and inelastic deformation limits.

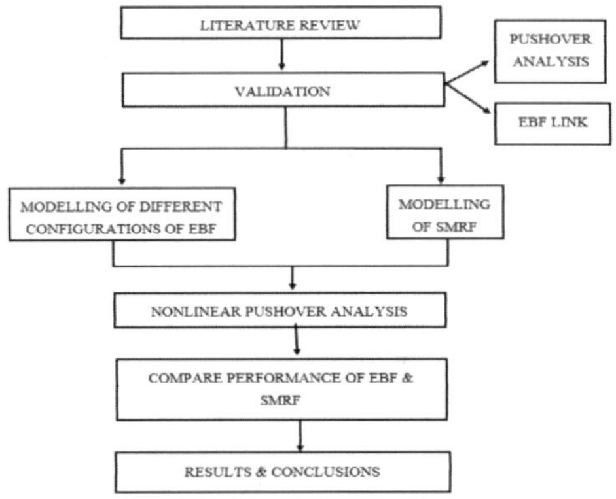

Figure 1. Methodology.

3.2 Pushover method

This nonlinear static procedure can be used to find the maximum displacement of the structure. By using the capacity spectrum method, the intersection point of the capacity (pushover) curve and a reduced response spectrum can be found out (ATC40 Vol. 1). In this method, the global force-displacement capacity of the structure can be plotted and is used to compare the response spectra representations of the earthquake demands. In this analysis, the structure is loaded in a specific predefined pattern and is boosted until the target displacement or before the collapse. SAP2000 have the ability to carry out nonlinear static pushover analysis for both two and three-dimensional structures as described in ATC-40 and FEMA-273 documents.

3.3 Hinge details

The inelastic behaviour of SMRF is obtained by the introduction of plastic hinges which estimates the additional capacity of the frames beyond the elastic limit. In this research, auto hinges are used for beam and column elements. From Tables in ASCE 41-13, moment M3 hinges are assigned at the beam endings and for columns, P-M2 hinges are provided at both ends. Nonlinear behaviour of EBF is captured by modelling the plastic hinges in the link member using the nonlinear modelling parameters as recommended by ASCE41-13 (ASCE, 2014). A deformation-controlled shear (V2) hinge is provided at the middle of the link where shear is expected to be maximum. All other elements are designed as elastic elements as they are expected to behave linearly. A default kinematics hysteresis model is considered in the study.

3.4 Modelling and Designing of frames

SAP2000 version 20 is the platform used for numerical modelling and analyses of EBFs. Both material nonlinearity and geometric nonlinearity such as P-Delta effects are considered for this study. Fixed base restraints are provided in all cases.

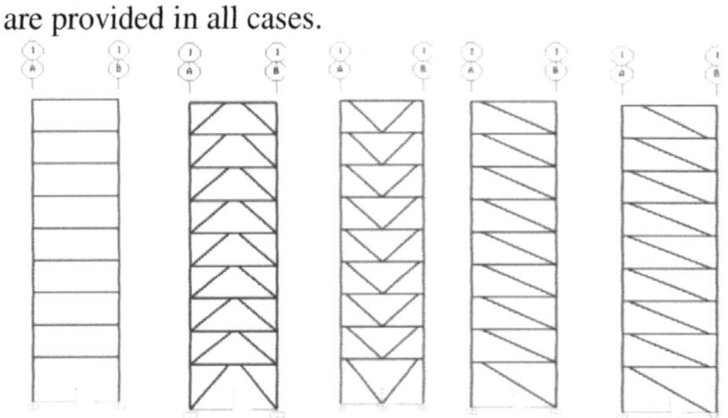

(a) SMRF (b) EBF 1 (c) EBF 2 (d)EBF 3 (e) EBF

Figure 2. Different configurations selected for the study.

The 9 storey 2D building SMRF and EBF designed as per ASCE provisions. Grade of steel used for the frame is A992 steel G50. The single bay 9storey 2D frame has a bay width of 9.15m, with a storey height of 3.96m. The soft storey of the frame have a height of 5.49m. The 2D frame is designed and is checked in SAP 2000. The designed frame section details are shown in Table 1 and Table 2. In this study, four different configurations of EBF frames are selected namely, Split V- Braced EBF, V-Braced EBF, D Braced EBF, 2D Braced EBF represented as EBF 1, EBF2, EBF3, EBF4 respectively.

Table 1. Section details of EBF frames.

Storey	Beam size	Column size	Brace size	Link size
1	W 150X24	W 250X131	HSS 335.6X335.6X12.7	W 150X24
2	W 150X22.5	W 250X131	HSS 406.4X406.4X12.7	W 150X22.5
3	W 150X22.5	W 250X115	HSS 406.4X406.4X12.7	W 150X22.5
4	W 150X18	W 250X101	HSS 406.4X406.4X12.7	W 150X18
5	W 150X18	W 250X89	HSS 406.4X406.4X12.7	W 150X18
6	W 150X13.5	W 250X80	HSS 406.4X406.4X12.7	W 150X13.5
7	W 150X13.5	W 250X73	HSS 406.4X406.4X12.7	W 150X13.5
8	W 150X13	W 250X67	HSS 406.4X406.4X12.7	W 150X13
9	W 150X13	W 250X58	HSS 406.4X406.4X12.7	W 150X13

Table 2. Section details of SMRF.

Storey	Beam Size	Column Size
1	W 150X24	W 250X131
2	W 150X22.5	W 250X131
3	W 150X22.5	W 250X115
4	W 150X18	W 250X101
5	W 150X18	W 250X89
6	W 150X13.5	W 250X80
7	W 150X13.5	W 250X73
8	W 150X13	W 250X67
9	W 150X13	W 250X58

Table 3. Link length of different configurations.

	Link Length			
Story	EBF 1	EBF 2	EBF 3	EBF 4
1	1.22	0.61	1.08	0.61
2	0.61	0.60	1.08	0.61
3	0.61	0.60	1.08	0.61
4	0.91	0.50	1.08	0.46
5	0.91	0.50	1.08	0.46
6	1.22	0.61	1.08	0.61
7	1.22	0.61	1.08	0.61
8	1.83	0.91	1.08	0.99
9	1.83	0.91	1.08	1.83

IOP Conf. Series: Materials Science and Engineering **989** (2020) 012021 doi:10.1088/1757-899X/989/1/012021

4. Results and discussions

4.1 Linear analysis

During earthquake shaking, building oscillates which induces inertia force in the building. During earthquake shaking, inertia force is induced in the building due to its oscillation. The magnitude of induced inertia force, its intensity and duration of oscillation depends on the dynamic features of a building. Dynamic characteristics of building includes modes of oscillation and damping. For steel buildings, the damping ratio is commonly adopted as 2%. The modal analysis gives a mathematical description of structures about its dynamic behaviour. The time period and frequency of different structural systems are shown in Table 4.

Table 4. Time period and frequency of different configurations.

Structural Systems	SMRF	EBF 1	EBF 2	EBF 3	EBF 4
Mode1					
Time Period	1.995	0.306	0.292	0.359	0.358
Frequency	0.501	3.268	3.425	2.788	2.795
Mode 2					
Time Period	0.611	0.133	0.102	0.133	0.158
Frequency	1.636	7.529	9.84	7.499	6.332
Mode 3					
Time Period	0.276	0.085	0.072	0.083	0.115
Frequency	3.62	11.790	13.825	11.988	8.719

From Table 4, it is clear that EBF has large stiffness as compared to a SMRF which owing to large time period to SMRF frames. SMRF undergoes large oscillations and takes more time to come back to its mean position. Introduction of braces in the structure improves the stiffness of the structure and reduced the time period. Among the four configurations, EBF 2 has shown the maximum stiffness and reduced the time period to 0.292seconds. EBF 2 is 85% stiffer than SMRF so that it can attract large forces into it. Mass participation of EBF frames in higher modes is negligibly small as compared to SMRF. Thus, lower modes are dominating the overall response of the structure in EBF systems. Introducing braces in EBF improves the in-plane stiffness in the vertical plane and increases the torsional stiffness of the building.

4.2 Pushover analysis

Earthquake-resistant design controls the type of damage and sequence of its occurrence in various structural elements of a building. In this design, some structural damage is allowed for normal buildings, but it prevents the collapse condition during a strong earthquake shaking. The nonlinear analysis makes it possible to understand the whole failure mechanism of the building. In this study, pushover analysis is used to study the nonlinear behaviour of the structure. The whole deformability of the building is assessed from the Pushover Analysis. The ideal lateral load-deformation curve of a building in pushover analysis reveals the linear, nonlinear and plastic behaviour under the monotonic lateral displacement loading.

Five different types of bare frames with the same cross-sections are analysed for the study and the undergo pushover analysis. The graph showing the variation of base shear with displacement is shown in Figure 3. EBF 4 is showing the maximum lateral strength and more ductile compared to other systems whereas EBF 2 is the stiffest configuration. EBF 2 frame has two links at a floor level which is connected to fixed support using braces. Additional support increases the redundancy of the structure and makes the structure stiffer. SMRF frames show poor resistance to lateral load and are more flexible compared to other systems. When the demand curve goes through the capacity curve of the structure makes the performance point.

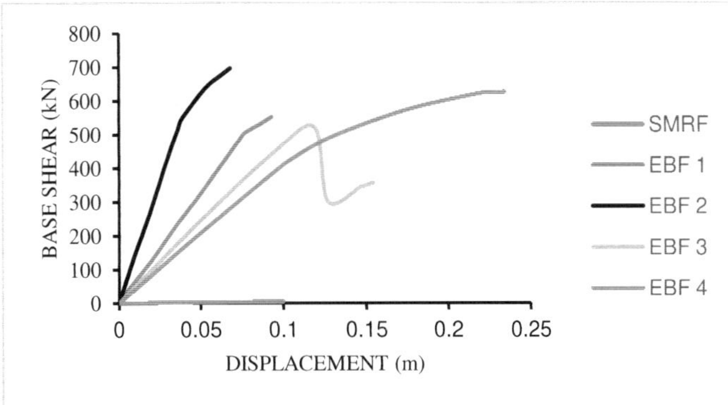

Figure 3. Graph showing variation of base shear with displacement for different structural systems.

Table 5. Performance points of different configurations.

Parameters	SMRF	EBF 1	EBF 2	EBF 3	EBF 4
Performance Point, V (K N)	16.223	218.967	213.75	191.572	186.494
Roof Storey Displacement (M)	0.3	0.033	0.029	0.038	0.044
Effective Time Period (S)	1.599	0.257	0.245	0.305	0.309

Table 5 illustrates the performance point of all five frames and is compared. EBF1attracts a large amount of lateral force compared to other frames. SMRF frames have larger roof displacement and base shear demand of magnitude 16.223k N. EBF2 frame reduces the roof storey displacement to 90% as compared to SMRF frames. SMRF resist lateral loads by using both beams and columns. The axial forces, bending moment and shear force generated in them will handle these forces. Braces and links in EBF frames diminish the overall lateral displacement, bending moment and shear force demands on beams and columns. Thus, EBF frames are able to resist more lateral loads as compared to SMRF and dissipate energy through shear deformations (shear link).

4.3 Post yeild mechanisms

Due to reversed cyclic loading, inelastic material steel will enter in to a stage beyond yield and eventually hysteresis will take place. EBF frames enter into the inelastic range after the yielding is much earlier than SMRF frames. EBF frames designed by the capacity design principles. Here links beam is the weak element which shows the inelastic response. In all the configurations of EBF, it is seen that nonlinear hinges are formed only in the link elements which are assigned with shear hinges and all other elements are in the elastic range. This behaviour of frames validates the capacity design principles. Figure 4 shows the collapse mechanisms of different frames. Uniformity in forming the hinges is not achieved in these frames. In EBF 3 formation of hinges is concentrated only on the first floor whereas in EBF 1 and 2 shows the formation of more hinges in two stories. Among all the four EBF frames, EBF 4 shows a better formation of hinges. Non-uniformity in the formation of hinges is due to the presence of the soft storey. The soft storey is subjected to large forces and is stressed more compared to other storeys resulting in the damage to be concentrated in the soft storey and it results in a collapse of link elements.

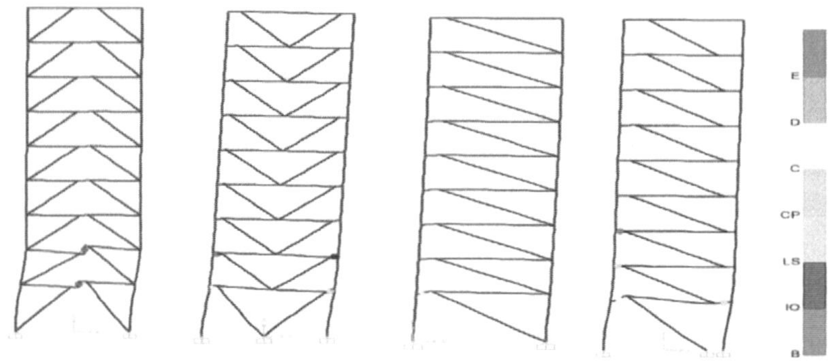

(a) EBF 1 (b) EBF 2 (c) EBF 3 (d) EBF

Figure 4. Collapse mechanisms of EBF frames.

From the analysis, the EBF 4 frame shows that up to a roof displacement of 0.02, the frame is in the collapse prevention stage. Table 6 shows the maximum roof displacement permitted to each frame before the collapse. EBF 4 frame shows the maximum displacement of 0.129 m before collapse confines that it is more ductile compared to other frames.

Table 6. Maximum displacement of different configurations.

Frames	Maximum Displacement (m)
EBF 1	0.101
EBF 3	0.094
EBF 4	0.129

5. Conclusions

Seismicity of the world is increasing day by day. This creates a new challenge in the construction industry to create more earthquake-resistant infrastructures having desired performance and economically viable structures. In this study, seismic performance of different configurations of EBF is analysed and is compared with the performance of SMRF. The main criteria selected for the study are target drift and post yield mechanisms. The frames under study are subjected to both linear and nonlinear analysis. Major observations and conclusions arrived from the study are:

1. Mass participation ratios of EBF frames are much more in lower modes than SMRF. Therefore, the lower modes predominate the overall response of the structure.

2. Modal analysis suggest EBF 2 is stiffer than all other frames, having a natural time period of 0.292 seconds.

3. Pushover analysis gives the performance point of various frames. Introduction of braces and links in EBFs makes the frame stiffer and reduces the displacement. EBF 2 reduces displacement to 90% as compared to SMRF whereas it enhances the demand of the structure to a magnitude of 213.75 k N.

4. SMRF show a maximum displacement of 300mm whereas EBF frames shows a maximum displacement of 44mm which is below the target displacement of 93mm.

5. Link beams dissipate energy through shear deformations which reduce the demand on beams and columns.

6. Post yield mechanisms of EBF 4 shows a better formation of hinges as compared to other EBF frames

7. Non uniformity in the formation of hinges is due to the presence of the soft storey. This can be improved by proper detailing of beam column joints and link connections.

8. EBF 4 allows a displacement of 0.129m before the collapse and is exhibiting more ductile behaviour.

9. Among all these frames, EBF 4 can be selected as the better configuration, whose displacement values are below the target drift, enough lateral strength, better post yielding mechanisms and is more ductile compared to other frames.

10. Stiffness of the structure can be improved by EBF frames results in minimum displacement as compared to SMRFs.

6. References

[1] Fujimoto M, Aoyagi T, Ukai K, Wada A and Saito K 1972 Structural characteristics of eccentric K-Braced frames *Transctions AIJ* **195** 39-49

[2] Tanabashi R, Naneta K and Ishida T 1974 On the rigidity and ductility of steel bracing assemblage *In Proceedings of the 5th World Conference of Earthquake Engineering* **1** 834-40

[3] Bruneau M, Uang C M and Whittaker A S 1998 *Ductile design of steel structures* (NewYork: McGraw-Hill)

[4] Popov E P, Kasai K and Engelhardt M D 1987 Advances in Design of Eccentrically Braced Frames *Earthquake Spectra* **3** 43-55

[5] Okazaki T, Arce G, Ryu H C and Engelhardt M D 2005 Experimental study of local buckling, overstrength, and Fracture of links in eccentrically braced frames *Journal of Structural Engineering* **131(10)** 1526-38

[6] Naji A and Zadeh M K 2019 Progressive collapse analysis of steel braced frames *Pract. Period. Struct. Des. Constr.* **24** 1-9

[7] Okazaki T, Engelhardt M D, Drolias A, Schell E, Hong J K and Uang C M 2008 Experimental investigation of link-to-column connections in eccentrically braced frames *Journal of Constructional Steel Research* **65** 1401-12

[8] Tande S N and Sankpal A A 2014 Study of inelastic behavior of eccentrically braced frames under non-linear range *IJLTET* **4** 273-86

[9] Hjelmstad K D and Popov E P 1983 Seismic Behaviour of Active Beam Links in Eccentrically Braced Frames *Report no. UCB/EERC-83/15. Berkeley (CA): Earthquake Engineering Research Institute*

[10] Engelhardt M D and Popov E P 1989 Behaviour of long links in eccentrically braced frames *Report no. UCB/EERC-89/01. Berkeley (CA): Earthquake Engineering Research Institute*

IOP Conf. Series: Materials Science and Engineering **989** (2020) 012022 doi:10.1088/1757-899X/989/1/012022

Empirical research on strength and durability properties of bottom ash for sustainable concrete practice

Ashwathi R [1,3], Sabarinathan K[2] and Velmurugan S[1]

[1]Department of Civil Engineering, Bannari Amman Institute of Technology, Sathyamangalam, Tamilnadu
[2] Department of Civil Engineering Excel Engineering College, Komarapalayam, Tamilnadu
[3]Email:ashwathisingh@gmail.com

Abstract: The prime constituents utilized in the construction field are concrete because of its easy availability and also by considering strength properties and cost-effective factors. It is composed essentially of fine aggregate, coarse aggregate and cement, where the fine aggregate manipulated during manufacturing is natural sand. The requirement for sand is thriving due to its abundant utilization in the construction industry, thereby the natural resource is becoming diminished. In sequence to overcome the depletion of sand, empirical research has been done on strength characteristics of concrete with bottom ash to partially counterbalance fine aggregate. Bottom ash is a residue obtained at the thermal power plant by combusting coal. It can be utilised in concrete as a partial replacing material for cement or fine aggregate or coarse aggregate. The practice of bottom ash in construction makes the disposal eco-friendly and hence reduces sand quarrying. In this paper, it is affirmed that bottom ash can partially (15%, 25%, 30%) replace fine aggregates without compromising the qualities of concrete.

1. Introduction

Natural river sand is more usually employed as fine aggregate which is one of the important components of concrete. Due to the wide usage of concrete around the world, the global consumption of natural river sand is increasing. The growth of the construction industry is meeting a great problem in many parts of the country because of the unavailability of sufficient quantity of ordinary river sand [1]. On the other side, the effective disposal of solid wastes from both municipal and industrial sector is a challenging task as it causes an impact on the environment. In those cases, the waste can be used in concrete making as a replacement to aggregate or cement to enhance the properties, which in turn reduces the wastage and its disposal [2, 3].Utilization of residues and waste in concrete production leads to green environment and those concrete manufactured are known as green concrete. When bottom ash is employed, the cost of manufacturing of concrete gets reduced to a certain scale. Carbon-di-oxide emission and landfill deposition is reduced and protects the environment from degradation [4].

The ultimate aspiration of this study is to examine the feasibility of bottom ash in concrete and to find the optimum value for replacing aggregate in concrete. Also, it aims at improving the compressive strength by partly displacing sand by bottom ash without negotiating the properties of concrete [5].

2. Materials and properties

2.1. Cement

Cement is a binding material utilised in concrete production with the property of hardening and setting. Ordinary Portland cement is highly durable and feasible. It is Limestone which is silicious in nature is burnt at 1450°C to manufacture OPC and is ground with gypsum. The optimum particle size distribution of OPC renders strength and durability to the structure [6].

Table 1. Physical properties of OPC cement.

Cement	CCN	Mass%
Calcium Oxide, CaO	C	61-67%
Silicon oxide, SiO2	S	19-23%
Aluminium Oxide, Al2O3	A	2.5-6%
Ferric Oxide, Fe2O3	F	0-6%
Sulphate	S	1.5-4.5%

Table 2. Chemical properties of OPC cement.

Name of the test	Value
Consistency	30%
Setting time - Initial	30 minutes
Setting time - Final	600 minutes
Fineness Modulus	3.5 %
Specific gravity	3.15

2.2. Coal Bottom Ash (CBA)

When the crushed coal is burnt in a boiler, 80% of unburnt material is collected. During the process of burning the crushed coal in a dry bottom boiler, nearly 80 percent of the unburnt material or ash is retrieved in the fly ash form while the rest 20% of the unburnt material is washed out at the bottom and recovered as bottom ash [7].

Table 3. Physical properties of coal bottom ash.

Property	Bottom ash
Specific gravity	2.7
Dry unit weight	15.72kN/m
Plasticity	None
Absorption	1.5%
Property	Bottom ash

Table 4. Chemical properties of coal bottom ash.

Composition	Percentage
SiO_2	33.5
Al_2O_3	15.8
Fe_2O_3	8.4
CaO	19.4
MgO	2.0
SO_3	9.3
ZnO	0.8
Na_2O	3.6
K_2O	1.9
TiO_2	1.5
Cl	1.1
SiO_2	33.5

2.3. *Fine Aggregate*

Riverbed sand available in the nearby area that is free from debris is used as fine aggregate. One of the most significant characteristics is its grading. The various test has been conducted for the fine aggregate conforming to IS: 2386 (Part III) - 1963 and has been used successfully [8].

Table 5. Physical properties of fine aggregate.

Property	Value
Specific gravity	2.44
Fineness modulus	2.87
Grading	Zone –II

2.4. Coarse aggregate

Aggregates used must be free from impurities and must be clean. Coarse aggregate of size 20mm size confirming to grading Zone-II has been used. The coarse aggregate of fineness modules 6.67 conforms to IS: 383-1970. The size of the coarse aggregate should range between 10mm and 20mm such that optimum compressive strength is offered with high cement content and low W/C ratio [9], [10].

Table 6. Physical properties of coarse aggregate.

Property	Value
Specific gravity	2.72
Fineness modulus	6.67
Grading	Zone –II

2.5. Water

Water is also a crucial ingredient in making concrete, as it helps to cause chemical reaction with cement. The durability of concrete is achieved due to the binding action. Water is an important constituent in attaining strength. The maximum strength is obtained by proper proportion and workability. When the quantity of water increases, thereby the strength of the concrete gets decreased. So it is inherent to maintain water-cement ratio throughout [11, 12, 13].

3. Tests and discussions

3.1. Strength test

3.1.1 Compressive strength

One of the most prominent and useful properties of concrete is its compressive strength which depends on various factors such as water-cement ratio, quality of materials used, quality control during production of concrete etc,. Concrete is strong in compression and thereby used to resist compressive stress in the structure. The compressive strength test on concrete is carried out to measure the properties of the compressive strength [14, 15].

The research has been carried out for finding the values of compressive strength for different replacement of Coal Bottom Ash at different curing ages. The test has been conducted and the results are given below.

Table 7. Compressive strength test result at 7th, 14th and 28th day.

	0% CBA	15% CBA	25% CBA	30% CBA
7th day	11.7	9.76	11.52	11.20
14th day	15.6	14.87	15.20	14.95
28th day	26	23.25	25.50	25.30

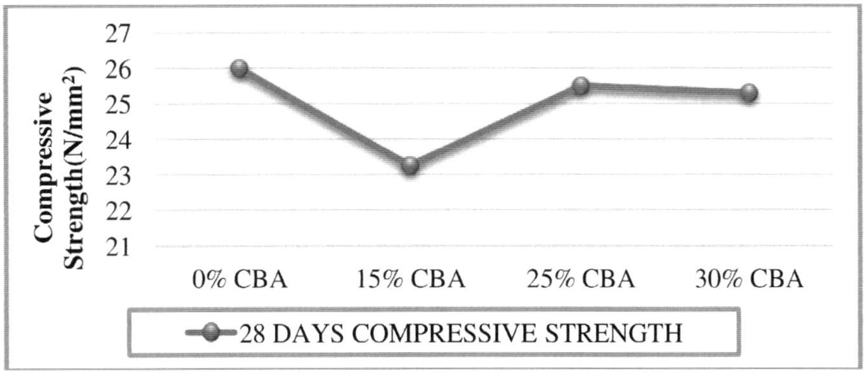

Figure 1. Compressive strength of concrete at 28 th day.

The results show the difference in compressive strength of concrete with and without replacement of CBA with fine aggregate at different curing ages. It is apparent from the figure, that compressive strength increases up to 25% replacement and on further replacement, it becomes nearly alike to that of conventional concrete [16, 17].

3.1.2. Split tensile strength
The split tensile strength is one of the most significant properties that considerably affect the size and depth of cracking in structures. As the concrete is brittle in nature, it is weak in tension and does not counter direct tension. So, when tensile forces exceed its tensile strength cracks develop [18].

Table 8. Split tensile strength test result at 7th, 14 th and 28 th day.

	0% CBA	15% CBA	25% CBA	30% CBA
7th day	1.17	1.02	1.22	1.15
14th day	1.93	1.68	2.03	1.90
28th day	2.55	2.25	2.50	2.30

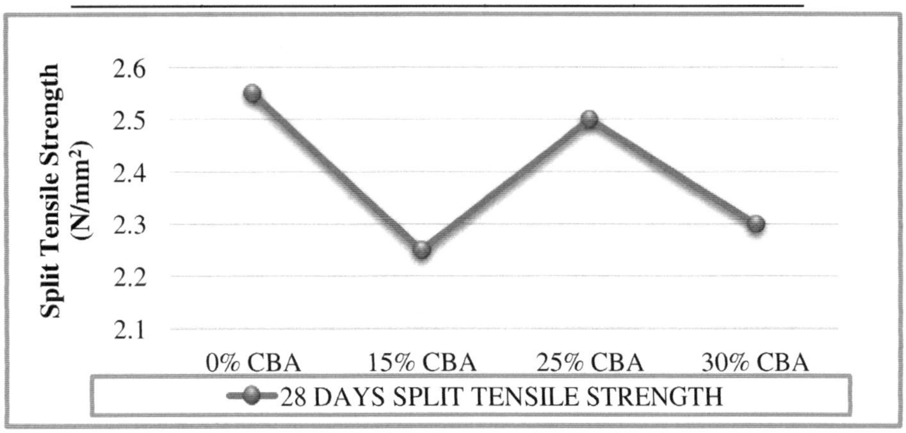

Figure 2. Split tensile strength of concrete at 28 th day.

The results show the contrast of the tensile strength of concrete with and without replacement of CBA with fine aggregate at different curing ages. The figure apparently shows that tensile strength increases up to 25% and on further replacement, it becomes almost equal to that of conventional concrete [19].

3.1.3. Flexural strength

The flexural strength of concrete is employed in resisting compressive stress. The dimension of the beam the pattern of loads has a significant impact one the value of modulus of rupture. The flexural strength of concrete with 0%, 15%, 25% and 30% of replacements were determined as per I.S. 516-1959 [19].

Table 9. Flexural strength test result at 7th, 14th and 28th day.

	0% CBA	15% CBA	25% CBA	30% CBA
7th day	2.17	2.12	2.31	2.35
14th day	3.12	3.06	3.21	3.26
28th day	3.45	3.37	3.48	3.44

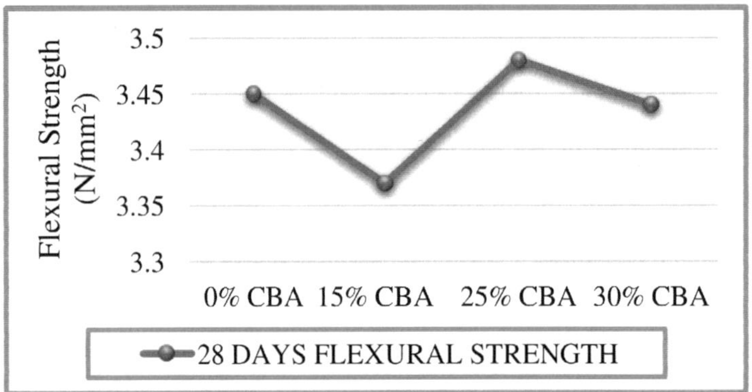

Figure 3. Flexural strength of concrete at 28th day.

The results show the variation of flexural strength of concrete with the replacement of CBA with fine aggregate at different curing ages. It is clear from the figure, that tensile strength increases upto 25% replacement and on further replacement, it becomes almost similar to that of conventional concrete.

3.2. Durability test

The durability of a structure is the ability to exhibit the life span of a structure which can withstand chemical attack and physical impairment. Mix proportion, selection and quality of materials used majorly influence the durability of the structure [20, 21].

3.2.1. Acid attack test

Concrete is generally susceptible to acid attack due to its alkaline nature. The cement paste normally breaks down when it reacts with acid. The cube is taken and casted. After 28 days curing, the specimens were taken out. For acid attack, sulphuric acid is diluted in water by 5 % of volume of water. After 45 days of immersion in acid, the cubes were taken out. The surface of cubes are cleaned and left for 24 hours and compressive strength test is done.

Table 10. Compressive Strength of cubes before and after Acid Attack

% Bottom ash	Compressive strength of concrete (N/mm²)	
	Before attack	After attack
0%	26	23.5
15%	23.25	21.4
25 %	25.5	21.2
30%	25.3	20.4

3.2.2. Sulphate attack test

Sulphate attack is a chemical breakdown mechanism where sulphate ions attack components of the cement paste. The cube is taken and casted. After 28 days curing, the specimens were taken. For sulphate attack, magnesium sulphate is diluted in water by 5 % of volume of water. After 45 days of immersion in acid, the cubes were taken out. The surface of cubes are cleaned and left for 24 hours and compressive strength test is done.

Table 11. Compressive Strength of cubes before and after Sulphate Attack.

% Bottom ash	Compressive strength of concrete (N/mm²)	
	Before attack	After attack
0%	26	24.2
15%	23.25	21.46
25 %	25.5	23.1
30%	25.3	23.2

3.2.3.Chloride attack test

The cube is taken and casted. After 28 days curing, the specimens were taken. The sodium chloride solution was prepared by adding 3.5% sodium chloride (by volume of water) to water. Then the concrete cubes are immersed in 3.5% sodium chloride (1N NaCl) solution. After 45 days of immersion in solution, the cubes were taken out. The surface of cubes are cleaned and left for 24 hours and compressive strength test is done.

Table 12. Compressive Strength of cubes before and after Chloride Attack.

% Bottom ash	Compressive strength of concrete (N/mm²)	
	Before attack	After attack
0%	26	25.2
15%	23.25	22.4
25 %	25.5	23.6
30%	25.3	23.45

3.2.4. Alkaline attack test

The cube taken was casted and cured for 28 days. Then 3.0% of sodium hydroxide was added with water and the solution was prepared. The concrete cubes were then immersed in the solution for a period of 45days. Then the specimens were taken out from the alkaline water (sodium hydroxide solution). After 45 days of immersion in acid, the cubes were taken out. The surface of cubes are cleaned and left for 24 hours and compressive strength test is done.

Table 13. Compressive Strength of cubes before and after Alkaline Attack.

% Bottom ash	Compressive strength of concrete (N/mm^2)	
	Before attack	After attack
0%	26	25.1
15%	23.25	22.1
25 %	25.5	23.4
30%	25.3	23.2

3.2.5. Water absorption test

The rate of absorption of water by hydraulic cement concrete can be determined by this method. This can be done by measuring the increase in the mass of a specimen resulting from absorption of water (1) as a function of time when only one surface of the specimen exposed to water. The exposed surface is immersed in water and water ingress of unsaturated concrete dominated by capillary suction during initial contact with water.

$$\text{Water absorption} = (\text{saturated weight} - \text{dry weight}) / \text{dry weight} \quad (1)$$

Table 14. Water Absorption on Concrete.

% Bottom ash	Initial weight in grams	Final weight in grams	% of water absorption
0%	2520	2551	1.23
15%	2347	2382	1.49
25 %	2431	2496	2.67
30%	2315	2366	2.2

4. Conclusion

From the research work carried over the strength and durability of concrete, the following observations are made:

The CBA is a low cost available industrial solid waste which on incorporation with concrete helps in attaining strength, thereby paving way for an eco-friendly environment. From the results, it's been inferred that 25% replacement of CBA is found to be expedient. Therefore the optimum percentage is found to be 25%. The durability properties also show optimal results which help in indulging CBA in concrete.

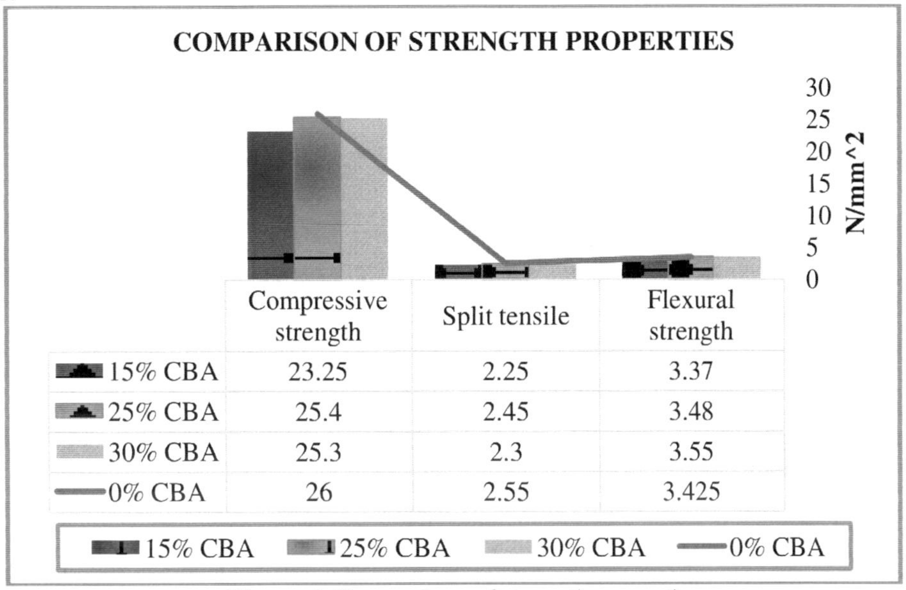

	Compressive strength	Split tensile	Flexural strength
15% CBA	23.25	2.25	3.37
25% CBA	25.4	2.45	3.48
30% CBA	25.3	2.3	3.55
0% CBA	26	2.55	3.425

Figure 4. Comparison of strength properties.

5. References

[1] Navdeep Singh, Mithulraj M., Shubham Arya 2018 Influence of coal bottom ash as fine aggregates replacement on various properties of concretes: A review *J.* Resources, Conservation and Recycling **138** 257-271

[2] Caprai V, Gauvin F, Schollbach K and Brouwers H J H 2019 MSWI bottom ash as binder replacement in wood cement composites *J. Construction and Building Materials* **196** 62-68

[3] Akkadath Abdulmatin, Weerachart Tangchirapat and Chai Jaturapitakkul 2018 An investigation of bottom ash as a pozzolanic material *J.* Construction and Building Materials **186** 155-162

[4] Jacques Rémy Minane, Frédéric Becquart, Nor Edine Abriak and Christophe Deboffe 2017 Upgraded mineral sand fraction from MSWI bottom Ash: An alternative solution for the substitution of natural aggregates in concrete applications *J. Procedia Engineering* **180** 1213-1220

[5] Diana Bajare, Girts Bumanis and Liga Upeniece 2013 Coal combustion bottom ash as microfiller with pozzolanic properties for traditional concrete *J. Procedia Engineering* **57** 149-158

[6] Sajedeh Sadat Ghazizadeh Hashemi, Hilmi Bin Mahmud, Jean Noël Yankwa Djobo, Chee Ghuan Tan, Bee Chin Ang and Navid Ranjbar 2018 Microstructural characterization and mechanical properties of bottom ash mortar *J. Journal of Cleaner Production* **170** 797-804

[7] Stabile P, Bello M, Petrelli M, Paris E and Carroll MR 2019 Vitrification treatment of municipal solid waste bottom ash *J. Waste Management* **95** 250-258

[8] Mahdi Rafieizonooz, Jahangir Mirza, Mohd Razman Salim, Mohd Warid Hussin and Elnaz Khankhaje 2016 Investigation of coal bottom ash and fly ash in concrete as replacement for sand and cement *J. Construction and Building Materials* **116** 15-24

[9] Wan Ibrahim M H, Hamzah A F, Jamaluddin N, Ramadhansyah P J and Fadzil A M 2015 Split tensile strength on self-compacting concrete containing coal bottom ash, *J. Procedia - Social and Behavioral Sciences* **195** 2280-2289

[10] Hongxu Zhou, Rabin Bhattarai, Yunkai Li, Shiyang Li and Youheng Fan 2019 Utilization of coal fly and bottom ash pellet for phosphorus adsorption: Sustainable management and evaluation *J. Resources, Conservation and Recycling* **149** 372-380

[11] Malkit Singh and Rafat Siddique 2016 Effect of coal bottom ash as partial replacement of sand on workability and strength properties of concrete *J. Journal of Cleaner Production* **112(1)** 620-630

[12] Yogesh Aggarwal and Rafat Siddique 2014 Microstructure and properties of concrete using bottom ash and waste foundry sand as partial replacement of fine aggregates *J. Construction and Building Materials* **54** 210-223

[13] Jawahar S, Magesh M, Jagen V and Vasugi V 2019 Performance evaluation of concrete using bottom ash as fine aggregate *International Journal of Innovative Technology and Exploring Engineering* **8(7)** 95-100

[14] Cristina Argiz, Amparo Moragues and Esperanza Menéndez 2018 Use of ground coal bottom ash as cement constituent in concretes exposed to chloride environments *J. Journal of Cleaner Production* **170** 25-33

[15] Jang J G, Kim H.J, Kim H K and Lee H K 2016 Resistance of coal bottom ash mortar against the coupled deterioration of carbonation and chloride penetration *J. Materials & Design* **93** 160-167

[16] Andrade L B, Rocha J C and Cheriaf M 2009 Influence of coal bottom ash as fine aggregate on fresh properties of concrete *J. Construction and Building Materials* **23(2)** 609-614

[17] Rafat Siddique 2013 Compressive strength, water absorption, sorptivity, abrasion resistance and permeability of self-compacting concrete containing coal bottom ash *J. Construction and Building Materials* **47** 1444-1450

[18] Mahdi Rafieizonooz, Jahangir Mirza, Mohd Razman Salim, Mohd Warid Hussin and Elnaz Khankhaje 2016 Investigation of coal bottom ash and fly ash in concrete as replacement for sand and cement *J. Construction and Building Materials* **116** 15-24

[19] Guodong Huang, Yongsheng Ji, Linglei Zhang, Jun Li and Zhihui Hou 2018 Advances in understanding and analyzing the anti-diffusion behavior in complete carbonation zone of MSWI bottom ash-based alkali-activated concrete *J. Construction and Building Materials.* **186** 1072-1081

Analytical study on the behaviour of beam-column joints made of electric arc furnace concrete

Gowri Krishna S R[1,2] and Rekha Ambi[1]

[1]Deptartment of Civil Engineering, T. K. M. College of Engineering, Kollam (Kerala), India

[2]Email:grikriss@gmail.com

Abstract: The advancement of the construction industry towards using recycled materials has substantially reduced the consumption of natural resources. Electric Arc Furnace (EAF) slag is the by-product obtained from the production of steel and is procured after the separation of molten steel from impurities. In this paper, the impact of EAF concrete on the structural behaviour of beam-column joints and the resisting mechanism of these joints are evaluated using ANSYS software. One joint made with conventional concrete and two joints made with EAF concrete were analysed subjected to incremented loading. The results indicate that EAF concrete exhibits better structural performance than conventional concrete in terms of the ultimate load attained by the joint in the failure condition.

1. Introduction

In Indian practice, a beam-column joint is not distinctively designed mainly due to economic considerations. Beam-column joints are indispensable zones in the analysis of moment resisting reinforced concrete frames, especially under combined loading condition [1]. The most important part of composite beam-column structures is the connection [2, 3]. This is better understood during earthquakes when the failure of reinforced concrete frames occurs primarily due to shear in joints resulting in the total breakdown of structure. Joint design can be conveniently avoided if the structure is solely subjected to gravity load. Adequate detailing and design of joints should be ensured to improve the strength and ductility to resist large deformations caused due to earthquakes. Since sufficient guidelines on joint design are not specifically provided in Indian codes of practice, adequate lapping of main reinforcement satisfies the strength criteria.

India is the world's second largest producer of crude steel. According to the World Steel Association, India's crude steel production in 2019 was at 111.2 million tons which accounted to 12 million tons of steel slag. This leaves a question as to what happens to this enormous quantity of slag produced. In contrast to certain other nations, steel slag is discarded and is becoming a problem due to paucity of land. Currently, there are no specific regulations or norms to control the slag generation or disposal. EAF slag is obtained as a by-product of steel production process when Electric Arc Furnace is used instead of Basic Oxygen Furnace (BOF) [4, 5, 6]. Although EAF slag has relatively low hydraulicity, its chemical composition can be controlled by adjusting the steel making process [7]. Correctly manufactured EAF concrete has good mechanical properties and high density [8]. It is desirable to use EAF instead of BOF due to the following reasons: (i) energy consumption of BOF process is 2.8 times greater than that of EAF process, (ii) material consumption of BOF process is 11 times greater than that of EAF process, (iii) carbon emission is low when steel is produced through EAF and (iv) cost of production is lower for EAF compared to the blast furnace route [4].

Worldwide industrial waste generation and quarry exploitation are becoming a major environmental problem. EAF is currently used for asphalt concrete pavements in many countries but however, huge quantities of this material are still land filled [4]. Reusing the slag as a recycled material in construction industry not only helps in the reduction of the amount of waste disposed off but it can also bring down the consumption of natural aggregates.

Current Indian codes like IS 1893:2002, IS 13920:1993 and IS 456:2000 is not equipped well enough to design a beam-column joint under seismic conditions. It fails to shed light on the assessment of shear demand and detailing of shear reinforcement. This lack of information may be due to the disparities among different countries in terms of the occurrence of earthquakes. In this study, an exterior beam-column joint is designed based on [10, 11] which provides codal recommendations on structural concrete and beam-column joint design in RC structures respectively. The designed joint is used for the analytical study on its behaviour under combined loading condition. Also, a comparative study is conducted on the joints made with EAF concrete and conventional concrete based on load-deflection plot, energy absorption capacity and crack propagation results. Analysis is done using ANSYS software. This study also brings to light the urgent need for updating the Indian codes of practice considering the critical joints in a moment resisting structure.

2. Design approach

An exterior beam-column joint is designed using ACI 352R-02 and ACI 318M-02 which apply only to structures using normal weight concrete with a compressive strength less than 100MPa. Based on the loading state and the expected deformation of the fastened frame members when resisting lateral loads, beam-column joints are classified into two categories – Type 1 and Type 2 respectively [11]. Type 2 connection is preferred for the joint since the frame members are required to dissipate energy and they have sustained strength when subjected to deformation reversals in the inelastic range [11]. Joint dimensions along with reinforcement details are given in Figure 1 below and all dimensions are given in millimeters. Clear cover of 25 mm is provided for beams and columns. The design shear force and the concrete shear capacity of the joint is obtained as 354 kN and 2.32 MPa respectively.

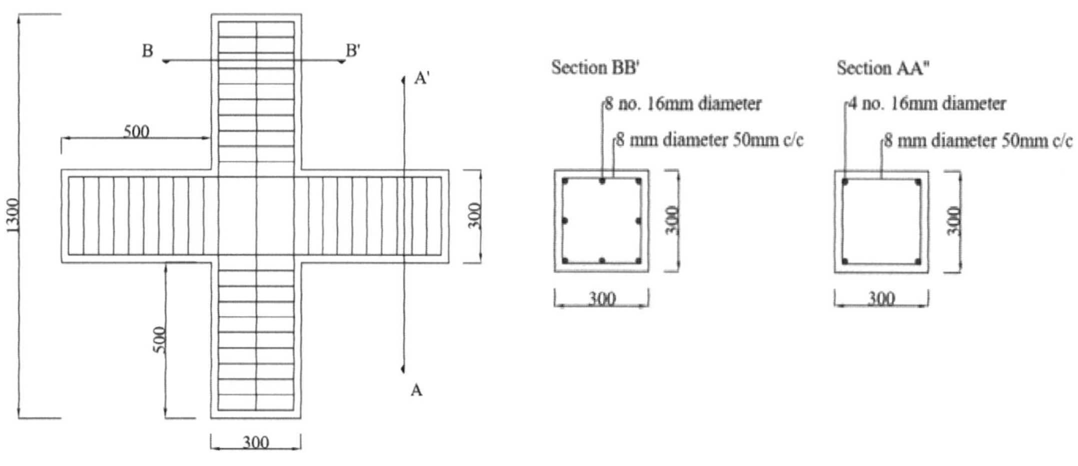

Figure1. Reinforcement details of the beam-column joint.

3. Finite element modelling and meshing

A finite element model is created as shown in Figure 2 using ANSYS, widespread Finite Element modelling and analysis software.

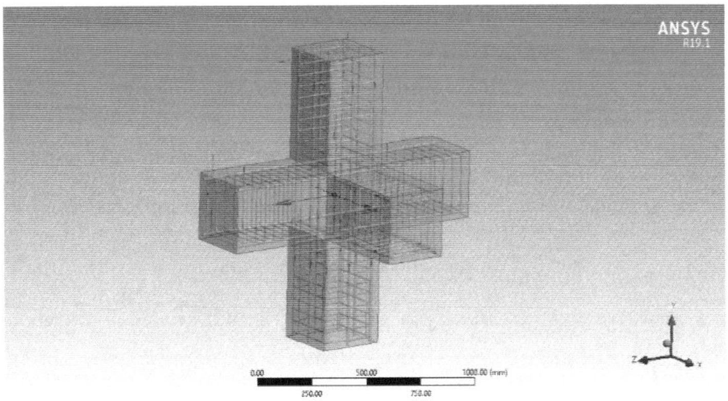

Figure 2. Model of the joint.

Solid 65 element is used for modelling concrete (Figure 3). Solid 65 is a 3D structural RC solid element which can indicate cracking and crushing [12]. It has eight nodes with three translational degrees of freedom at each node. It also exhibits non-linear material properties.

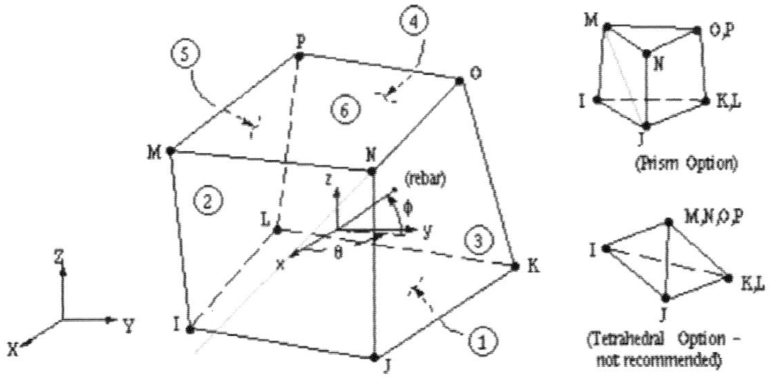

Figure 3. Solid 65 element.

Reinforcements are modelled using bilinear kinematic hardening properties where translation of the yield surface is allowed. Excessive number of stirrups is provided in beams and columns to initiate more shear cracks within the joint area than along the beam or column. This helps in the study of beam-column joint mechanism. Since the beam-column joint region is given utmost importance, full length of a beam or column is not considered. Hence beam or column is considered mid-length from the support. Due to this reason, all beam ends and bottom region of the column are provided with simple supports. Top portion of the column is allowed to move freely in the direction of applied loads and constrained in the remaining directions.

For meshing, hex dominant method is adopted for the entire body by using all quadrilateral elements. Body sizing and face meshing is also done with an element size of 30mm.

4. Material properties

Concrete is a quasi-brittle material and shows different behaviour in tension and compression. The mechanical properties of both concrete materials are provided in Table 1. The reinforcement is characterized with a yield stress of 250MPa and tangent modulus of 1450.

Table 1. Material properties of concrete (Rajasekhar et al. 2017) [13].

Type of concrete	Conventional	EAF
Open shear transfer coefficient	0.3	0.3
Closed shear transfer coefficient	1	1
Uniaxial cracking stress	3.35 MPa	3.77MPa
Uniaxial crushing stress	27.85 MPa	33.11MPa
Poisson's ratio	0.2	0.2
Elastic modulus	26386MPa	29155MPa

5. Non-linear static analysis

- A constant vertical load of 400kN is applied on top of the column corresponding to the gravity load induced by upper floors.
- Horizontal load is applied on top of the column corresponding to the quantified movement of the joint to the inter-story drift of a whole frame. A 2mm displacement is provided in the first step which lies within the elastic range. The successive displacements have values between 1.25 and 1.5 the preceding one.
- According to ACI 318-2011, allowable story drift is 3.5% for joints [14].
- Maximum principal stress, total deformation and force reaction in the direction of deformation are obtained as results from Ansys. Crack and crushing results are obtained from Ansys Parametric Design Language commands.
- Maximum principal stress theory is used for the analysis of brittle materials. In this case, failure occurs when maximum principal stress for complex loading in the joint at any point reaches equal or more than the maximum principal stress of simple tension test.

6. Results and discussion

Non-linear static analysis is used as an efficient tool to assess the performance of EAF concrete and conventional concrete. Non-linear static methods are improving day to day for evaluating the seismic analysis with more accuracy in a simple way. Load deflection graphs are developed out of the results obtained from non-linear static analysis. In Finite Element Analysis (FEA), it is recommended to use nonlinear static analysis as nonlinear dynamic time history analysis is very complex and time consuming. The main objective of this study is to evaluate the effect of EAF concrete on an exterior beam-column joint and to assess the performance by developing force-displacement graph. A non-linear quasi static analysis study of the considered joint model is conducted to assess the effect of constant axial load and incremented displacement on the top of column. Non-linear studies are conducted mainly to determine the yield point and the ultimate load of the joints. The significance of beam-column joints becomes more evident especially in high seismic prone areas. Most of the building collapse during earthquake is majorly caused due to the shear failure occurring in joints.

6.1. Study on failure criterion and load-deflection curve

The main purpose of this analysis is to evaluate the effect of concrete properties on the overall performance of the joint. Joint shear cracks are developed at the ultimate drift ratio (θ = 3.5%) from the crack and crush results using APDL commands. Cracks also developed along the beam near the column. This failure mode is known as *BJ* type failure where a mixed cracking is observed in which shear cracks in the joint occurs along with flexural damage in the beam [14]. Figure 4 shows the results referring to the force-displacement envelope curves of EAF and conventional concrete joints.

Joints made with EAF concrete is marked by higher ultimate loads achieved at lesser displacement than in conventional concrete. Hence, they showed greater stiffness than the joint made with conventional concrete due to higher elastic modulus values. As a result of *B-J* failure mode, deformation capacity and strength reduction of the elements occurred as an outcome of cracking of joints.

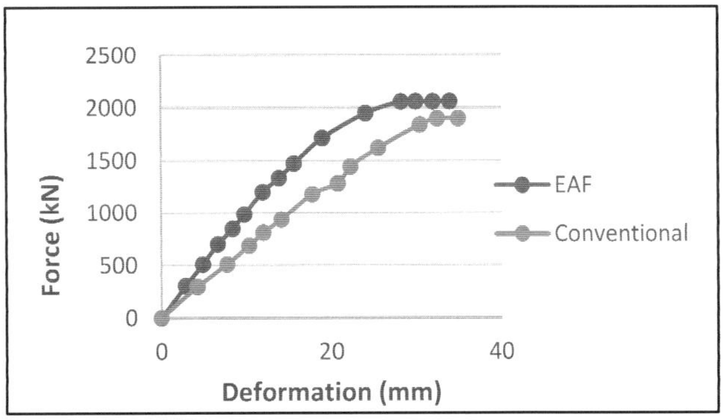

Figure 4. Load-deflection plot obtained from analysis.

The conventional concrete joint is less stiff compared to the EAF joint. Ultimate load achieved by the joint made of EAF concrete is 2057.93 kN and that of conventional concrete is 1897.08 kN. Maximum drift value attained by the joint is about θ = 3.7% which is greater than the allowed limit θ = 3.5%. After reaching the ultimate load, the stiffness of joints is seen to be suddenly reduced which typically characterizes a shear failure mechanism [14].

According to [15], there are three major points in the load-deflection graph of exterior beam-column joints failing by *B-J* failure mode representing the major stiffness changes. These key points (Figure 5) are obtained as follows: (i) a tangent line is drawn from origin until a point A that sets off another tangent line, (ii) another tangent line is drawn from A until it reaches a point B that sets off another line, (iii) point C is placed at the ultimate load attained by the joint.

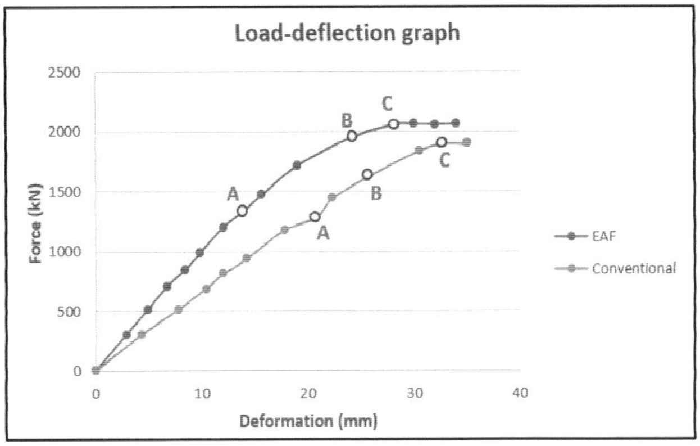

Figure 5. Plot showing the key points of stiffness change.

IOP Conf. Series: Materials Science and Engineering **989** (2020) 012023 doi:10.1088/1757-899X/989/1/012023

Point A stands for first stiffness change due to shear cracks in the joint. Point B stands for reinforcement yielding and point C stands for concrete crushing within the joint. It can be concluded that EAF concrete attains maximum load compared to conventional concrete and also at reduced applied displacement.

The displacement is 28.24 mm for EAF concrete at ultimate load whereas the displacement for conventional concrete is 32.52 mm at ultimate load. The shear values at point A are 1331.39 kN and 1280.66 kN for EAF and conventional concrete at 13.93 mm and 20.807 mm displacement respectively. The first stiffness change for EAF takes place at greater force and smaller displacement. The shear values at point B are 1947.68 kN and 1618.09 kN for EAF and conventional concrete at 24.06 mm and 25.55 mm displacement respectively. In the second stiffness change, an increase of total load attained by EAF concrete is observed. Also, the increase in shear force and displacement for EAF concrete in the stiffness change is much greater than that for conventional concrete.

6.2. Crack results

Crack results can be used to identify damage progression during the loading. In RC structures, the rigid joint assumption is a coarse concept and since concrete has low tensile strength, cracks are developed even at the early stages of loading. Crack occurs when the stress evolving on a portion of the structure match up with a point in the principal stress space situated outside the surface which defines the failure criterion and this, in turn results in localized material failure. This failure is evidenced by crack formation which results in the loss of load carrying capacity in the direction normal to the plane of crack. Crack results for EAF and conventional concrete in step 2 of loading are shown in Figures 6-9.

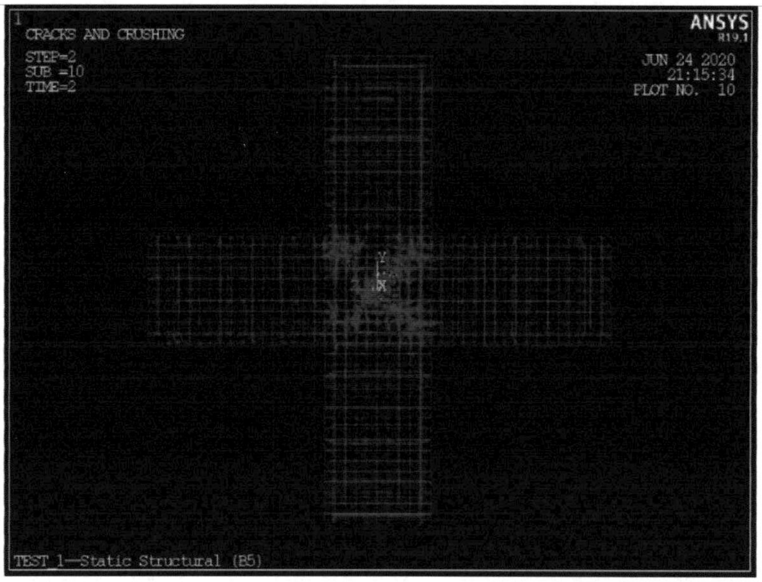

Figure 6. Front view of crack results for conventional concrete.

IOP Conf. Series: Materials Science and Engineering **989** (2020) 012023 doi:10.1088/1757-899X/989/1/012023

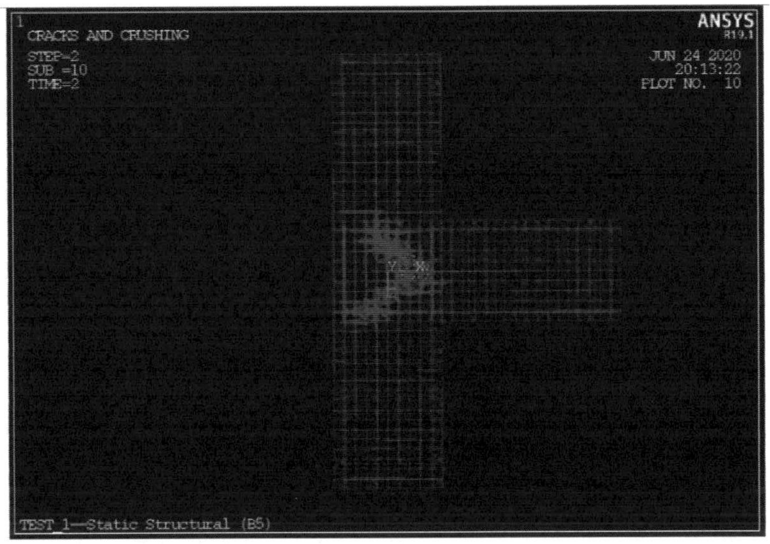

Figure 7. Side view of crack results for conventional concrete.

Figures 6 and 7 show the front and side view of conventional concrete joint. It can be seen that first crack and second cracks are formed at the joint portion. As each step of loading progresses, cracks continue to grow in number. But no sign of cracks were found along the beam length or column length. Hence, in this case, the majorly affected area is the beam-column joint region. Figures 8 and 9 show the front and side view of crack and crush results obtained for EAF concrete. Here, same as the conventional concrete, second step observation is taken in order to make comparison more solid. Crack has not yet developed in the joint region in this step for EAF concrete. But first and second cracks are developed in the top half portion of the column. As the load step progresses, crack continues to grow in number along the column portion as well as in the joint region.

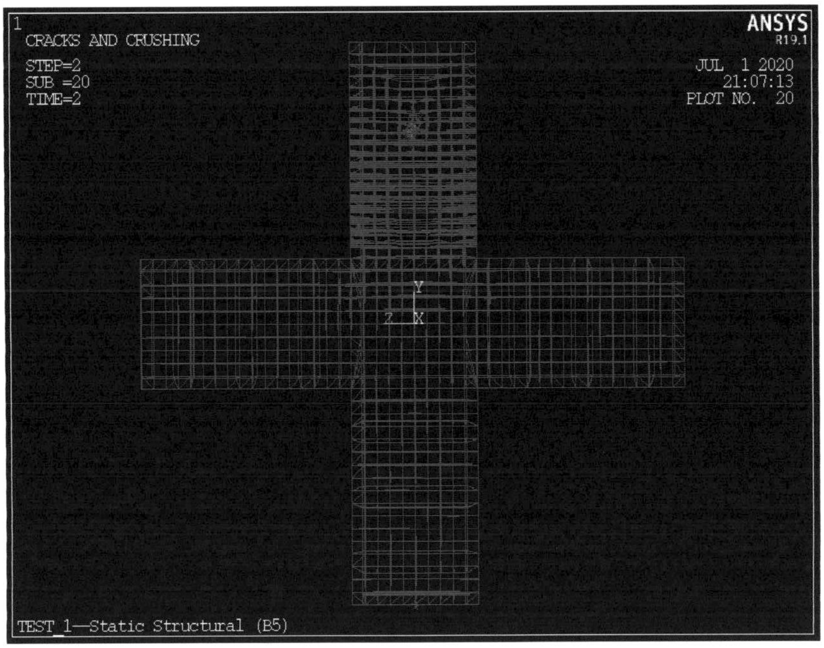

Figure 8. Front view of crack results for EAF concrete.

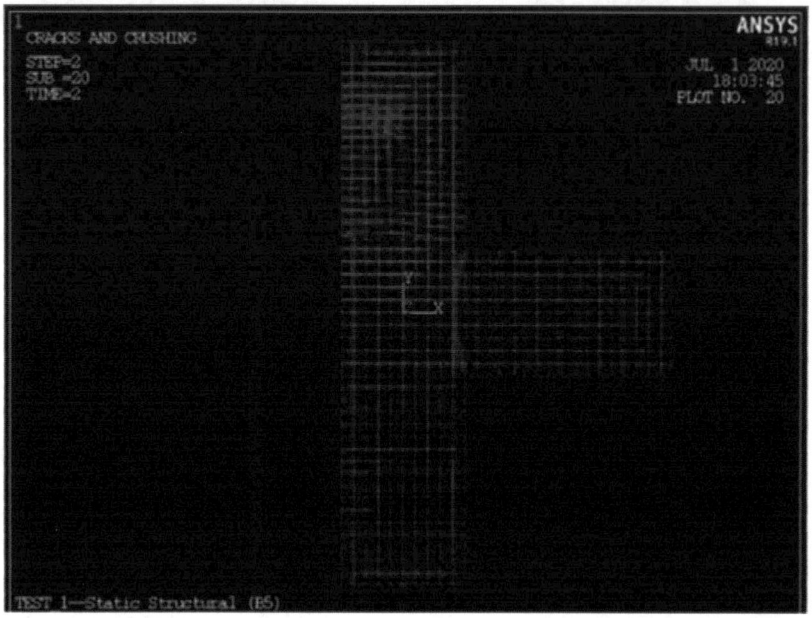

Figure 9. Side view of crack results for EAF concrete.

There are 3 directions of principal stresses for finding cracks in a 3D solid 65 element. If the value of the maximum principal stress attains the value of the material strength , the first crack forms; if the value of the average principal stress attains the value of the material strength , the second crack forms and if the value of the minimum principal stress attains the value of the material strength , the third crack forms. Last load step showed all three cracks in the beam-column region as well as along the length of column. But the number of cracks developed is less in EAF compared to the conventional concrete even during the last step of loading.

6.3. Energy absorption capacity

Load-deflection plot of both the concrete materials are given in the previous session. The area under the load–deflection plot characterizes the energy absorption capacity of the joint [16]. Considering the limitations in analysis running condition, the deflection after ultimate loading could not be tracked and hence the area under the load–deflection plot considered in this study constitutes the area under the upward slope portion up to the ultimate load. The energy absorption capacity was calculated for both joints and is shown in Table 2. It may be noticed from Table 2 that the energy absorption capacity of EAF is about 22.8% higher than that of conventional concrete.

Table 2. Comparison of the energy absorption capacity.

Joint type	Energy absorption capacity (kNm)	% increment in EAC
Conventional concrete	38.266	---
EAF	47.001	22.8

7. Conclusions

From the analysis results obtained in this study, the following conclusions can be drawn:

- Failure mode of both the materials showed that an increase in shear force and displacement for EAF in the stiffness change is much greater than that for conventional concrete.
- Load attained by EAF concrete is greater than that attained by conventional concrete.
- Cracks are developed in the joint region for conventional concrete in the early stages of loading whereas for EAF concrete joint cracks are developed in the final steps of loading.
- Energy absorption capacity (EAC) for EAF concrete is 22.8% higher than that of conventional concrete.

These results give an indication about the capacity of EAF slag to be used in structural concrete members as a replacement of aggregates. This, in turn will help to reduce the overexploitation of quarries.

8. References

[1] Sarkar P, Agrawal R and Menon D 2007 Design of RC beam-column joints under seismic loading- A Review *Journal of Structural Engineering* **33(6)** 449-457

[2] Mirza O and Uy B 2010 Finite element analysis of the behaviour of composite beam-to column connections subjected to seismic loading *In the Proceedings* of *Structures Congress* 3431-3442

[3] Ponmani K P, Sugumar K and Vasugi V 2019 Analytical and experimental study of RC beam-column joint with steel slag aggregate *International Journal of Innovative Technology and Exploring Engineering* **8(4)**1382-1386

[4] Faleschini F, Brunelli K, Zanini M A, Dabala M and Pellegrino C 2016 Electric arc furnace slag as coarse recycled aggregate for concrete production *Journal of sustainable metals and materials* **2** 44-50

[5] Rondi L, Bregoli G, Sorlini S, Cominoli L and Collivignarelli C 2015 Concrete with EAF steel slag as aggregate: A comprehensive technical and environmental characterization *Composites* **90** 195-202

[6] Sekaran A, Palaniswamy M and Balaraju S 2015 A Study on Suitability of EAF Oxidizing slag in Concrete: An Eco-Friendly and Sustainable Replacement for Natural Coarse Aggregate *The Scientific World Journal* **15** 1-8

[7] Lee J, Choi J, Yuan T, Yoon Y and Mitchell D 2019 Comparing properties of concrete containing electric blast furnace slag and granulated blast furnace slag *Materials* **12(9)** 1371-1382

[8] Manso J. M., Polanco J. A., Losanez M., Gonzalez J. J. 2006 Durability of concrete made with EAF slag as aggregate *Cement and Concrete Composites* **28(6)** 528-534

[9] Li X, Sun W and Zhao L 2018 Material metabolism and environmental emissions of BF-BOF and EAF steel production routes *Mineral Processing and Extractive Metallurgy* **39(1)** 50-58

[10] ACI 318M-11 2011 Building code requirements for structural concrete and commentary Reported by ACI Committee 318, American Concrete Institute, Farmington Hills, Michigan

[11] ACI 352R-02 2002 Recommendations for design of beam-column connections in monolithic reinforced concrete structures, Reported by Joint ACI-ASCE Committee 352, American Concrete Institute, Farmington Hills, Michigan

[12] Elmasry M I S, Abdelkader A M and Elkordy E A 2018 An analytical study of improving beam-column joints behaviour under earthquakes *Conference: International Congress and Exhibition on Sustainable Civil Infrastructures: Innovative Infrastructure Geotechnology* 487- 500

[13] Rajasekhar K and Praveenkumar M 2017 Stress-strain behaviour of confined normal grade concrete *International Journal of Professional Engineering Studies* **8(3)** 271-282
[14] Faleschini F, Hofer L, Zanini M A, Benetta M and Pallegrino C 2017 Experimental behavior of beam-column joints made with EAF concrete under cyclic loading *Engineering Structures* **139** 81-95
[15] Kim J and LaFave M J 2007 Key influence parameters for the joint shear behaviour of reinforced concrete (RC) beam-column connections *Engineering Structures* **29** 2523-2539
[16] Sheela S and Geetha A B 2012 Studies on the performance of RC beam-column joints strengthened using different composite materials *Journal of Institution of Engineers (India)* **93(1)** 63-71

An experimental study on corrosion initiative of infused reinforcing bar in concrete

Pa Velci Shridevi[1,3], Ashok M[2], Daniel Ronald Joseph J[2] and Dhanalakshmi A[1]

[1]Department of Civil Engineering, P.S.R. Engineering College, Sivakasi, Tamilnadu, India
[2]Corrosion and Materials Protection Division, CSIR-Central Electrochemical Research Institute, Karaikudi, Tamilnadu, India.

[3]Email:velcipaulraj@gmail.com

Abstract: Durability of concrete is the essential property for the lifespan of concrete structures. Rusting of steel is the major factor responsible for the loss of durability property. In order to study the corrosion behaviour of three types of reinforcing bars like Thermo-mechanically treated bar(TMT), Corrosion resistant steel (CRS), Epoxy coated rebar(ECR) were experimentally studied. The reinforcing bars having 12 mm diameter were infused in 300×300×75 mm size of concrete slab samples were kept under room temperature and certain samples are spraying with 3% of NaCl solution to attain a simulated marine exposure condition. There are various techniques including Open circuit potential (OCP), Linear polarisation resistance (LPR), Electrochemical impedance spectroscopy (EIS) are used to monitorize the corroding performance of samples. On analysing the results of the electrochemical studies, it could be inferred that the efficient rebar to withstand the corrosive nature by improving the life expectancy of the concrete structure.

1. Introduction

Fortified structures have been disintegrated by corrosion which is considered as risky that lead to undesirable occurrences. The two parts of a concrete contains strong compressive capacity and less tensile capacity which is repay by steel having solid rigidity. Rusting might be named generalized and localized corrosion. In Generalized, it forms uniformly in all surface of the rebar called carbonation. For localized, it incorporates with both the chloride penetration results in pits and the pressure consumption leads to break [1].

The rust development in steel breaks the connection between the steel and concrete prompts parting and spalling. The pattern of corrosion develops rapidly which changes the organization and properties of the metal surfaces and the local condition, for example, development of oxides, changes in nearby pH level and so on. The damaged concrete having high permeable breaking point prompts a fast infiltration of chemicals. The calcium hydroxide in concrete contains an alkaline area with a pH scope of 12 to 13 which makes a slender film around the steel known as the passivating layer which keeps from erosion.

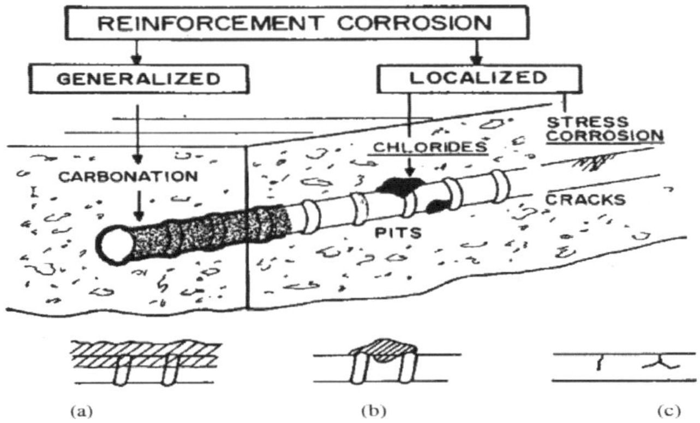

Figure 1. Types of corrosion of reinforcement: (a) carbonation, (b) chloride attack and (c) stress corrosion cracking.

Chloride fixation and carbonation causes the loss of passive layer affected by the outside condition and the nature of the concrete. The chloride particles infiltrate to concrete through splits and by scattering that are accessible in the water. When it enters the steel surface, they infiltrate the inactive film and respond with iron to form a solvent iron chloride [2]. Maximum chloride level in concrete at the strengthening steel significance is around 0.71 kg/m^3. The most extreme permissible estimation of 0.4% of Cl^- per weight of concrete was resolved just as the Cl^-/OH^- proportion of 0.6 [3].

When the concrete is presented to the air, the cycle of carbonation can start by carbon dioxide which gradually infiltrate and shows up the depth of infused steel, the passivating layer around the rebar is separated by the pattern of carbonation and makes the rebar to open to the destructive impacts of air and water. As carbon dioxide enters when the steel fortification consumes, it grows in volume prompts create power on the concrete to form breaks and spalls, causing the pace of disintegrate. As per D.A. Jones [4], it is commonly considered for rebar without surface coatings, erosion causes because of the failure of passive film on the rebar. Within the sight of a detached film, the erosion cycle may start from the metal surface and chloride particles. It is accepted that chloride particles respond at the specific spots where the inactive film is irregular or harmed on the steel surface.

Apparently, corrosion prompts the collapse of structures about 40%. It is prevented using different strategies by corrosion inhibitor, protected by cathode, use of polymer based material as a coated rebar for corrosion resistance.It can be postponed by the usage of inhibitors, which lessens either anodic oxidation or cathodic decrease, or the two responses [5].To beat the corrosion issues, the successful path is the utilization of covered rebar by coatings which frames a physical boundary from the avoidance of chloride particles that defers the inception of rust. Alblas and Van London [6] states that the impact of chloride sullying on the consumption of covered steel surfaces lesser contrasted with others.

The corrosive nature in fortified structures is incredibly high in the marine condition. A few components including high-temperature, moderately high dampness and high NaCl fixation have an impact in the corrosion of structures in the marine condition [7] and [8]. Additionally, the tedious substituting wet–dry introduction at high-recurrence in flowing zones will prompt a quickened corrosion of the strengthened structures [9]. As per Castro [10], A chance of getting quick data about the material crumbling if there should arise an occurrence of reenacted marine presentation is the utilization of quickened tests. These sort of tests should be possible through various strategies, for example, use of electrical potential estimations, cyclic wetting and drying in 3.5% sodium chloride arrangement which is a choice to do fast studies gave when the rusting kinetics do not change.

A few field and research center investigations completed to assess the corrosion execution. Electrochemical techniques are usually applied to evaluate and screen the rust execution of fortified concrete. The estimation of the corrosion potential decides the seriousness of the rust state of the fortified concrete [11]. Ha-Won Song and VeluSaraswathy [12] recommended that, Electrochemical impedance spectroscopy (EIS), a compelling strategy for depicting the highlights of different electrochemical instruments, can be utilized to figure out what commitments anode measures have made in these systems.

In this paper the corrosion conduct of three distinct sorts of rebar i.e., Thermo - Mechanically Treated Rebar (TMT), Epoxy Coated Rebar (ECR) and Corrosion Resistant Steel Rebar (CRS) which are implanted in concrete leveled out and reproduced situations. For quicken the corrosion execution the samples were showering with 3% of NaCl arrangement and drying at room temperature. The solid samples are then normally observed through the electrochemical investigations starting with the corrosion potential estimations with Open Circuit Potential (OCP), Linear Polarization Resistance (LPR) and Electrochemical Impedance Spectroscopy (EIS).

2. Experimental program

2.1. Materials

2.1.1. Concrete mixture proportions

As indicated by IS 1489 (Part I) – 1991, the Portland Pozzolana concrete of 53 grade and the properties of PPC utilized is introduced in Table 1, Coarse aggregates under Zone II of 12mm and 16mm size by extents of 40 % and 60 % and fine total under 4.75 mm size of Zone I adjusting to IS 383 – 2016 alongside proportion of 1 : 2.02 : 3.5 with water concrete proportion of 0.5 were used to make the samples. The 28 days compressive quality of cement was got as 36.9 MPa. The particle size distribution is introduced in Fig. 2. The concrete mixture proportions used to cast the samples are given in Table 2.

Table 1. Properties of PPC.

Consistency	Initial setting time	Final setting time
32%	45 mins	8 hours

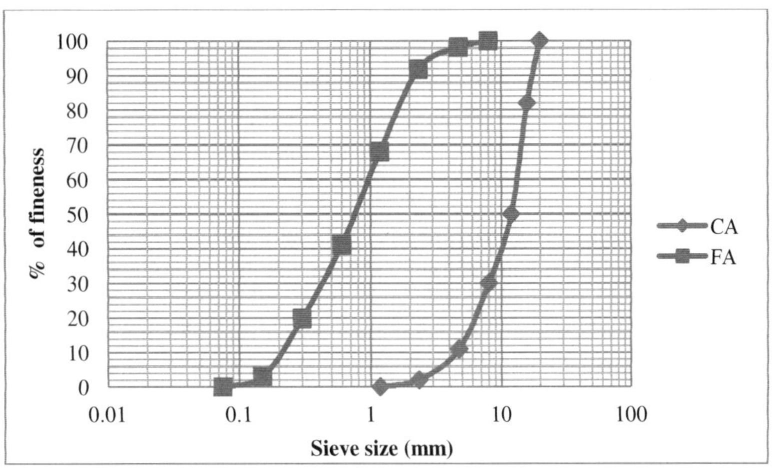

Figure 2. Fineness modulus of aggregates.

Table 2. Concrete mixture proportions.

Constituents	Units	Quantity
Cement	Kg/m^3	340
Coarse aggregate	Kg/m^3	1185
Fine aggregate	Kg/m^3	687
Water	Kg/m^3	180.2
Water cement ratio	-	0.5

2.1.2. Steel rebar

The rebar of TMT, CRS and ECR was adjusted to ASTM A615/A615M-18 [17], ASTM A1035/A1035M – 19 [18], ASTM A775/A775M – 17 [19] having 12mm measurement of Fe500D are utilized. The rebar of 250 mm length was bound to a copper wire to get further investigations. Initial lengths and weights of steel rebar were estimated using a balance with 0.01 g sensitivity.

2.2 Sample Geometry

The slab of 300 x 300 x 75 mm size is cast alongside steel bar infused from the depth of 25mm from the top. Each rebar is ought to be cleaned by pickling using sulphuric acid to eliminate the rust on the rebar. The isometric and top view of sample specimen is given as figure 2.

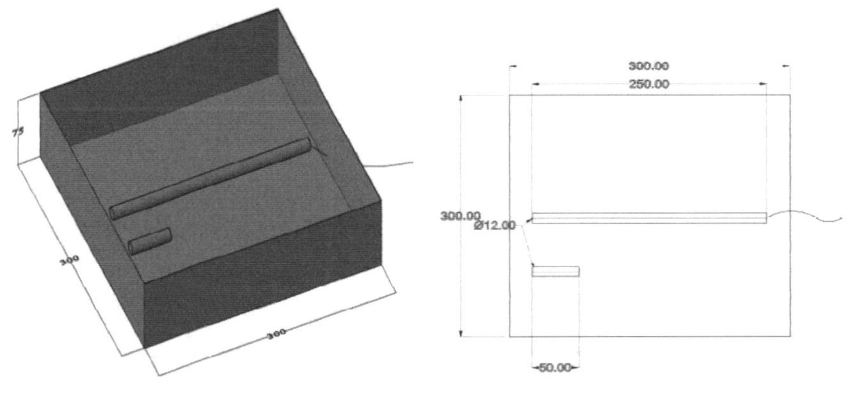

Figure 3. Isometric view and Top view of specimen.

2.3 Exposure Conditions

The samples with infused steel bars were exposed to room temperature and also some more samples are kept to a cyclic state of four days of wetting by showering 3% NaCl arrangement followed by three days of drying in the outside. This redundant exchanging wet–dry introduction at high-recurrence will quicken the erosion commencement which is fundamental to screen just as to comprehend the corrosion condition of the concrete structure upkeep.

3. Assumptions of corrosion of rebar in concrete

3.1 Open Circuit Potential (OCP)

The values had been recorded using an electrode called saturated calomel electrode (SCE) along with a multimeter. Then the potential values were compared with the ASTM C 876 [15] guidelines as follows:

IOP Conf. Series: Materials Science and Engineering **989** (2020) 012024 doi:10.1088/1757-899X/989/1/012024

Table 3. Potential values.

(mV vs. SCE)	Corrosion condition
Less than -426	Extreme
Less than -276	High
-126 to -275	Moderate
Greater than -125	Low

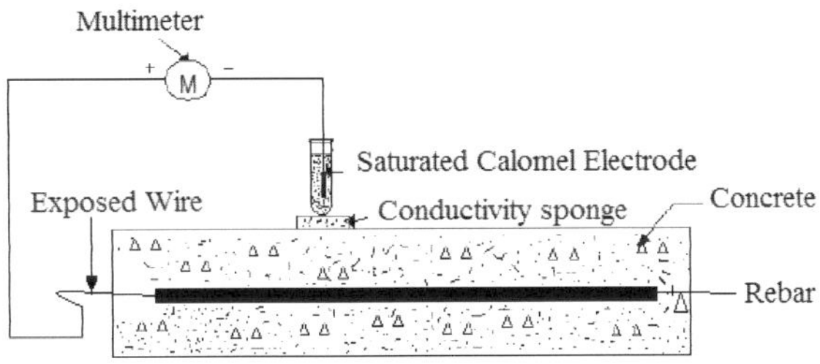

Figure 4. Open circuit potential apparatus setup.

3.2 Linear Polarization Resistance (LPR)

It is a basic dc process to get a quick estimate of corrosion rate having a sweep normally goes in between ±20 mv at a scan rate of 0.2mv/s for which the curve turns out to be practically straight. the icorr values can be attained by using the formula(1):

$$Icorr = \frac{B}{Rp} \quad (1)$$

Table 4. Corrosion current versus state of the rebar.

Corrosion current (Icorr) μA/cm^2	state of the rebar
Less than 0.1	Passive
0.1 to 0.5	Low – moderate
0.5 to 1.0	Moderate –high
Greater than 1.0	High

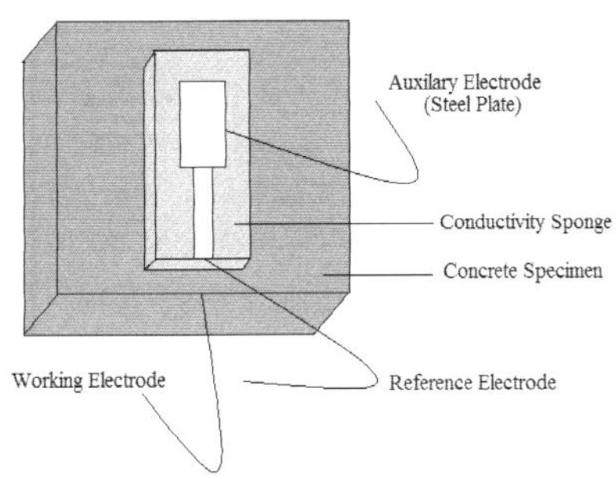

Figure 5. Electrode alignment for electrochemical study.

3.3 Electrochemical Impedance Spectroscopy (EIS)

EIS is a method for resolving the resistive capacity of steel which take into account for forecast the corrosion rates. An A.C voltage of around 10-20 mV is supplied to the rebar inserted in concrete. The reaction to an A.C input is a perplexing impedance that has both real (resistive) and imaginery(capacitive or inductive) segment z and z" as shown in fig 6. The imaginery impedance with the real impedance gives a semicircle with a diameter equivalent to Rt.

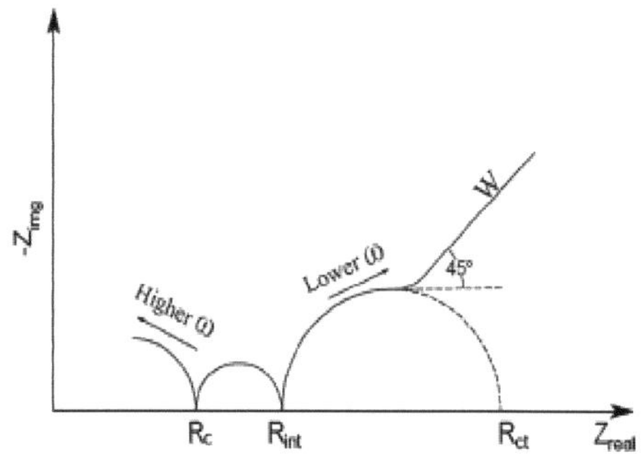

Figure 6. Nyquist Plot for steel in concrete.

4. Results

4.1 Corrosion Potential Measurements

The variety of the corrosion possibilities with no of days is graphically given in fig 7. At first the potential qualities are not firm, by following barely any weeks it dropped to more electronegative qualities. The likely readings under controlled condition for the TMT, CRS, Epoxy covered rebar is in the scope of - 133.39, - 100.21 and - 344.9 mV/SCE and for alternative condition for the rebar are in the scope of - 199.67 and - 189.8 mV/SCE individually.

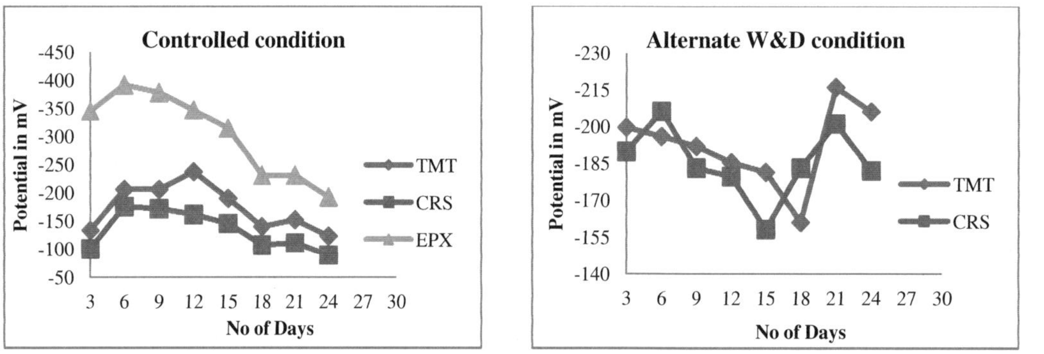

Figure 7. Corrosion potential versus No of Days.

4.2 Linear Polarization Resistance Study

A Nonlinear reaction began close to - 20 mV polarization for all the samples. Typically the bend shows a straight conduct having polarization scope of 5mV to 20mV. If the resultant Icorr value is less than 0.1μA/cm^2, then the concrete is in passive condition when it reaches more than 1.0 μA/cm^2, then the specimen undergoing high corrosion rate. Nearly the acquired outcome from the estimations, Icorr values portrays that the samples are in detached condition which is shown in table 5. On comparing the normal condition the corrosion rate and current for both the TMT and CRS rebar are lesser than Epoxy covered rebar.

IOP Conf. Series: Materials Science and Engineering **989** (2020) 012024 doi:10.1088/1757-899X/989/1/012024

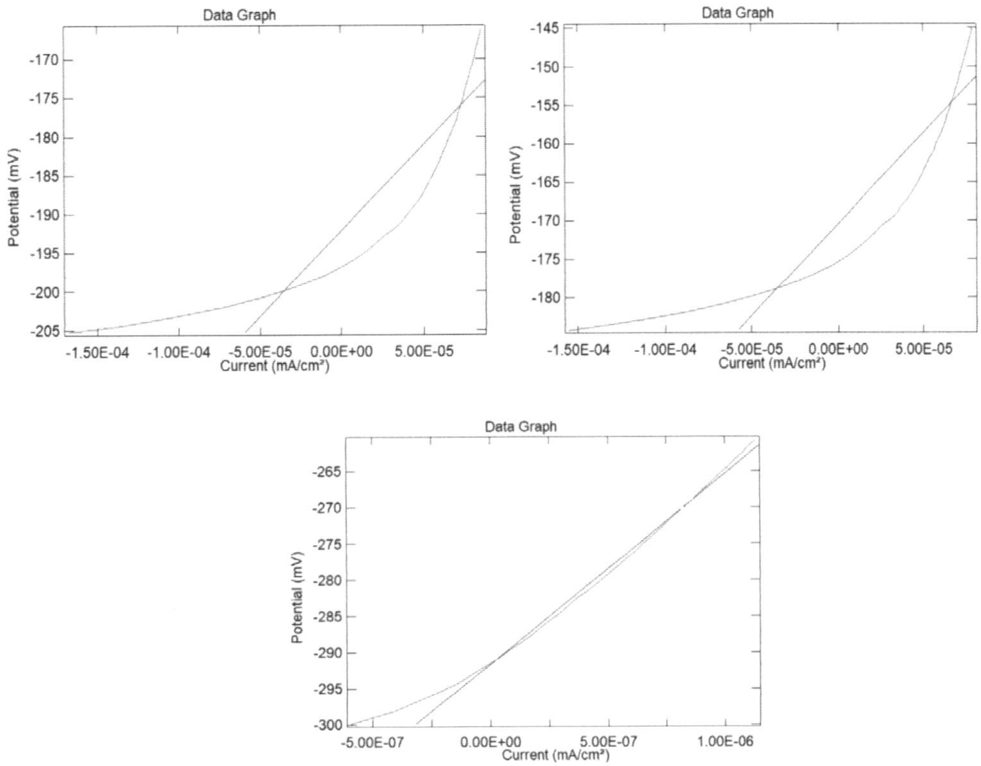

(a)

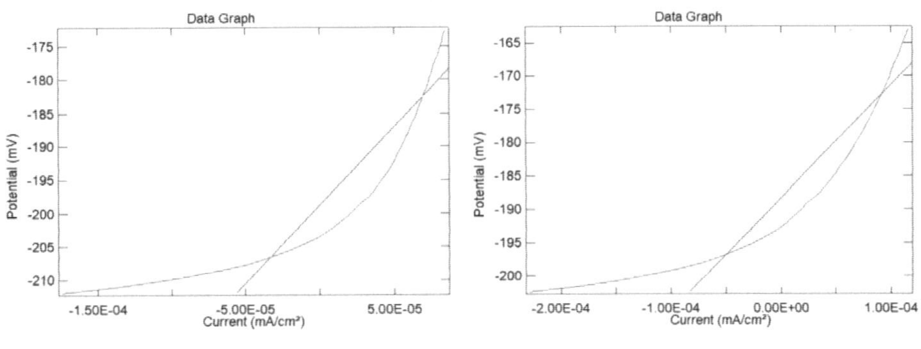

(b)

Figure 8.LPR Curve (a) Controlled condition (b) Alternate wet and dry condition.

Table 5. Corrosion current from EIS study.

CONTROL SPECIMEN	
Rebar	Icorr (mA/cm^2)
TMT	8.94E-05
CRS	4.85E-04
ECR	2.66E-06
ALTERNATE WET & DRY SPECIMEN	
Rebar	Icorr (mA/cm^2)
TMT	4.08E-04
CRS	5.66E-04

4.3 Electrochemical Impedance Spectroscopy study

The arrangement of bends in EIS results turned out to be progressively perplexing coming about because of the heterogeneity of the samples. The distinguished bends were detached and identified with each of the phenomena and a local investigation was performed, which improves the exactness. From this Electro chemical examination the charge transfer resistance (Rct) and double layer capacitance (Cdl) was obtained as given in table 6. From the EIS study, it shows that the estimations of TMT and CRS are lesser when contrasted to the Epoxy covered rebar. Subsequently the coated rebar has less inclinations to corrode quickly than the other kind of rebar. The highest passivation capability of epoxy coated rebar is lower than the other two kinds of rebar.

Table 6.EIS analysis.

Control Specimen				
Bar	Corr. rate (mm/year)	Corr. rate (mils/year)	Rct (ohms.cm²)	Cdl (F)
TMT	1.04E-03	4.08E-02	2.92E+05	3.30E-05
CRS	5.62E-03	2.21E-01	5.38E+04	3.28E-05
ECR	3.08E-05	1.21E-03	9.80E+06	4.62E-08
Alternate Wet & Dry Specimen				
Bar	Corr.rate (mm/year)	Corr.rate (mils/year)	Rct (ohms.cm²)	Cdl (F)
TMT	4.73E-03	1.86E-01	6.39E+04	2.16E-05
CRS	6.56E-03	2.58E-01	4.61E+04	2.38E-05

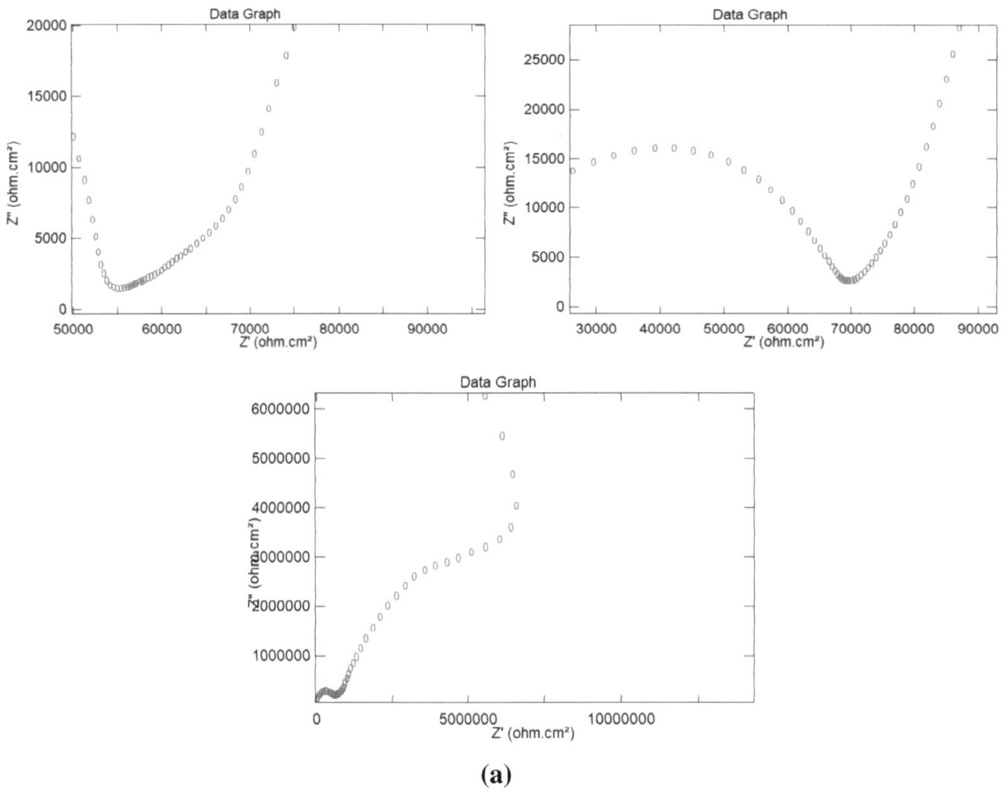

(a)

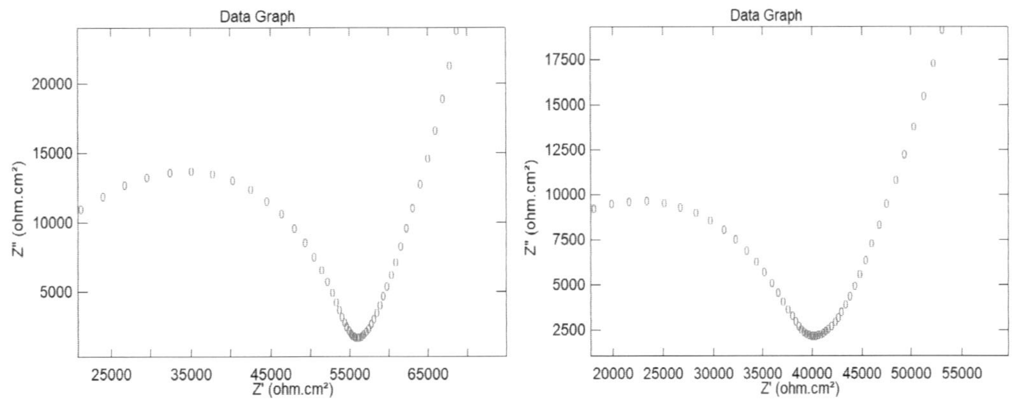

(b)

Figure 9. AC Impedance curve (a) Controlled condition (b) Alternate wet and dry condition.

5. Conclusion

The sturdiness execution of cement under controlled with alternate wet and dry condition by showering with 3% of NaCl arrangement regarding starting the consumption cycle utilizing electrochemical method was tentatively examined and the conduct of epoxy rebar under wet and dry couldn't assess inside the particular time frame. The quickened corrosion method is a proficient and solid approach to achieve corrosion of steel inside a limited capacity to focus time. Following are the outcomes acquired,

i) The corrosion execution of the rebar doesn't give a consistent rate in electrochemical cycle with fluctuating in rate alongside time. This is depicted by high starting changes in the open circuit potential readings which inevitably balanced out to a steady rate. The OCP drop from latent to dynamic state was most minimal for ECR in examination with the TMT and CRS. The total example of the corrosion potential clarifies the consumption rate which is marginally expanding for certain time.

ii) On comparing the rebar corrosion current worth ECR is less contrasted with other rebar that it will take time to attain corrosion than others. Polarization Resistance can be decided by the Icorr value including Stern – Geary constant [12]. The LPR shows that Rp range for ECR is higher than TMT and CRS.

iii) The EIS strategy was utilized to evaluate the corrosion states of solid samples regarding Icorr in mm/yr and mils/yr. The outcomes in EIS study expresses that Rct and Cdl estimations of ECR is vary from other rebar by increment in Rct qualities and abatement in Cdl values compared with the TMT and CRS rebar.

6. References

[1] Andrade C 2007 Corrosion of steel reinforcement *State of the Art in Science and Engineering* **28** 185

[2] John Fraczek1987 A review of electrochemical principles as applied to corrosion of steel in a concrete or group environment *Corrosion, Concrete and Chlorides - Steel Corrosionin Concrete: Causes and Restrains* American Concrete Institute

[3] Dillard JG, JD Glanville, WD Collins, RE Weyers and IL Al-qadi 1993 Concrete bridge protection and rehabilitation *Strategic Highway Research Program* National research council Washington DC

[4] DA Jones 1995 Principles and prevention of corrosion (Macmillan : New York)

[5] Schiessl P and R Bakker 1988 Measures of protection corrosion of steel in concrete *The International Union of Testing and Research Laboratories for Materials and Structures* **5** 73
[6] BP Alblas and AM Van London 1997 Protective coating (Europe)
[7] Cao M, Liu L, Yu Z, Fan L, Li Y and Wang F 2019 Electrochemical corrosion behaviour of CaO_2al alloy under an accelerated simulation marine atmospheric environment *J. Mater. Sci. Tech.* **35** 651–659
[8] Lu S and Ba H 2010 Monitoring the service condition of chloride contaminated cement mortar – *Sensors* **10** 4145 – 58
[9] Mu X, Wei J, Dong J and Ke W 2014 In situ corrosion monitoring of mild steel in a simulated tidal zone without marine fouling attachment *J. Mater. Sci. Tech.* **18** 1043 – 50
[10] Castro P, Weva L and Balancim M 1997 Corrosion of reinforced concrete in a tropical marine environment and in accelerated tests *Construction and Building Materials* **11** 75-81
[11] Khaled KZ, Vaca – cortés E, Jirsa JO, Wheat HG and Carrasquillo RL 1998 Corrosion performance of epoxy-coated reinforcement – beam tests (Texas USA) **3** 126-135
[12] Ha – Won Song and Velu Saraswathy 2007 Corrosion monitoring of reinforced concrete structures – a review *International J. Electrochem. Sci.* **2** 1- 28
[13] Moreno M, Morris W, Alvarez M G and Duffo G S 2004 Corrosion of reinforcing steel in simulated concrete pore solutions: Effect of carbonation and chloride content *Corr. Sci.* **46** 2681– 99
[14] Cheng A, Huang R, Wu JK and Chen CH 2005 Effect of rebar coating on corrosion resistance and bond strength of reinforced concrete *Construction and Building Materials* **19** 404–412
[15] ASTM C876-09 2009 *Test method for corrosion potentials of uncoated reinforcing steel in concrete* http://www.astm.org/standards/c876.htm
[16] Sathiyanarayanan S, Panjali Natarajan, Saravanan K, Srinivasan S and Venkatachari G 2006 Corrosion monitoring of steel in concrete by galvanostatic pulse technique *Cement and Concrete Composites* **28** 630 – 637
[17] ASTM A615/A615M – 18 2018 *Standard specification for deformed and plain carbon steel bars for concrete reinforcement* http://www.astm.org/standard/a615.htm
[18] ASTMA1035/A1035M – 16b *Standard specification for deformed and plain, low-carbon, chromium, steel bars for concrete reinforcement* https://www.astm.org/standards/a1035.htm
[19] ASTMA775/A775M–19 2018 *Standard specification for epoxy coated reinforcing bars* http://www.astm.org/standard/a615.htm
[20] Vedalakshmi R, Rajagopal K and Palaniswamy N 2008 Long term corrosion performance of rebar embedded in blended cement concrete under macro cell corrosion condition *Construction and Building Materials* **22** 186 – 199
[21] H-W Song, Saraswathy V, Muralidharan S, C – H Lee and Thangavel K 2009 Corrosion performance of steel in composite concrete system admixed with chloride and various alkaline nitrites *Corrosion Engineering Science and Technology* **44** 408-415
[22] Shigang Dong, Bing Zhao A, Chang Jian Lin AC, Rong Gui Dua, Rong Gang Hua and Gregory Xiaoge Zhang D 2012 Corrosion behavior of epoxy/zinc duplex coated rebar embedded in concrete in ocean environment *Construction and Building Materials* **28** 72–78
[23] Daoming Shen 2017 Electrochemical impedance spectroscopy study on corrosion inhibitor for reinforced concrete *Inter. J. Electrochem. Science* **12** 4183 – 92
[24] Wei J, Dong JH and Ke W 2013 Corrosion evolution of scaled rebar in concrete under dry and wet cyclic condition in 3.5% NaCl solution *Inter. J. of Electrochem. Sci.* **8** 2536 – 50
[25] Bayezid Baten, Jihan Hasan MD, Baishakhi Bose and Tanvir Manzur 2017 An efficient experimental setup for rapid assessment of rebar corrosion in reinforced concrete structures *International Journal of Advances in Mechanical and Civil Engineering* **4** 83-88

IOP Conf. Series: Materials Science and Engineering **989** (2020) 012025 doi:10.1088/1757-899X/989/1/012025

An experimental study on assessing the corrosion performance of steel reinforcement for the durability of concrete

Naveen Kumar V[1,3], Daniel Ronald Joseph J[2], Ashok M[2] and Suresh Kumar MP[1]

[1] Department of Civil Engineering, K.S. Rangasamy College of Technology, Tiruchengode-637 215, Tamil Nadu, India

[2] Corrosion and Material Protection Division, CSIR - Central Electrochemical Research Institute, Karaikudi-630 006, Tamil Nadu, India

[3]Email: narayansrinivasan.structures@gmail.com

Abstract. Durability of structures made of steel-reinforced concrete is greatly to scale back the speed of moisture transport or the ingress of aggressive ions. An experimental study to assess the initial corrosion behaviour of Thermo-Mechanically Treated rebar (TMT), Corrosion Resistant rebar (CRS) and Fusion bonded Epoxy Coated rebar (EC) under controlled condition (CC) and in simulated marine environment condition (SME). Rebars of 12 mm dia with a span of 250 mm are enclosed in specimens having and length and breadth of 350 mm and a thickness of 138 mm then initial study was carried out. The corrosion potential is noted throughout the specified duration and once in three days of observing the corrosion potential gives the likelihood of corrosion rise. The A.C Impedance and Polarization resistance methods are used to study the corrosion rate in the surface of the rebar. The polarisation resistance (R_p) of TMT rebar is found to be 1.8 times but that of CRS rebar. A.C Impedance used for EC showed that epoxy coating is harmed and affected because of chloride attack. Initial quantitative study from this research gives clear interpretation of rust production which can be utilized in fragility of the structure.

1. Introduction

Early negligence of reinforced concrete assembly happens fundamentally because of premature corrosion of rebar. The two most persistant action occurring foremost in rebar corrosion are effervescence of concrete and chloride intrusion which supply an extremely decaying environment for the implanted steel. The corrosion by-product so found are further more generous besides the rear constituents. The progress of corrosion in reinforcing steel, expands and occupies a volume of about 6–10 times greater than that of steel which results in the formation of cracks leads to failure of structures. As, when prevailing corrosion rate quantification methods are practical for speedy monitoring of structures, which enhance fairly high cost and if inappropriate examining is carried out over prolonged duration. The corrosion of rebar in concrete is a significant issue from the outlook of protection and economic restraint which influences the feasibility of the framework configuration.

It is noticed that constituents of reinforcement bar plays a remarkable role in critical chloride effectiveness that commences the decay of rebar. The amount of chloride ions need to commence the decay of rebar in concrete matches with 0.10% dissolved chloride ion on cement weight [1,2] some corrosion resisting agents influences the protective film on the enclosed rebar which commences the decay of rebar in place effervescence starts which eventually decreases the decay of rebar. This phenomenon can be prevented by a number of techniques by providing sufficient cover to concrete, use of migratory corrosion inhibitors such as calcium nitrate, sodium nitrate, sodium monoflurophosphate etc, using concrete sealers as silane sprays, acrylic sealers and linked oil emulsions and by adding silica fume to concrete which shows carbonation and keeps chlorides out which reduces rebar corrosion and it is obviously detected that addition of silica fume uncertainly imparts to lower the porosity of concrete to chloride ions [3]. Daoming Shen suggested that steel becomes inactive when chloride to nitrite ratio should not be equal to greater than 1 [4-6]. It is established that decay of rebar on the anode area mostly rely on the oxygen supply and chloride content and results showed that proliferation under macro cell corrosion is due to the consequence of w/c ratio, cover depth etc [7]. Vedalakshmi et.al suggested that decay of rebar in PPC and PSC had more decaying resistance in 30 MPa concrete [24]. In the vigorous corrosion condition, the high chromium steel normalized beneath ASTM A1035/A1035M-16b having a mild carbon content of 0.1% and chromium content of 9-11% [9] provided to have 1.5-2 times higher confrontation to charge transfer in vital condition to corrosion. The durability of RC structures in zones were moisture and chloride mass are excessive like marine environment, the utilization of EC and HC is anticipated [25]. An imperfection less overlay is extremely impenetrable to the dispersing of chloride and other ions. Condensation yet, infiltrates past coating but the amount of corrosion rate irritates only when chloride ions are nearer to the junction of covering [11] and visual inspection reveals that no decay of rebar was found below galvanized coating when pH is greater than 13.3 and found more active to rear a long time shielding [12-14] were the epoxy/galvanized binary cover over the rebar possibly flatters a capital approach in corrosive habitat construction [15]. Erdogdu et al. concluded that EC when subjected to couple of years under chloride habitat, 1% and 2% impairment experiences minor decay of rebar and no severe symptoms of corrosion is detected [16]. Interpenetrating Polymer Network-coated reinforcement possess sustainable firmness, more extended life on concrete which experiences greater corrosion resistivity [17]. Venkatesan et al. concluded that at different environmental conditions CPCC coated rebar performed well when tests conducted at atmospheric level, high tide level and at sea floor level [18].Song et al. studied that by mixing of PPC+PSC along with 0.3% NaCl and with 0.2% of alkaline nitrates showed that the emergence of this kind intricate much lower permeability and decay of rebar [19], the corrosion rate of the reinforcement steel is inversely proportional to the polarization resistance [20] were existing A.C Impedance is a well-founded technique to evaluate the corrosion rate of rebar beneath epoxy coating steel which showed that R_p for TMT, HC is lower than that of EC [25].

In the following study, three types of concrete reinforcements, Thermo-Mechanically Treated Bars (TMT), High Chromium Steel Bars (CRS) and Fusion Bonded Epoxy Coated Reinforcements (FBECR) are implanted in the concrete specimens and corresponding corrosion rates are monitored in a confined environment as well as in environment supposed to high salinity by sprinkling the surface with 3% of NaCl solution. Then the specimens are subjected to Electrochemical Techniques, for corrosion monitoring by Open-Circuit Potential (OCP), Linear Polarisation Resistance (LPR), A.C Impedance. and by Gravimetric loss of the reinforcement, however the A.C Impedance which demonstrates to be the utmost acceptable application to monitor the corrosion effect of EC, were the corrosion of elemental steel (Iron), was utilized.

2.Experimental program

2.1. Materials and Mix proportions

The considered Bare TMT, CRS and Epoxy Coated confirming to A615/A615M – 18 [8], A1035/A1035M – 16[9], A775/A775M – 19 [10] respectively. Incipient weight of rebar is deliberated with the help of physical balance having a sensitivity of 0.01g which when done after treating the rebar with pickling solutions. Concrete specimens were casted for a mix proportion having a characteristic compressive strength of 30MPa after 28 days of curing. Ingredients of mix proportion are given in Table 1. Type I Portland Pozzolana Cement of 53 grade confirming to IS 1489(Part I):1991, Coarse aggregates with maximum sizes of 16mm and 12mm are used in fractions of 60%,

40% with a fineness modulus of 5.62 and Fine aggregates passing through 4.75mm sieves having a fineness modulus of 4.82, both confirming to IS 383:2016 is used for concrete specimens. The grading of coarse and fine aggregate are shown in figure1. Potable water was used for casting the concrete.

Table 1. Details of concrete mix proportions.

Concrete Grade	w/c Ratio	Cement (kg/m^3)	Water (kg/m^3)	Fine Aggregate (kg/m^3)	Coarse Aggregate (kg/m^3)
M30	0.5	340	180.2	687	1185

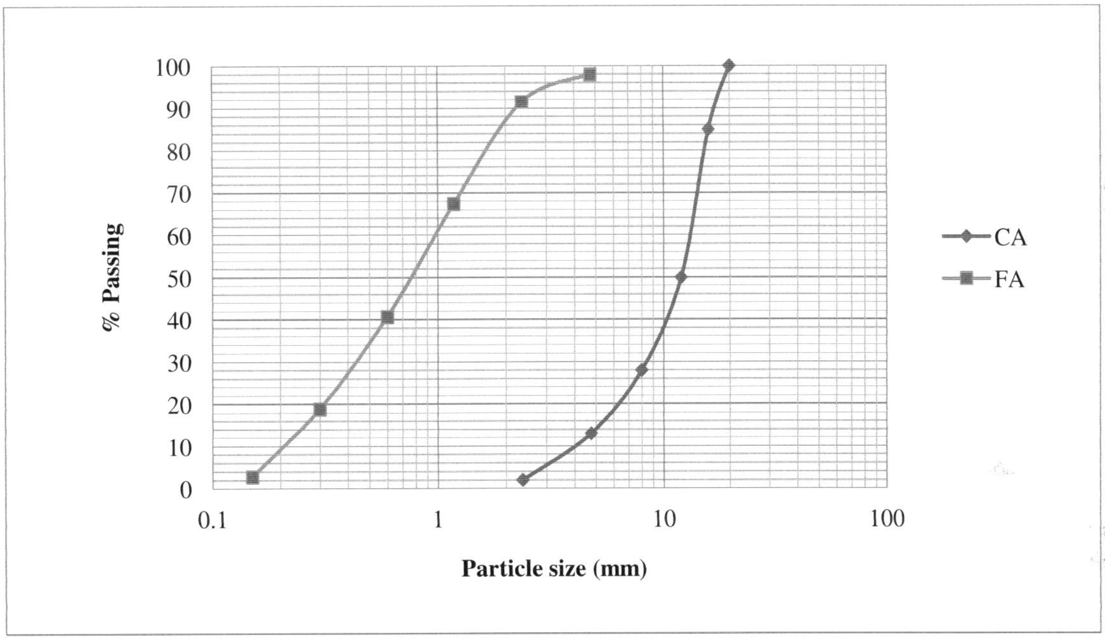

Figure 1. Grading of coarse and fine aggregate.

2.2 Sample configuration

Twenty-four concrete slabs with dimensions of 350mm×350mm×138mm were cast. Reinforcement bars of 12 mm diameter were cut at 250mm length and 50mm length, is implanted in centre of concrete slabs. Cover concrete thickness of 50 mm was provided from the top of each specimen. Sample of three numbers in each type of rebar is casted to make sure the fickleness in ingredient effects and evaluating methods is ensured properly. Specimens casted is identified by means of punching number, i.e., TMT, Corrosion Resistant Steel and Epoxy coated steel followed by a number from 1 to 3, e.g., first TMT specimen is identified as 1, and along with others. To setup the interdependence for electrochemical measurements a copper wire is affixed. To setup electrical connection in EC rebar, a 2mm wide open of epoxy is eradicated in the circumference for suppressing moisture entry and chloride admittance M-Seal is pertained over the rebar surface. The rectangular configuration permits constant dissolution on rebar by exterminating the mistakes and in assessing of polarised steel area throughout the electrochemical study. Rebar is perfectly implanted in specimens to ignore the effect of surroundings which causes corrosion. In advance of implanting the rebar, the oxidised lamina is detached from all the rebar by scouring the surface with dil. Sulphuric acid and then allowed to dry. After measuring the incipient weight, electrical leads made of copper is winded on top of rebar which enables the rebar perfectly implanted in the concrete specimen and only the electrical leads is unveiled externally through which electrochemical measurements are done. Equivalent slab samples are casted. The specimens are cured for 28 days after demoulding.

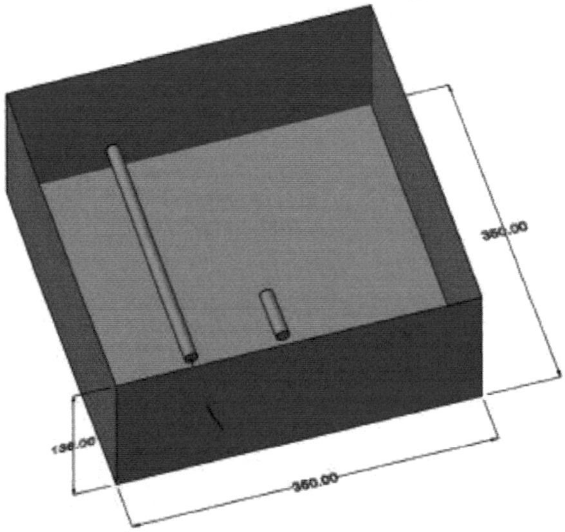

Figure 2. Concrete test slab for corrosion monitoring.

2.3 Method of Exposure

2.3.1 Controlled Environment Exposure

The slab samples are influenced to normal environment condition, having a humidity of $27^0 \pm 2^0$C. The surface of the specimens are wetted with potable water atleast 1 hour prior to the observations taken, so that the passivity if the steel is removed considerably with the variation in time. Observations are taken once in three days for the specimens and the loss in passivity of the specimens are observed.

2.3.2 Simulated Marine Environment Exposure

The specimens were subjected to alternate wetting and drying in 3% NaCl solution. The NaCl solution was sprayed over the surface of the specimens at an interval of eight hours, and the readings are observed once in three days. Throughout the elucidation of salt infusion, chloride ions tends to spread past the concrete cover and motivates the rebar to corrode.

3. Testing methods

3.1 Open Circuit Potential (OCP) measurement

The steel reinforcement implanted in the specimen is observed periodically in case of saturated calomel electrode (SCE) as reference. OCP measurements were made using a multimeter, were the positive terminal was connected to the reference electrode and three readings are taken very close to the rebar and the variation should not be more than ±10 mV. OCP estimates only the probability of corrosion and it does not specify the corrosion rate. ASTM C876 relates the OCP to the likelihood of the rebar corrosion in concrete specimens. Table 2 relates the diversity correlated with lowest relevancy to extreme corrosion risk.

IOP Conf. Series: Materials Science and Engineering **989** (2020) 012025 doi:10.1088/1757-899X/989/1/012025

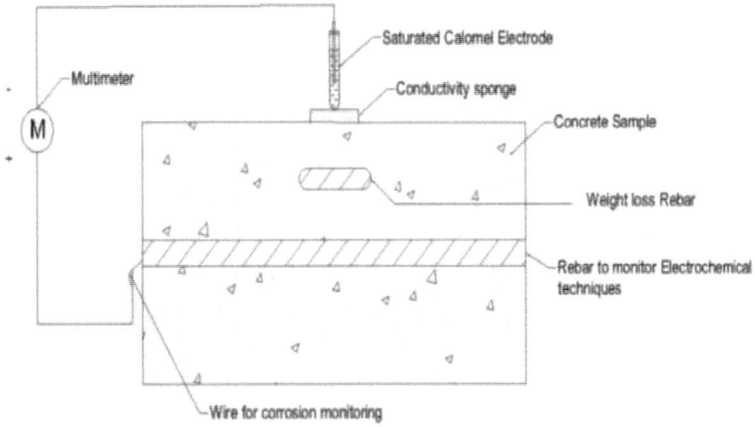

Figure 3. Schematic representation of Open circuit potential (OCP) measurement.

Table 2. Corrosion condition related with OCP measurements.

Open circuit potential (OCP) values		Corrosion condition
mV vs. SCE	mV vs. CSE	
≯ -426	≯ -500	Extreme corrosion
≯ -276	≯ - 350	Highest relevancy (90%)
< -126 to -275	< -350 to -200	Intermediate relevancy
> -125	> -200	Lowest relevancy (10%)

3.2 Linear Polarization Resistance (LPR) technique

Linear polarisation resistance measurements are taken at regular intervals once corrosion initiation was determined based on OCP values. LPR quantification is directed with the help of ACM field machine (UK). The probe has a three electrode fabrication were rebar implanted in concrete which functions as working electrode (WE), a reference electrode (RE) is the saturated calomel electrode is used and a stainless steel plate which acts as a counter electrode (CE) were used. During testing the samples are properly wetted with potable water, this was to make sure the accessibility of oxygen in the junction of steel and concrete.

LPR in this study is obtained, when the rebar is polarized in the potable scale of ±20 mV from OCP with a sweeping frequency of 0.2mV/s is applied on the steel rebar and the current retaliation is recorded. The polarisation resistance R_p, from the rebar implanted in concrete is acquired from the slope of potential curve. With the help of the following relationship (1) in the corrosion current Icorr (μA/cm^2) and rate of corrosion is acquired.

$$Icorr = \frac{B}{Rp} \tag{1}$$

where B is the Stern-Geary constant = 52 mV

3.3 Electrochemical Impedance Spectroscopy (EIS) technique

The guideline of this method is to put in an interchanging indicator of lower extent (5-20 mV) to the electrode (rebar) implanted in the electrolyte (concrete). The EIS works in the frequency domain (resistance, capacitance and inductance), in which impedance, $Z = \frac{\Delta E}{\Delta l}$, is measured using Ohm's law by applying the selected sine curve, $V(t) = V_0 \sin(\omega t)$at a stipulated sweep of recurrence, $f = \frac{\omega}{2\pi}$ is related to the rebar and the transient response, I□ $= I_0\sin(\omega t+\theta)$, which has a phase shift of θ from the applicable electric potential difference is observed.

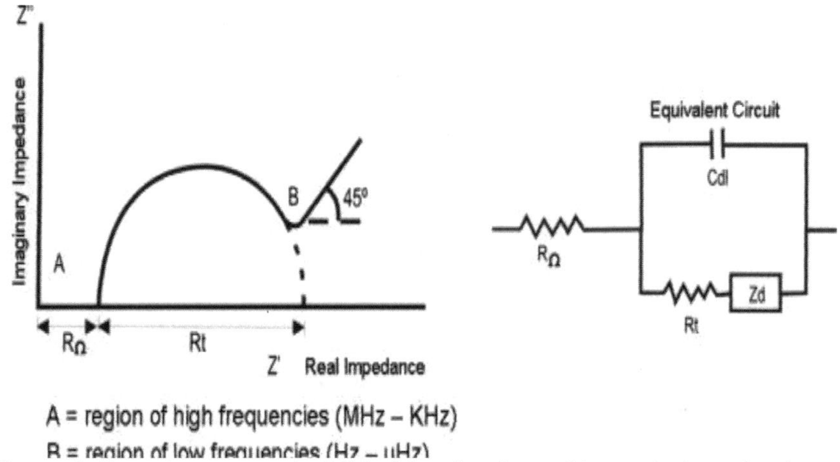

Figure 4. Corrosion process in RC by Nyquist plot and its equivalent circuit.

In the present study, the EIS sweep is conferred in Nyquist plot, showing the electrical resistance in argand plane of cartesian co-ordinates, i.e., Z = Z′+ jZ′′, where Z′ and jZ′′ show the existent and non-existent segments strained on the polar co-ordinates and abscissa of the graph appropriately. Further description called as Bode diagram, showing the approximation of the electrical resistance modulus and phase shift as a sweep of recurrence approximation. Depending upon the configuration of corrosion process in concrete, the illustrative of classic Nyquist plot for the EIS sweep in reinforced concrete is given in fig. 4. To acquire the EIS sweep, an electric potential difference of 10 mV is pertained with a sweep of recurrence of 0.1 Hz to 30kHz. Electrochemical Impedance spectroscopy is examined as a practical non-corrosive method for measuring the corrosion of reinforcement implanted in concrete. This method is extremely appealing since, worn in an extensive sweep of recurrence which gives elaborate details regarding the appliance energy of electrochemical responses.

4. Results and discussion

4.1 Open Circuit Potential

The OCP of TMT, CRS and EC under controlled condition (CC) and simulated marine environment condition (SME). OCP decreased during the first few days before the concrete sample becoming stable in case of both CC and SME, then later on dripped to increase the electronegative values. In the LPR study, the OCP sample for both CC and SME varied from -193mv/SCE to -224.31 mV/SCE for TMT, from -164.23mV/SCE to -213.68mV/SCE for CRS and from -345.07mV/SCE to -415.42mV/SCE for EC. The pattern of corrosion potential for CC is influencing lesser corrosion susceptibility when compared with SME as per fig. 5 (a) and (b). The sensitivity of corrosion on rebar is assumed from OCP values in the succeeding manner from high level to low-level: EC, TMT and CRS.

The standard drip in OCP estimates from non-resistant to resistant state is less in case of CRS rebar among the three types of rebar used. This is in fact shows that samples under SME have the ability to undergo an excessive level of corrosion rate after termination of investigation. The effect of perviousness due to corrosion sensitivity of concrete samples under CC and SME is explained in former segments.

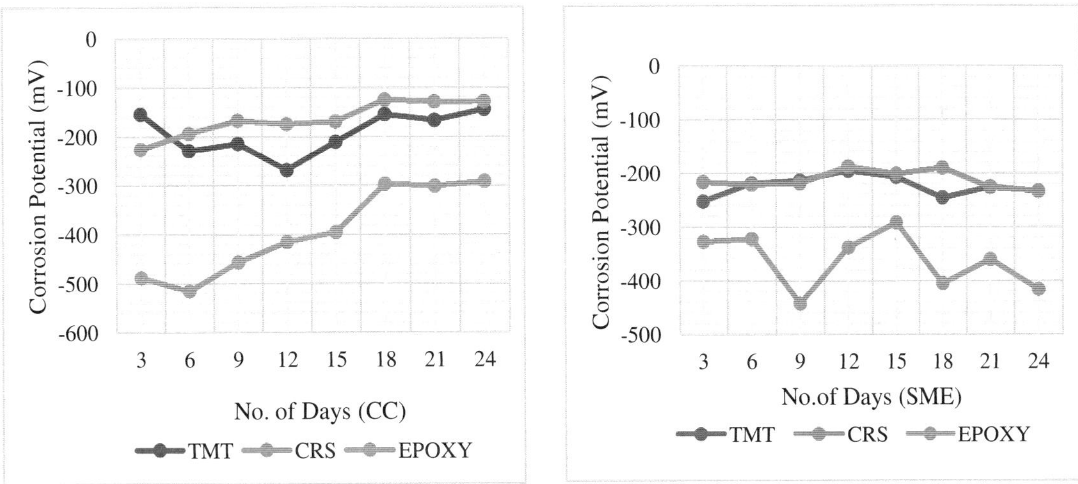

Figure 5. OCP in Controlled and Simulated Marine Environment Condition.

4.2 Linear Polarization Resistance

The LPR curves shown in the fig. 6 shows an apparently lined behaviour in the complete dissolution sweep; were, a divergent reaction of -20 mV dissolution is given at the start of the experiment for all the three types of rebar embedded specimens. In case of LPR study, the unique line distribute of the curve is analysed to find the dissolution resistance R_p. For LPR study, a 5 mV to 20 mV dissolution is endorsed and the lined distribute in this range can be analysed to acquire the dissolution resistance.

Since the sweeping frequency commences at 0.2 mV/sec, there can be a little variation of current connecting the stainless steel plate which acts as counter electrode and water around this voltage acts as a capacitor. The average values of R_p over three specimens be 92.2 kΩ.cm^2, 33.76 kΩ.cm^2 and 20.9 kΩ.cm^2 for TMT, CRS and EC accordingly, during the initiation of the study.

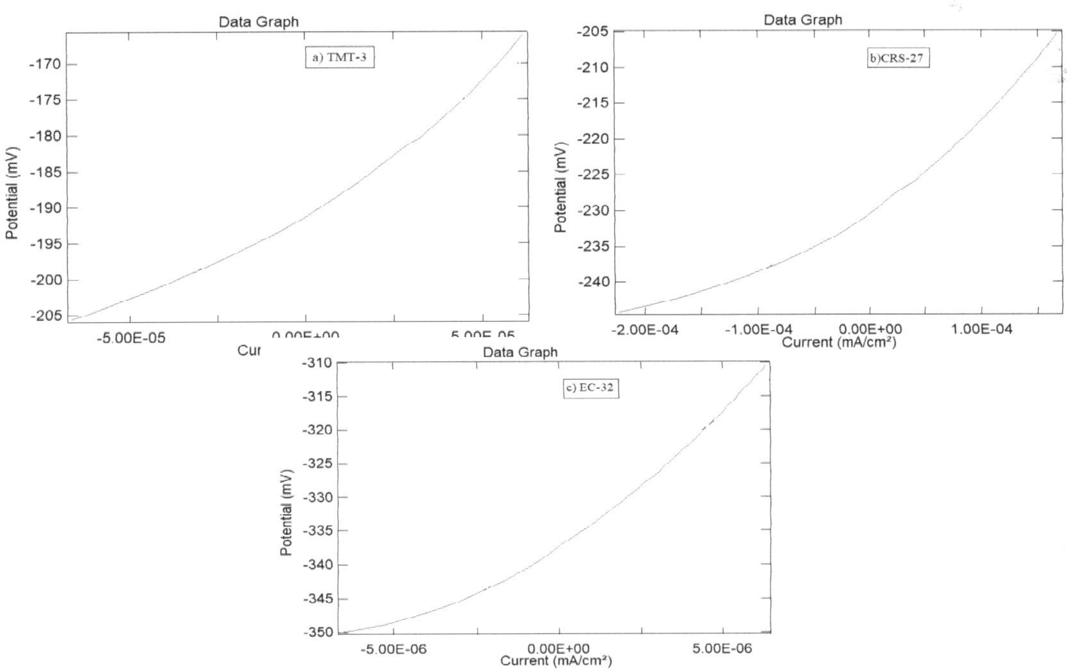

Figure 6. Linear Polarization Resistance curves for TMT, CRS and EC.

4.3 Electrochemical Impedance Spectroscopy

The Nyquist plots of the studied rebar in fig.7 shows the initial corrosion study. The TMT and CRS rebar showed a similar behaviour, in case of EC it showed a bit dissimilar Nyquist plot formation when compared with coating free rebar at greater sweep of recurrences. The complete resistance accounts for charge transfer resistance (R_{ct}) and concrete resistance (R_c). Generally, the concrete and the steel-concrete junction are impure condensors, in which these condenser curve as an alternative existence of an ideal half circle found over the existent axis (as shown in fig. 7), they are impaired with centres beneath the existent axis.

An unresisting steel below the EC showed a greater charge transfer resistance than TMT and CRS. The standard parameters of the electrical resistance limit acquired by connecting the plot is shown in Table 3. The concrete resistance (R_c) was acquired to be within 22.9 kΩ.cm^2, and 38.5 kΩ.cm^2, while the capacitance of bulk concrete was between 10^{-4} mA/cm^2 and 10^{-6}mA/cm^2. The charge transfer resistance (R_{ct}) was observed to be 105 kΩ.cm^2, 17 kΩ.cm^2, and 46 kΩ.cm^2 for EC, CRS and TMT accordingly. The higher steel resistance designates that the inner curve acquired from Nyquist plot emerged from the surface of the electrode. However, the LPR and EIS methods recommend that R_p values exists in similar manner from (fig.7), the EIS method shown a bit less R_p for coating free rebar. Considering the EIS method which applies alternating current (AC) which passes pass the epoxy coating is declared that the entire span of EC is dissoluted; then the range of electrochemical impedance sweep is in allowable reach as the rebar in corrosion offers a lower dissolution resistance under CC and SME.

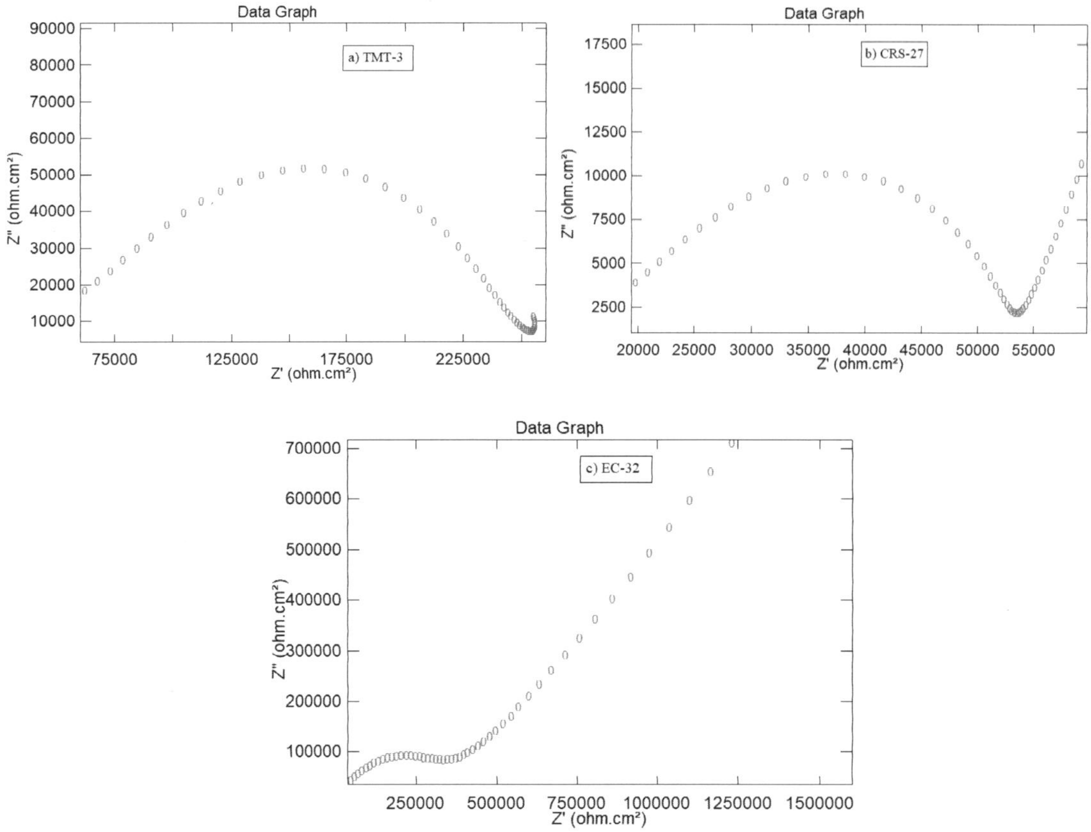

Figure 7. Electrochemical Impedance Spectroscopy curves for TMT, CRS and EC.

Table 3. Average values of LPR and EIS parameters.

Type of Rebar	R_c (kΩ.cm^2)	R_{ct} (kΩ.cm^2)	C_{dl} F/cm^2	Icorr mA/cm^2	Corrosion rate (mm/year)
TMT	22.9	46	1.36×E-04	5.64×E-04	6.53×E-03
CRS	26	17	3.43×E-05	1.54×E-03	1.78×E-02
EC	38.5	105	3.51×E-07	2.48×E-05	2.88×E-04

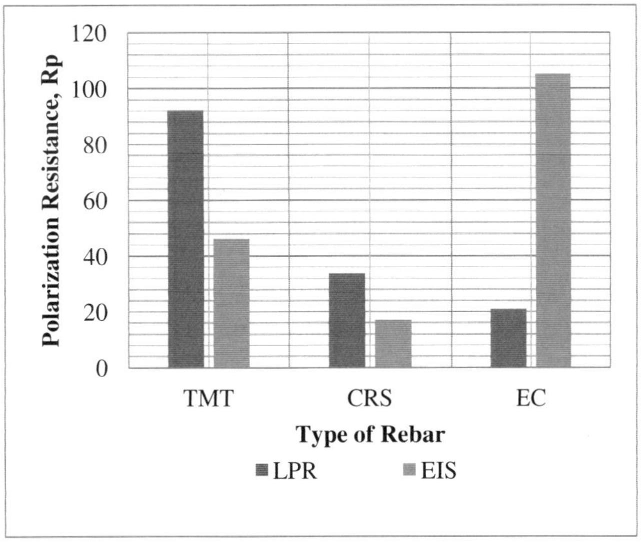

Figure 8. Charge transfer resistance and Polarization resistance calculated by LPR and EIS.

5. Conclusions

In the modern development of infrastructures, stability and resilience is given the utmost importance, this consideration is essential in assessing and resembling the effect of corrosion in concrete. Corrosion study have become a requirement of periodic life support as insistent circumstances arise in concrete because of chloride ingress or by carbonation. The resilience effect on three types of rebar embedded in concrete with reference to defiance of corrosion in controlled as well as in simulated marine environment condition was evaluated for the initial study. The specimens in CC subjected to a room temperature of 27±2^{0}C and in case of SME the specimens are sprayed with 3% NaCl over the surfaces at an interval of eight hours.

The OCP values suggested the probability of corrosion of the steel rebar during the study. During the study, the OCP values under CC and SME were in the range of -193 mV/SCE to -224.31 mV/SCE for TMT, from -164.23 mV/SCE to -213.68 mV for CRS and from -345.07mV/SCE to -415.42 mV/SCE for EC. These values are observed in the initial condition of the study were more electronegativity will be observed in later durations.

The dissolution defiance of rebar is acquired by LPR and EIS. The polarization resistance measured from LPR were 92.2 kΩ.cm^2 and 33.76 kΩ.cm^2 and for TMT and CRS appropriately which shows the more resistance than EIS. In case of EIS study for EC, it showed more resistance when compared to LPR i.e., 105 kΩ.cm^2 by EIS and 20.9 kΩ.cm^2 by LPR; which shows that for uncoated rebar LPR will be more useful and EIS is utilized to monitor the well-being of epoxy and to monitor the corrosion of steel below epoxy coating. The benefit of EC is anticipated to increase the durability of RC structures, particularly when periodic decay is extended in the regions where chloride accumulation is heavy and at that point deterioration of RC structures will become a considerable fit out.

6. References

[1] Beeby AW 1978 Corrosion of reinforcing steel in concrete and its relation to cracking *The Structural Engineer* **56 A(3)** 77–81.

[2] Feliu S, Gonzalez JA, Andrade C and Feliu V 1988 On-situ determination of the polarization resistance in a reinforced concrete beam. *Corrosion Science* **44(10)** 761–5

[3] Stern M and Geary AL 1957 Electrochemical polarization a theoretical analysis of the shape of polarization curves *J. Electrochem Soc* **104** 56–63

[4] Ahmad S and Bhattacharjee B 1995 A simple arrangement and procedure for in-situ measurement of corrosion rate of rebar embedded in concrete *Corrosion Science* **37(5)** 781–91

[5] Güneyisi E M, Mermerdaş K, and Gesoğlu M 2015 Numerical modeling of time to corrosion induced cover cracking in reinforced concrete using soft-computing based methods *Materials and Structures* **48** 1739–1756

[6] Dagher HJ and Kulendran S 1992 Finite element model of corrosion damage in concrete structures *ACI Struct J* **89(6)** 699–708

[7] Suzuki K, Ohno Y, Praparntanatorn S and Tamura H 1990 Some phenomena of macro cell corrosion of reinforcement in concrete *London: Elsevier Publishers* 87–95

[8] ASTM A615/A615M – 18:2017 *Standard Specification for Deformed and Plain Carbon-Steel Bars for Concrete Reinforcement*

[9] ASTM A1035/A1035M-16:2016 *Standard Specification for Deformed and Plain, Low -Carbon, Chromium, Steel Bars for Concrete Reinforcement*

[10] ASTM A775/A775M–19:2019 *Standard Specification for Epoxy-Coated Steel Reinforcing Bars*

[11] Panjali Natarajan et al 2005 A high current RCC monitor for corrosion studies *J. Corrosion Reviews* **23** 4-6

[12] Tan ZQ and Hansson CM 2008 Effect of surface condition on the initial corrosion of galvanized reinforcing steel embedded in concrete *Corrosion Science* **50** 2512–22

[13] Blanco MT, Andrade C and Macias A 1984 SEM study of corrosion products of galvanized reinforcement immersed in solution on the range 12.6–13.6 *Br Corrosion Journal* **19** 41–48

[14] ShiGang Dong, Bing Zhao, ChangJian Lin, RongGui Du, RongGang Hu and Gregory Xiaoge Zhang 2012 Corrosion behavior of epoxy/zinc duplex coated rebar embedded in concrete in ocean environment *Construction and Building Materials* **28** 72–78

[15] Erdogdu S, Bremner T and Kondratova I 2001 Accelerated testing of plain and epoxycoated reinforcement in simulated seawater and chloride solutions *Cement Concrete Composites* **31** 861–867

[16] Asthana K K, Aggarwal L K and Lakhani R 1999 A novel interpenetrating polymer network coating for the protection of steel reinforcement in concrete *Cement Concrete Composites* **29** 1541–1548

[17] Venkatesan P, Palaniswamy N and Rajagopal K 2006 Corrosion performance of coated reinforcing bars embedded in concrete and exposed to natural marine environment *Progress in Organic Coatings* **56** 8-12

[18] Song H W, Saraswathy V, Muralidharan S, Lee C H and Thangavel K 2009 Corrosion performance of steel in composite concrete system admixed with chloride and various alkaline nitrites *Corrosion Engineering, Science and Technology* **44** 408-415

[19] Panjali Natarajan and Ramakrishnan KR 2005 LPR meter for concrete corrosion studies *J. Instrum. Soc. India* **37(1)** 1-6

[20] Paul J 1996 Inspecting and repairing concrete before lining Protection Coating Europe **16**

[21] Moreno M, Morris W, Alvarez M G and Duff G S 2004 Corrosion of reinforcing steel in simulated concrete pore solutions Effect of carbonation and chloride content *Corrosion Science* **46** 2681-2699

[22] Sathiyanarayanan S, Panjali Natarajan, Saravanan K, Srinivasan S and Venkatachari G 2006 Corrosion monitoring of steel in concrete by galvanostatic pulse technique *Cement and Concrete Composites* **28** 630-637

[23] Vedalakshmi R, Rajagopal K and Palaniswamy N 2008 Longterm corrosion performance of rebar embedded in blended cement concrete under macro cell corrosion condition *Science Direct Construction and Building Materials* **22** 186–199

[24] Muazzam G Sohail, Ramazan Kahraman, Nasser A Alnuaimi, Bora Gencturk, Wael Alnahhal, Mina Dawood and Abdeldjelil Belarbi 2020 Electrochemical behavior of mild and corrosion resistant concrete reinforcing steels *Construction and Building Materials* **232** 1-14

IOP Conf. Series: Materials Science and Engineering **989** (2020) 012026 doi:10.1088/1757-899X/989/1/012026

A parametric study on fatigue life of plate girder railway bridge with varying velocity of rolling stock

Ramzi Nasar K[1,3], Althaf M[1] and Hisham Ajmal P C[2]

[1]TKM college of Engineering, Kollam, Kerala, India.
[2] Junior Engineer, Kerala Public Work Department, Kozhikode, Kerala, India.

[3]Email: ramzinazark@gmail.com

Abstract. It is important to measure the fatigue life of a structure as it helps in finding the probable life of a structure before its failure. It also helps in making cost effective decision on existing structure regarding replacement or rehabilitation in time. In this study deterministic approach is used for the estimation of fatigue life of a plate girder railway bridge. Finite element analysis is carried out using software to find the critical region of fatigue failure and the stress developed there. Number of cycles of different stress range during the movement of train is evaluated with the help of S-N curve. The S-N curve is selected based on type of welds and type of stress considered. Rain flow counting method is used for evaluating the number of cycles corresponding to each stress range. The fatigue damage is then evaluated by using Palm Miner damage accumulation rule and corresponding fatigue life is evaluated. The fatigue life corresponding to different velocity of train ranging from 30 kmph to 140 kmph is generated similarly and compared to find the influence of velocity of moving load on fatigue life of structure.

1. Introduction

The majority of structures involve parts that are subjected to fluctuating or cyclic loads. In such loading cases fluctuating or cyclic stresses induced can often result in failure by fatigue. The damage occurred during the fatigue loading is cumulative and is unrecoverable in nature. Same is the case with plate girder railway bridges. It is not economic and is time consuming to build a new structure when the structure starts to crack or fail. Finding the fatigue life helps us get an idea when the structure should be retrofitted to make maximum utilization of the structure and increase the net life span of the structure. Also, considering the need for faster transportation in the current state points out the relevance of the study.

The increased service loads and speed indicates the requirement to find how long the structure would sustain (Pipinato *et al.,* 2012). A study by Zhao *et al.* (1994) found that 80% to 90% of steel structure fails by fatigue.

Among various approaches that exist to analyse fatigue damage of welded joints nominal stress concept is least time consuming where as effective notch stress concept is most time consuming (Pettersson and Barsoum, 2012). The structural hotspot stress method and Linear Elastic Fracture

Mechanics (LEFM) lies in between them. According to the study conducted by Aygül *et. al.,* (2013) in comparing all the methods mentioned before, it was found that in spite of the simplicity the nominal stress approach gave fairly accurate result. But, considering how complex it is to create finite element for numerical analysis, it is difficult to distinguish nominal stress near the welded joints. One of the available standards by Hobbacher (2009) conveys how to use Hot spot stress for finite element analysis in welded joints. Thus, Hot spot stress approach is used in this study.

S-N curve-based design approach as mentioned in Euro Code is conveniently adopted to find the design life depending on the types of weld, location of joint, loading type, number of lanes etc. The Euro Code defines the effect due to above in based on loading condition at Euro Nations. But it can be adopted with slight variations according to the practical conditions that prevail in certain other country (Goel, 2005). Euro Code also give guidelines on hot spot stress method in addition to nominal stress method.

From a comparative study conducted by Goel and Kumar (2006) on existing provisions of the BS-5400, Part-10 and Indian Railway Standard Steel Bridge Code with respect to fatigue it was concluded that BS-5400 takes in heavier loads compared to IRS Modified Broad Gauge (MBG). This puts the loading condition in Indian aspect at a conservative spot.

Allowable working stress was adopted from the Appendix G of IRS Steel Bridge Code. This is depending on the ratio of minimum stress and maximum stress, also number of stress cycle repetition and the type of connection is a factor. All the details were designed such a way that the stress induced under the influence of design loads are within the allowable limits (Goel and Kumar, 2006). Goel and Kumar (2006) points out the limitations of IRS approach as follows:

- There is no clear basis for adopting counts of 10 million number of cycles to determine allowable stress levels.
- The cumulative phenomena of Fatigue are not clearly reflected in the IRS procedure.
- No S-N curves are used in the current procedure.
- The standard train load can be represented as an equivalent uniformly distributed load. Thus, the actual variation of stresses in members due to the passage of rolling stock is not counted.

Considering the Indian scenario IS 800:2007 fatigue provision is based on the nominal stress approach. There is no indication of finite element modelling or other approaches. The International Institute of Welding (IIW) gives recommendations on fatigue of welded components and structures in their comprehensive code IIW-1823-07 (2008). It gives design S-N curves and clear guidelines on how to determine hotspot stress compared to other design codes.

This study is focused on finding the effect of variation in velocities of the rolling stock on a plate girder railway bridge in Indian scenario using finite element analysis. For this study IRS plate girder bridge of span 12.2 m is adopted

2. Methodology

The deterministic approach of fatigue analysis based on the stress cycle is adopted in this study. The stress cycle was counted using the principle of Palm-Miner damage accumulation and it was implemented using rain flow counting method with respect to S-N curve. The stress type considered for obtaining the count is hot spot stress. Fig. 1 shows the methodology adopted for this study.

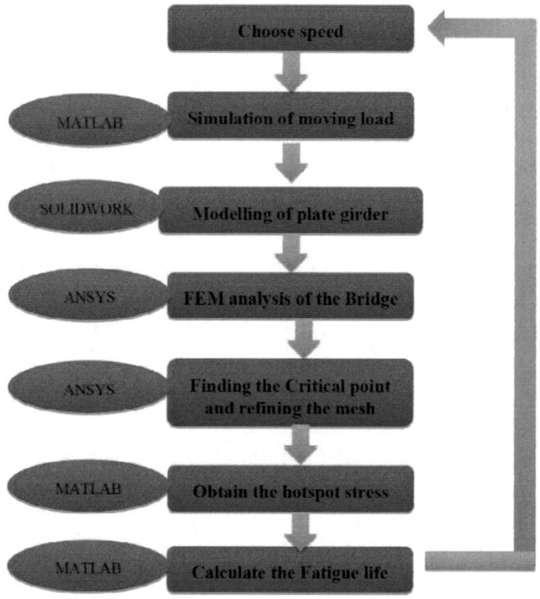

Figure 1. The methodology adopted for fatigue assessment.

With the help of a MATLAB program time histories were developed corresponding to the axle load, distance between loads and velocity of rolling stock. These time histories were then given a load in ANSYS on the imported model of bridge that was modelled using Solid works. Most critical train load was selected to find critical region of the structure, so that that meshing at that region can be refined, the hot spot stress was obtained by standard surface stress extrapolation technique.Then obtained stress history was used as input to the rain flow cycle counting algorithm where number of cycles to failure of different stress range in context to S-N curve defined by IIW-1823-07 (2008) to obtain number of cycles of each stress range. Fatigue damage was then computed with the help of Palm-Miner damage accumulation concept, inverse of which gives the fatigue life.

3. Modelling of plate girder bridge

The bridge used for this study is the IRS welded plate girder bridge as per the Research, Design and Standards Organization (RDSO) drawing no. B16003. The span of the bridge is 40 feet (12.2 m) having a total weight of 10.26 t. The plate girder was modelled with the help of Solid Works and the imported to ANSYS to perform finite element analysis.

The bridge consists of two built up I girders that are placed parallel to each other and is connected using lateral bracings. The web of the girders is stiffened with the help of intermediate and end stiffeners. The intermediate stiffeners and bracings are Indian Standard Angles (ISA) while the web, flange and end stiffener are made up of Indian standard plate sections. The Material that the girder is made is the Structural steel conforming to the Indian standard IS 2062: 2011. It has a density of 7850 kg/m ^{3}, yield strength 250 MPa, young's modulus 200 GPa and poison's ratio 0.3.

3.1Meshing and boundary conditions

While modelling plates were modeled using the shell elements and the intermediate stiffeners and bracings were modelled as beam elements. If any one of the dimensions of the geometry that is to be meshed is very small compared to the other two, then 2D elements are preferred for finite element modeling (Gokhale, 2008). By iteration the element size for mesh was found to be around 200 mm which is when the frequency of structure converged. According to Ajmal and Mohammed (2018) the bridge is supported on piers with bearing plate placed between them. The bearing plates are then welded to the bottom flange of the girder and rigidly bolted into abutment concrete. The bearing plate were modelled using shell elements and the welded connection with the flange were represented by bonded type contact in ANSYS. The support is taken as fixed. The train load acting on the rail is transferred to the bridge through two channel steel sleepers placed back to back. Bearing plates are then sandwiched between the sleepers and top flange of the bridge. The axle loads are applied as pressure on the bearing plate. The standard sleeper density of M+7 was taken which gives 20 sleepers on the bridge span (Saxena and Arora, 2012). Figure. 2. shows the bridge model after initial meshing.

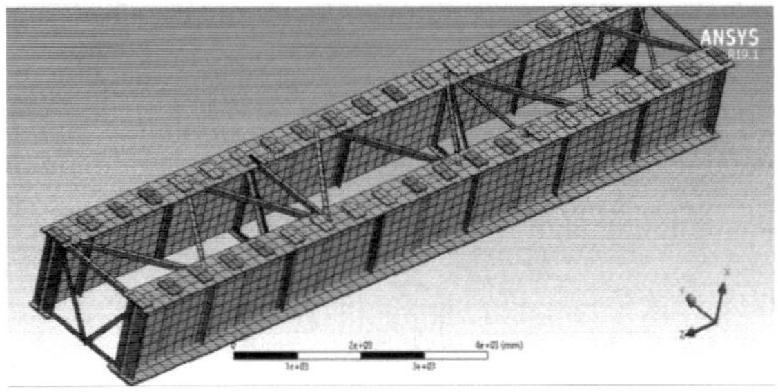

Figure 2. The bridge model after initial meshing

3.2 Fatigue loading

This study mainly considered axle load due to loco and EMU according to values given by IRS MBG loading. The fatigue load model developed by Goel (2007) for Indian railway bridges was used in this study. Goel (2007) classified the Indian railway traffic into four: heavy freight traffic, mixed traffic lines with heavy traffic, suburban traffic and mixed traffic lines with light traffic.

The suburban traffic was chosen for this study as it has the least variety of trains. The most heavily used suburban line in India is at Western Railway, Bombay-Virar, which currently has a traffic of 200 suburban trains in a day and in addition to which 10 other passenger trains also runs in the same route. The two type of passenger trains are composed of 2 locomotives each with 22 and 26 coaches respectively. Table 1 gives an outline of the load model for various types.

Table 1. Fatigue load model of suburban traffic (Goel, 2007).

Train	Train type no. (As per Goel, 2007)	No. of trains per day
2 Loco_1 + 22 Coach_2	2	5
2 Loco_1 + 26 Coach_3	3	5
EMU	4	200
Total		210

The fatigue strength of the bridge is mainly governed by the suburban train consisting of Electric Multiple Units (EMU). Table 2 shows axle loads in tons and spacing between them in millimetres for each type of rolling stock units. Since we assume linear damage accumulation, the total fatigue damage caused by these units will be the sum of individual blocks. The stress developed corresponding to each combination is assessed and the total repetition of load is computed over a year while calculating fatigue life.

Table 2. Axle load diagrams for various train units.

Block name	Load diagram
Loco_1	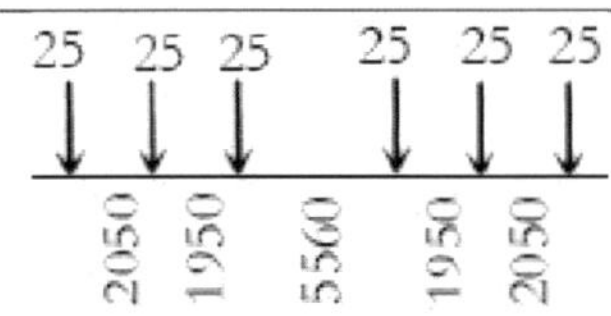
Coach_2	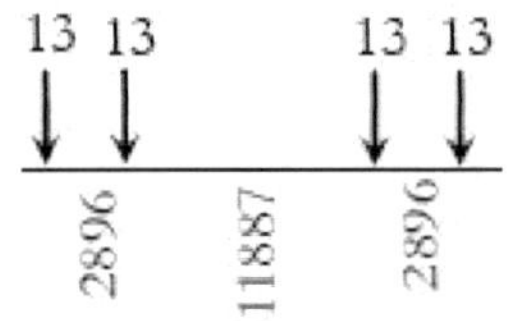
Coach_3	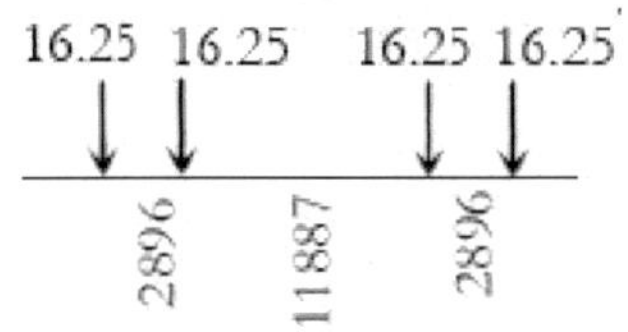
EMU	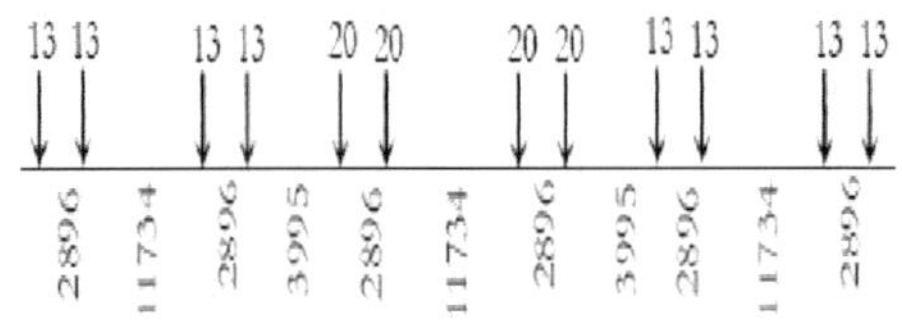

4. Moving load analysis

According to Imam *et. al.* (2007) principal stress histories can be used to apply miner rule for fatigue damage accumulation of steel railway bridge. IIW recommendations and also the Eurocode 3 advise us to use maximum principal stress as the more appropriate stress component (Heshmati, 2012). Considering that the maximum positive tensile stress is given by the maximum principle stress, it could be used to get more reliable results.

Hot spot stress can be obtained from point of maximum principle stress by refining the mesh size. It was observed that as we refine the mesh size, the point of maximum principle stress tends to change. The refining of maximum point is done until it stays intact. The point is identified by applying the most critical load that could act on the structure, here the load of EMU. It is also considered that instead of loading the full axle load of the whole train, the axle load of the repetitive sets needs to be considered. It is such that the effect of repeated axle load sets imparts same dynamic behavior on the structure.

5. Determination of hot spot stress

The Recommendations for fatigue design of welded joints and components (IIW-1823-07) shows how to determine the hot spot stress by the extrapolation of surface stress, particularly in certain reference points to the weld toe. It was found that the mesh size had great influence on the finite element analysis results. Fig. 3 shows the variation of stresses near weld toe.

As per IIW recommendation there are two types of hotspots according to their location and orientation- *Type a* in which the weld is located on plate surface and *Type b* in which the weld is located at plate edge. The procedure is first to identify the type of hotspot. Then establish the reference points required and determine the structural hot spot stress by extrapolation to the weld toe from the stresses of those reference points using the extrapolation equations given by the code. The hotspot in this problem is identified as *Type a.* IIW recommendations provide linear extrapolation equations for both cases.

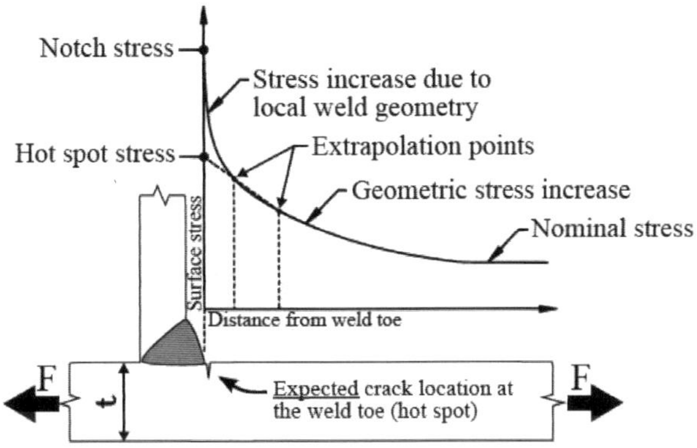

Figure 3. Vector plot of maximum principal stress at the critical node.

5.1. Surface stress extrapolation

IIW recommendations suggest fine or coarse meshing densities for shell and solid elements. It is a confirmed fact that the finite element model that have finer mesh size generally give more accurate results. Whereas relatively coarse mesh can be used for determining the structural hot spot stress. The coarse mesh enables easier computation. When coarse mesh is used for the FE model, the recommended element size should be strictly followed and no alterations are permitted. These recommendations are as shown in Table 4. The thickness of the web plate was 12 mm and that of stiffeners are 10 mm. The below gives the expression (1) for surface stress extrapolation by IIW-1823-07.

$$\sigma_{hs} = 1.5\sigma_{0.5t} - 0.5\sigma_{1.5t} \tag{1}$$

Where, σ_{hs}: Hot spot stress, $\sigma_{0.5t}$: Stress at a distance of 0.5t from the weld toe, $\sigma_{1.5t}$: Stress at a distance of 1.5t from the weld toe and t: element size

To satisfy IIW guidelines, the plates were refined near the hotspot area. In order to use the above relation for stress extrapolation, stress values at midpoint of the element is needed. Since ANSYS does not allow to directly extract stress time history from mid node points. It gives only the time history at nodes. Hence it becomes essential to interpolate the stress values at the extraction points *0.5 t* and *1.5 t* (where t refers to the thickness) from the nodal values. The shape function for the 4 nodded quad shell element shell181 in ANSYS is given as below.

$$u = \frac{1}{4}\left(u_i(1\text{-s})(1\text{-t}) + u_j(1+\text{s})(1\text{-t}) + u_k(1+\text{s})(1+\text{t}) + u_l(1\text{-s})(1+\text{t})\right) \tag{2}$$

Where u is the value at any point inside the element, u_i , u_j , u_k and u_l are the values at nodes i , j , k and l respectively and s, t are the natural coordinates. The interpolation point lies at the midpoint between the nodes i and l. As told by Ajmal and Mohammed (2018) the natural coordinates of this point are s = -1 and t = 0 respectively. The stress values have to be interpolated for all time points in the time history. The time histories obtained from ANSYS results for each train load cases with the help of MATLAB coding. This time histories were then used to compute the fatigue damage using a suitable S-N curve. Table 3 show IIW recommendations for meshing and extrapolation.

Table 3. IIW recommendations for meshing and extrapolation (IIW-1823-07).

Type of model		Relatively coarse mesh		Relatively fine mesh	
		Type a	*Type b*	*Type a*	*Type b*
Element size	Shells	t x t	10 x 10mm	≤ 0.4 t x t	≤ 4 x 4 mm
	Solids	t x t	10 x 10mm	≤ 0.4 t x t	≤ 4 x 4 mm
Extrapolation points	Shells	0.5 t & 1.5 t (mid-side points)	5 & 15 mm (mid-side points)	0.4 t & 1.0 t nodal points	4, 8 & 12 mm nodal points
	Solids	0.5 & 1.5t (surface center)	5 & 15 mm (surface center)	0.4 t & 1.0 t (nodal points)	4, 8 & 12 mm (nodal points)

6. Calculation of fatigue life

The determination of stress ranges (Δσ) and corresponding number of cycles (n) from the structural hot spot stress time histories was carried out using rain flow cycle counting algorithm. The guidelines for cycle counting are provided by ASTM standard E 1049-85: Standard practices for cycle counting in fatigue analysis (1997). The MATLAB algorithm prepared by Nieslony (2003) is adopted for rain flow counting.

To choose a proper S-N curve is an important part of the fatigue life estimation welded joints. The basic principle of S-N curves has a fatigue limit, below which fatigue failure will not occur and the curve becomes horizontal, named a knee point. It is mostly depicted in terms of the number of cycles on the S-N curve, $N=10^7$ being the most common assumption. The slope of the curve is taken as 3 below this point. From further study it was found that a constant amplitude fatigue limit does not exist and the S-N curve should continue on the basis of a further decline in stress range at a slope of m= 5. Fig. 4 gives the Modified resistance S-N curves of steel defined by IIW-1823-07 for Palm - Miner summation.

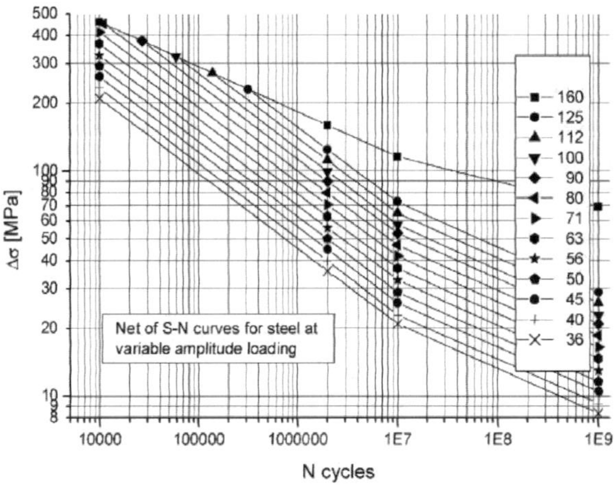

Figure 4. Modified resistance S-N curves of steel for Palm - Miner summation (IIW-1823-07).

IIW recommendations defines nine weld types for fatigue life estimation using structural hot spot stress approach. The most suitable fatigue design class for this type of joint as per IIW recommendations is FAT 100. For this FAT class the knee point value is given as 58.5 MPa. The fatigue resistance is limited by the fatigue resistance of the base material for welded joints. The FAT class of the base material here is FAT 160 and is expected to give higher fatigue resistance than the joint. The equation (3) of an S-N curve is given by,

$$N = C.\Delta\sigma^{-m} \tag{3}$$

Where Δσ = Stress range and m is the slope of the curve. C is a constant found experimentally which accounts for different factors affecting the fatigue. The S-N curve used for this problem is the two-slope curve proposed by IIW recommendations. The Table 4 gives the design values for the S-N curve for FAT 100.

Table 4. Design values for the S-N curve.

FAT class		100
Stress at 2 x 10^6		100 MPa
Stress at knee point (@ 10 7 cycles)		58.5 MPa
Constant, *C*	For stress ranges that are above knee point	2.000×10^{12}
	For stress ranges that are bellow knee point	6.851×10^{15}
Slope, m	For stress ranges that are above knee point	3
	For stress ranges that are bellow knee point	5

The expressions (4-6) for the IIW S-N curve for FAT 100 are given first two below. The N values from the S-N curve and the no of cycles n from the cycle counting results were used in following equations for linear damage accumulation according to Palm-miner rule.

$$N_i = 2.000 \times 10^{12}\ \Delta\sigma^{-3}\text{, for } \Delta\sigma \geq 58.5\,\text{MPa} \tag{4}$$

$$N_i = 6.851 \times 10^{15}\ \Delta\sigma^{-5}\text{, for } \Delta\sigma \leq 58.5\,\text{MPa} \tag{5}$$

$$D = \sum_{i=1}^{j} \frac{n_i}{N_i} \tag{6}$$

The fatigue damage accumulation was done for the structural hotspot stress time histories for all train loads. The damage caused by each train loads were added to get the total damage due to all trains. The fatigue life of the plate girder bridge was then calculated by taking the inverse of the total damage.

7. Fatigue life behaviour for varying velocity

It was found from the analytical analysis using ANSYS that there is mainly a total of 6 critical location for various velocities ranging from 30 kmph to 140 kmph. These locations are considered for the structural hotspot stress extraction as mentioned earlier. When the frequency of loading varies with speed, the location of hotspot stress tends to vary. Even though most of the time most of the same spot repeats as critical for various speed. This makes that spot most vulnerable part in the structure. The critical locations are as shown in the Table 5.

Table 5. Location of Maximum principal stress.

	Overall View	Isolated View of Critical point
Location 1	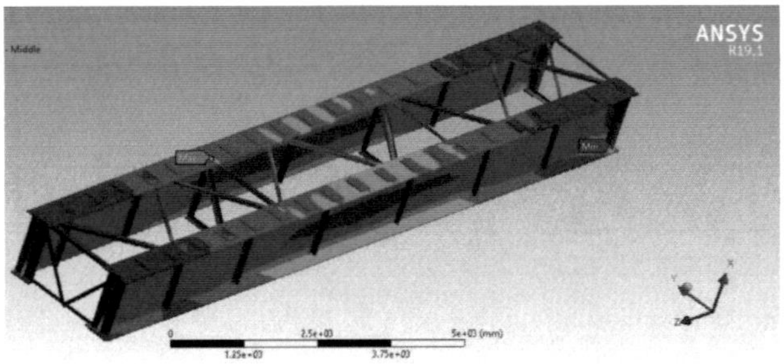	
Location 2	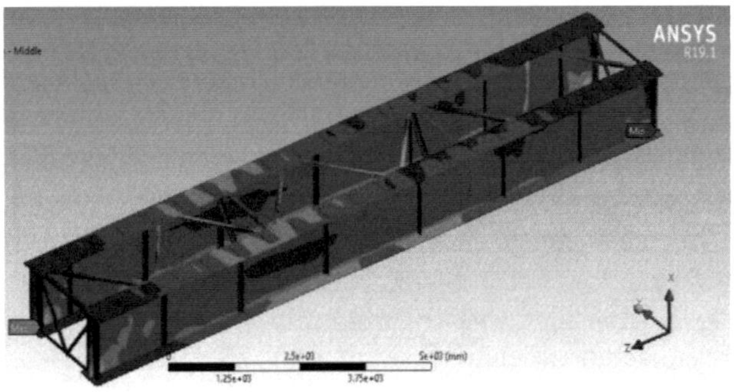	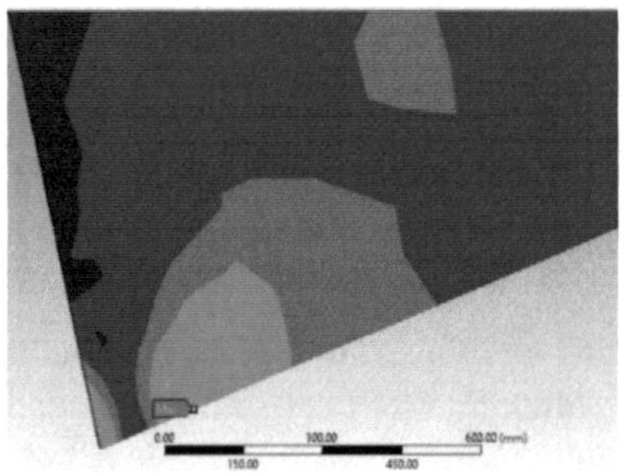
Location 3		
Location 4	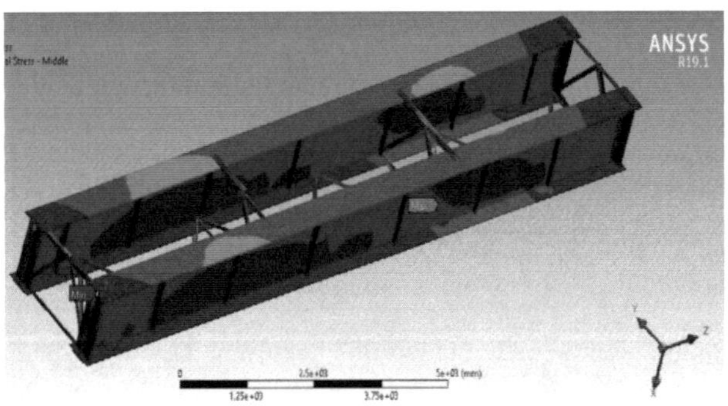	

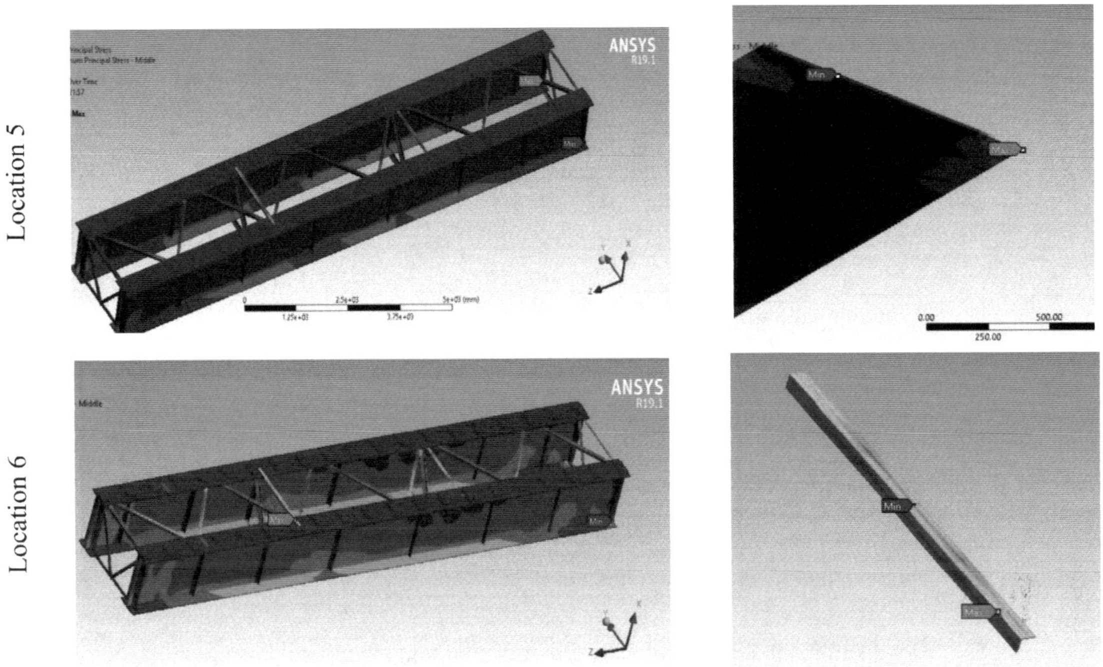

The fatigue life corresponding to each velocity are calculated using a MATLAB program by giving the hotspot stress history as input. The corresponding damage is calculated and its inverse is found to get the fatigue life of structure at that velocity. The results obtained are shown in the Table 6.

Rao *et al.* (2013) conducted fatigue life analysis of a railway bridge made of plate girders designed as per Indian railway standards using S-N curve approach in frequency domain. The fatigue life estimate of 12.2m span plate girder bridge by Rao *et al.* (2013) is found to be 45 years. "It is reported that the MBG loading specified by IRS is comparable to RU loading specified in BS 5400: Part 10" (Rao *et al.*, 2013). Hence Rao *et al.* (2013) had used the RU loading in their study. The fatigue load model used in the present study is based on IRS MBG loading. The geometric properties of the bridge analysed by Rao *et al.* (2013) are slightly different from the bridge analysed in the present study. Even though the loading conditions and bridge model are not exactly the same for both works, a reasonable comparison may be acceptable.

According to the experience of railway bridge engineers, the time taken for the first fatigue crack to develop could be possibly less than the obtained results. This could be due to many other factors like corrosion that affects the fatigue phenomenon. In Indian railways the toilet wastes splash out of the trains that are highly corrosive in nature. Due to this problem the bridge parts especially, the horizontal bracings corrode and needs replacement after a few years. Other factors like over loading, braking load effects, variations in train traffic, variations in train composition etc. are also not considered in the present study.

Table 6. Fatigue life obtained corresponding to each speed and their critical location.

Speed (kmph)	Fatigue life	Critical location
30	62 Years 173 Days	1
40	32 Years 154 Days	1
50	25 Years 28 Days	2
60	45 Years 4 Days	1
70	54 Years 162 Days	3
80	45 Years 4 Days	1
90	45 Years 4 Days	1
100	26 Years 109 Days	4
110	45 Years 4 Days	1
120	45 Years 4 Days	1
130	10 Years 112 Days	5
140	58 Years 277 Days	6

8. Conclusion

The fatigue life of a plate girder bridge can be calculated using Parlmgren-Miner damage accumulation model through MATLAB coding. The design S-N curve proposed by IIW recommendations was used in the analysis. The results obtained in the present study were found satisfactory.

It is clear from the study that most of the time the bracing tends to have the critical locations, thus periodic repair of the bracing is recommended for healthy service of the plate girder railway bridge. Also considering the unpredictable behaviour of fatigue phenomenon, it is quite hard to find a pattern of effect of velocity on the structure. Though we could accept that for a bridge on which the rolling stock moves at a velocity range between 30 kmph and 140 kmph the theoretical retrofitting should be conducted by 10 years and 112 days. That is taking apart all other factor's corrosion, unpredicted loads etc. The current study is very complex and sensitive to every parameter considered; thus, a numerical study needs to be validated through actual field observations.

9. References

[1] Ajmal P C and Mohammed A 2018 Finite element analysis-based fatigue life evaluation approach for railway bridges: a study in Indian scenario *Structural Monitoring and Maintenance* **5(4)** 429-443

[2] Alavala C R 2008 *Finite Element Methods: Basic Concepts and Applications*, PHI Learning Pvt. Ltd., New Delhi, India.

[3] ANSYS Mechanical User's Guide, ANSYS Inc.

[4] ASTM: E1049-85 2011 Standard practices for cycle counting in fatigue analysis, ASTM Standards, Philadelphia, USA.

[5] Aygül M, Al-Emrani M and Urushadze S 2012 Modelling and fatigue life assessment of orthotropic bridge deck details using FEM *International Journal of Fatigue* **40** 129-142

[6] Aygül M, Bokesjö M, Heshmati M and Al-Emrani M 2013 A comparative study of different fatigue failure assessments of welded bridge details *International Journal of Fatigue* **49** 62-72.

[7] BS 5400 1980 Steel, concrete and composite bridges - Part 10: Code of practice for fatigue, British Standards Institution, London, UK.
[8] Dieter G E 1986 *Mechanical Metallurgy (Vol. 3)*, McGraw-Hill, New Delhi, India.
[9] Downing S D and Socie D F 1982 Simple rainflow counting algorithms *International Journal of Fatigue* **4 (1)** 31-40
[10] Eurocode 3 2005 *Design of steel structures- Part 1-9: Fatigue*, European Standard, The European Union
[11] Goel R K 2005 Advancements in codal provisions for fatigue design of railway steel bridges *Proceedings of the National Conference on Advances in Bridge Engineering,* IIT Roorkee, Roorkee.
[12] Goel R K 2007 *Emerging Load Model for Fatigue Assessment of Railway Steel Bridges*, Golden Jubilee Special Issue of Indian Railway Technical Bulletin, 183-190.
[13] Goel R K and Kumar D 2006 Fatigue life assessment of steel bridges *Proceedings of The Institution of Permanent Way Engineers(IPWE) Seminar on Bridge Design, Construction, Rehabilitation & Maintenance,* Chennai. India.
[14] Gokhale N S 2008 *Practical Finite Element Analysis*, Finite To Infinite, Pune, India.
[15] Heshmati M 2012 Fatigue life assessment of bridge details using finite element method *Master's Thesis in the Structural Engineering and Building Performance Design* , Chalmers University of Technology, Göteborg, Sweden.
[16] Hobbacher A F 2009 The new IIW recommendations for fatigue assessment of welded joints and components – a comprehensive code recently updated *International Journal of Fatigue*, **31** 50-58
[17] IIW-1823-07 2008 Recommendations for fatigue design of welded joints and components, International Institute of Welding, Paris, France.
[18] Imam B M, Righiniotis T D and Chryssanthopoulos M K 2007 Numerical modelling of riveted railway bridge connections for fatigue evaluation *Engineering Structures* **29** 3071–3081
[19] Indian Railway Standard 2003 Steel bridge code, Research Designs & Standards Organisation, Ministry of Railways, Lucknow, India.
[20] Indian Railway Standard 2008 Bridge Rules, Research Designs & Standards Organisation, Ministry of Railways, Lucknow, India.
[21] IS 2062 2011 Hot rolled medium and high tensile structural steel-Specification, Bureau of Indian Standards, New Delhi, India.
[22] IS 800 2007 General construction in steel - Code of practice, Bureau of Indian Standards, New Delhi, India.
[23] MATLAB Documentation. https://in.mathworks.com/help/matlab/
[24] Miner M A 1945 Cumulative damage in fatigue *Journal of Applied Mechanics*, **67**, 159–164.
[25] Niemi E, FrickeW and Maddox S J 2006 Structural hot spot stress determination using finite element analysis *Fatigue Analysis of Welded Components - Designer's Guide to the Structural Hot-Spot Stress Approach*, Woodhead Publishing, Sawston, Cambridge, 14–22.
[26] Pettersson G and Barsoum Z 2012 Finite element analysis and fatigue design of a welded construction machinery component using different concepts *Engineering Failure Analysis* **2** 274–284
[27] Pipinato A, Pellegrino C and Modena C 2012 Fatigue damage estimation in existing railway steel bridges by detailed loading history analysis *ISRN Civil Engineering* **2012** 1-13
[28] Rao K B, Anoop M B, Raghava G, Prakash M and Rajadurai A 2013 Probabilistic fatigue life analysis of welded steel plate railway bridge girders using S-N curve approach", *Proceedings of the Institution of Mechanical Engineers, Part O: Journal of Risk and Reliability* **227**(4) 385-404
[29] RDSO/B-16003 (2004), MBG Loading-1987 Plate Girder Welded Type 12.2m Span, Drawing of Research Designs & Standards Organisation, Ministry of Railways, Lucknow, India.
[30] RDSO/T-5155 to T-5155 1999 Rail Fastening Arrangement on Girder Bridges with Steel

Channel Sleepers, Drawing of Research Designs & Standards Organisation, Ministry of Railways, Lucknow, India.

[31] Saxena S C and Arora S P 2010 *A Textbook of Railway Engineering,* Dhanpat Rai Publications, New Delhi, India.

[32] Zhao Z, Haldar A and Breen F L 1994 Fatigue - reliability evaluation of steel bridges *Journal of Structural Engineering* **120**(8) 1608-1623

IOP Conf. Series: Materials Science and Engineering **989** (2020) 012027 doi:10.1088/1757-899X/989/1/012027

Optimization of high strength concrete with construction and demolition waste

Geetha S[1,2], Selvakumar M[1] and Muthulakshmi S[1]

[1] Rajalakshmi Engineering College, Thandalam, Chennai, India.

[2]Email:geetha.s@rajalakshmi.edu.in

Abstract. The development in the Infrastructure Sector has led to a revolution in the Construction Industry with fast track projects. Hence the use of prefabricated construction is gaining importance in India in the recent days. The structural elements like beams, columns and wall panels are fabricated in a factory and assembled at the site. The concrete used in the fabrication of these members are conventional concrete with density range 2400-2500 kg/m^3. The transport, handling and erection of these members involve large machineries. Moreover due to the use of conventional concrete the self weight of the structure also increases leading to large column size and foundations. Hence this work is based in optimizing reactor powder with mineral admixtures, construction and demolition waste and steel fibres that can be used in precast construction. The compressive strength achieved is 63.12 Mpa, flexural strength is 6.21 and tensile strength is 6.38MPa.Rapid chloride penetration test (RCPT) was done to test the durability of the concrete and it was found to be 469 coulombs which proved it to be resistant to chloride penetration.

1. Introduction

The current trend of construction Industry is witnessing fast track projects with the use of precast structural units. But this technique has not gained popularity due to bulky structural elements resulting in difficulty in transportation and erection in the site. Also Lightweight concrete is available in construction practice but the strength attained so far is not sufficient to take up the structural loads. Reactive powder concrete (RPC) is a high strength concrete comprising powder admixtures and eliminating the coarse aggregate [1].RPC is not conventionally used due to its high cost. RPC possess excellent strength capacities, fracture energy and can be used in any type of severe exposure conditions [2–4]. The strength development in RPC is mainly based on particle packing and the strength achieved so far is 200 to 800 MPa [5–11].The application of RPC is mainly in bridges, nuclear power plants, tall structures and impact resistant buildings [12–14]. In spite of its many advantages RPC has not been used in the construction practices due to high cost and energy demand [12,15,16].

The components required for proportioning RPC are Portland cement, fine quartz sand, low water - binder ratio, chemical and mineral admixtures and discrete fibres [17]. The high cost of RPC is due to more cementitious binders in the matrix, fine quartz sand and steel fibres. Due to the advancement in concrete technology several mineral admixtures have been proven to perform better that ordinary Portland cement in terms of strength and durability. The mixture can be proportioned economically by the use of these mineral admixtures as replacement to cement. Also in this paper an attempt is made to use the construction and demolition waste (C&D) as replacement to the fine quartz sand. The material is pulverized and the fraction retained in 600 micron sieve is used as fine aggregate in this concrete mix. Obviously, these expensive raw materials are responsible for the high production cost. Moreover not all structural elements require high compressive strength of 200 MPa. The maximum strength that is required in high rise buildings range from 60 to 80MPa. As the strength of the concrete increases it becomes more brittle. Hence a balance in the strength and flexural toughness is needed for concrete used in structural members to prevent the cracking in concrete due to excessive load.

RPC is a good choice in casting precast structural elements as the size of the member can be reduced due to high compressive strength which reduces the dead weight of the member also facilitating ease in transportation and handling these structural members. The flexural toughness is increased as steel fibres are included in the matrix. Accelerated curing methods like autoclaving or steam curing are generally adopted for producing RPC which normally results in high energy consumption and as a result the cost of production of RPC gets increased [18,19]. Also accelerated curing methods facilitate the structural components to be ready in 24 hours which will facilitate fast track construction of the projects.

2. Materials and methodology

Portland Pozzolano cement of 53 grade was used in this work. There are several components involved in the making of reactive powder concrete like fly ash, steel fibres, silica fume and fine quartz sand. In this work fly ash F has been used as cement replacement upto 20%. And to reduce the usage of quartz sand pulverized construction and demolition (C&D) waste has been used as replacement to quartz sand upto 20%. The mixture proportion was done as follows: 60% of the binding material was Portland cement in which the replacement with flyash was varied from 0 to 20% of cement, 10 % was silica fume, 10% was metakaolin and 20% was ground granulated blast furnace slag (GGBS). The other modification made in this project is the use of CNC waste steel fibres instead of commercial available steel fibres. These fibres were available as fine coiled springs. They were cut to a length of 10 mm. The W/C ratio was maintained as 0.35%. Superplasticiser (commercially available Glenium-53) was used as 1% by weight of cement. The individual components are dry mixed in Hobart mixer for 3 min. Then water and superplasticiser are added to the mix and mixed for 5 minutes approximately. The mixture is then poured in moulds for the appropriate testing. Since many components are involved in the making of reactive powder concrete (RPC), the traditional methods of experimentation will lead to number of trials.

In this work, the experimental trials were designed using design of experiment software in order to minimize the trial runs and also to get meaningful conclusions from the experimental results. The range of factors that were varied was entered in the software and combination for each trial was designed by the software. Specimens for testing the compressive strength (15 cm cube), flexural strength (beam of size 10 cm x 10 cm x 500 cm) and split tensile strength (cylinders of 10 cm diameter and 20 cm height) were casted in the laboratory based on the designed trial mixes. The specimen were subjected to steam curing at 90°C for 24 hours and tested on the next day. The test results which are the responses were entered in the software. The significance of the test results were analyzed by the software. Thus the advantage of using design expert software is that, apart from minimizing the trials it also facilitates to verify the reliability of experimental results by statistical analysis (ANOVA). The interaction effect between the individual factors can also be inferred from the response surface graphs [20]. The major factor and the range of the factor can be identified by using this tool and based on the results the reasons for decrease or increase in the responses can be evaluated from the available literature and reactions of the components.

Durability of concrete was tested using Rapid Chloride Penetration Test (RCPT). The test set up is as shown in figures1-2. The quality of concrete can be assessed based on the coulombs passed. The lesser the charge passed, concrete is durable.

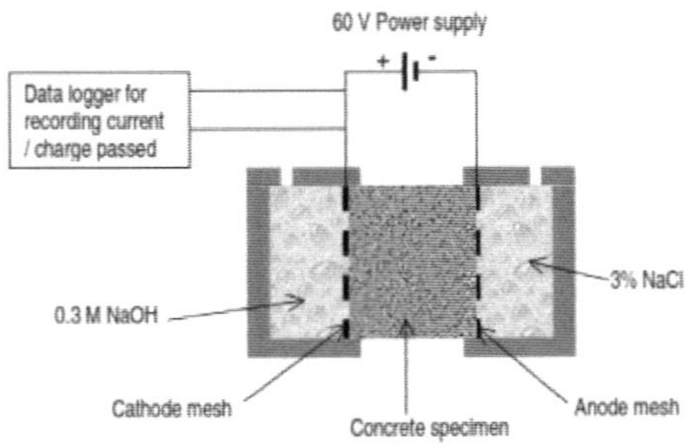

Figure 1. Details of the test.

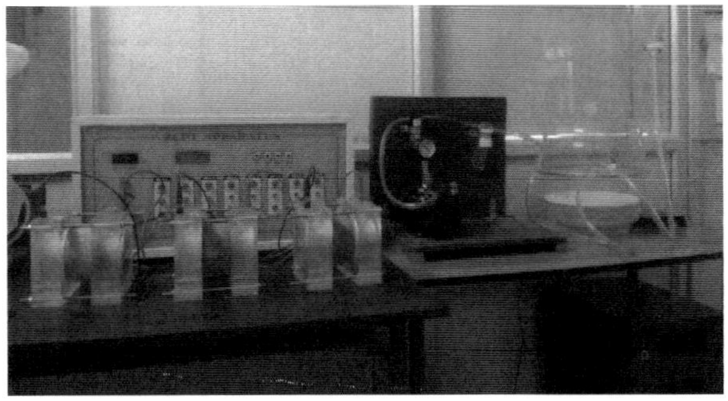

Figure 2. Test set up in the laboratory.

3. Results and discussions

The concrete specimens are to be tested for compressive, tensile and flexural strength. The results of the response surface graph are given in the graphs. Fig-3 shows that compressive strength increases as the fly ash and C& D waste content increased. Similarly as the steel fibres content increased there was increase in strength (Figure 4). The same trend was observed with respect to flexural strength and tensile strength also as given in graphs (5 to 8). The improvement in strength parameters is mainly attributed to the use of 60% mineral admixtures (fly ash, silica fume, metakaolin and GGBS) which enhances the filler effect and improves the particle packing of the matrix resulting in dense microstructure. The density of the concrete varied from 1680 to 1852 kg/m^3, which is less when compared to the density of conventional concrete (2400 kg/m^3).

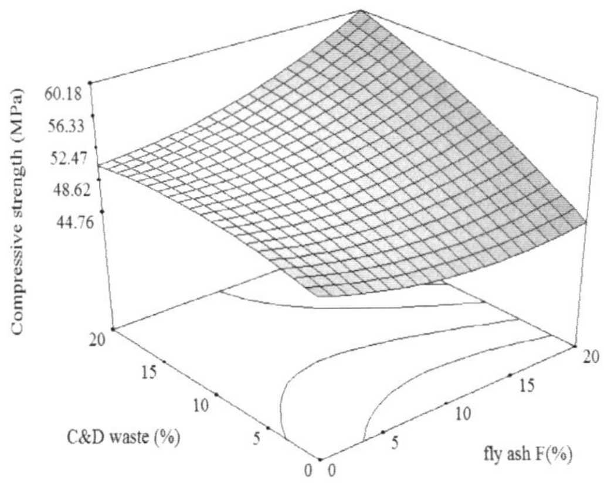

Figure 3. Response surface with fly ash and C& D waste for compressive strength.

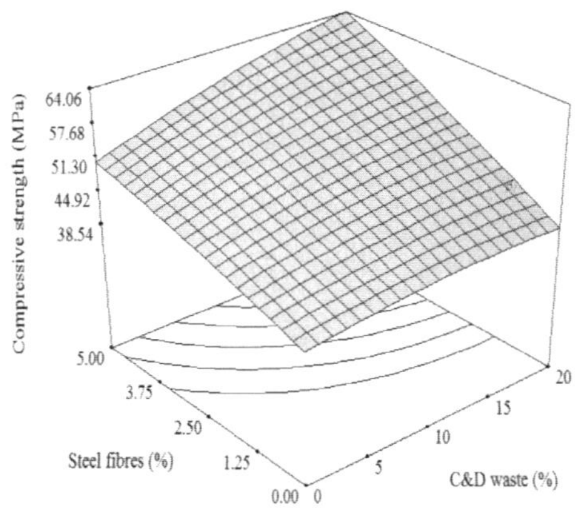

Figure 4. Response graph with C& D waste and steel fibres for compressive strength.

IOP Conf. Series: Materials Science and Engineering **989** (2020) 012027 doi:10.1088/1757-899X/989/1/012027

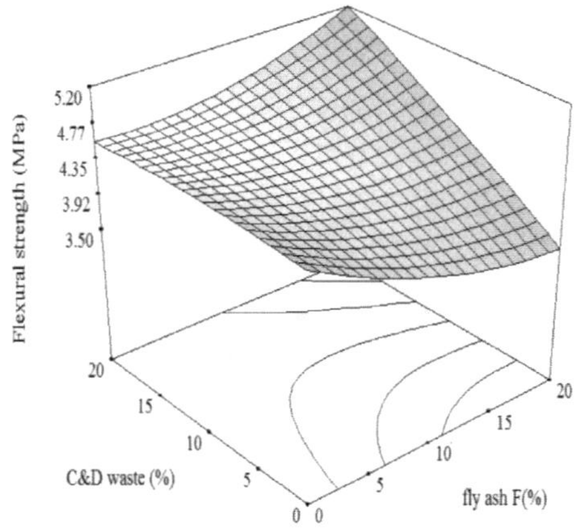

Figure 5. Response surface with fly ash and C& D waste for flexural strength.

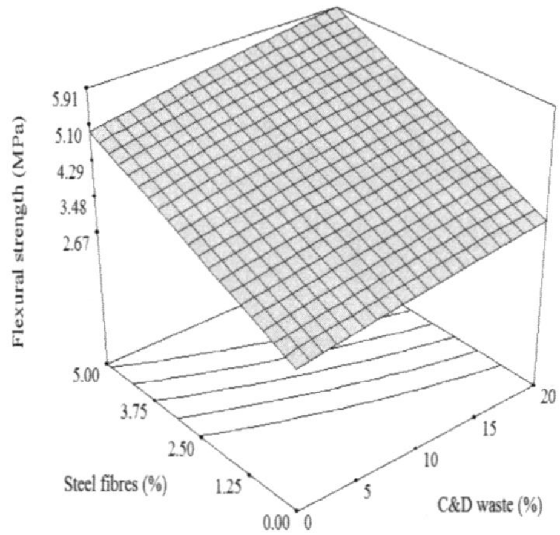

Figure 6. Response surface with C&D waste and steel fibres for flexural strength.

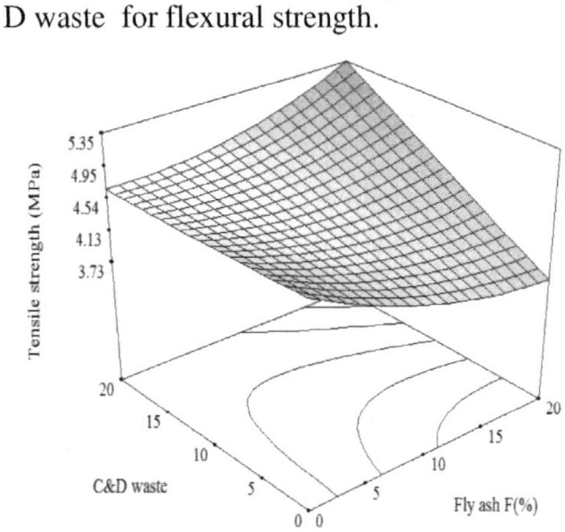

Figure 7. Response surface graph with fly ash and C& D waste for tensile strength.

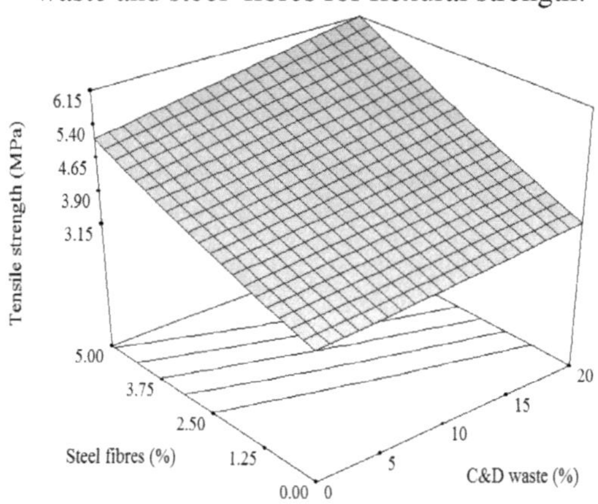

Figure 8. Response surface graph with C& D waste and steel fibres for tensile strength.

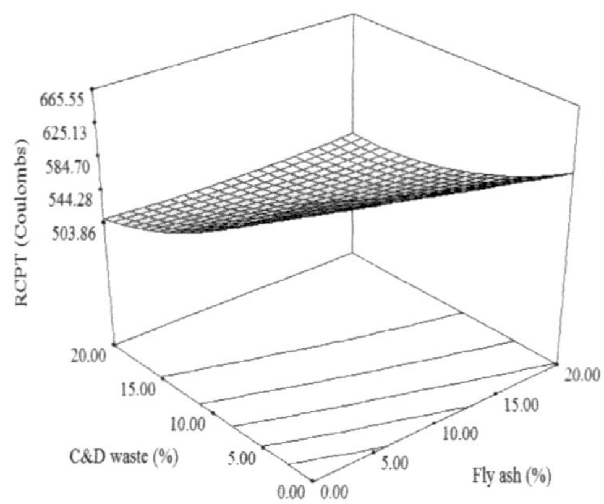

Figure 9. Response surface with fly ash and C& D waste for RCPT .

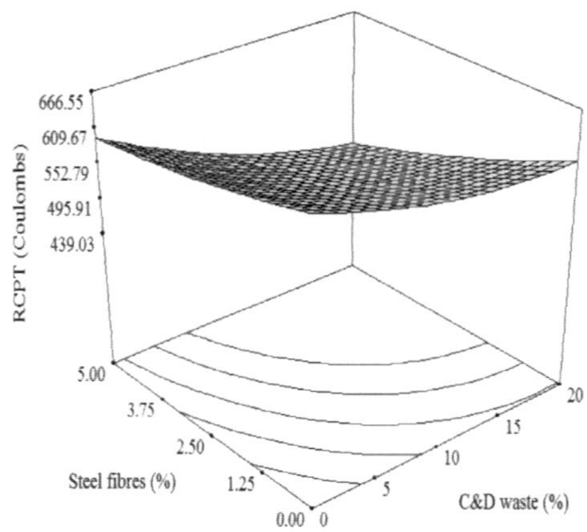

Figure10. Response surface with C& D waste and steel fibres for RCPT.

The RCPT results given in fig 8 and 9 shows that the value decreases with increase in fly ash and C&D waste which implies that the concrete is durable. The range of RCPT is from 500 to 650 coulombs.

The Optimization was done for high strength within the range of the parameters used. The optimized mix was casted and tested to check the predicted values as given in Table 1 and 2. The observed values were closer to the predicted values.

Table1. Optimized Values.

Fly ash (%)	C& D waste (%)	Steel fibres(%)
17.87	18.92	5

Table 2. Confirmatory tests.

	Compressive strength(MPa)	Flexural strength (MPa)	Tensile strength (MPa)	RCPT (Coulombs)
Predicted	62.90	6.10	6.28	458.92
Observed	60.6	5.89	6.12	478.32

4. Conclusions

Based on the experimental results the following conclusions were derived:

- Construction and demolition waste (C&D waste) finds a potential use as replacement for sand in the production of reactive powder concrete which could reduce the cost of this concrete.
- The addition of supplementary cementitious materials and admixtures contributed to the strength enhancement of the RPC thereby making it a suitable material that can be used in construction of structural elements.

5. References

[1] Richard P and Cheyrezy M 1994 Reactive powder concrete with high ductility and 200–800 MPa compressive strength *J. ACI SP* **144** 507–18

[2] Bonneau Olivier, Lachemi Mohamed and Dallaire Eric 1997 Mechanical properties and durability of two industrial reactive powder concretes *J. ACI Materials Journal* **94(4)** 286–90.

[3] Matte V and Moranville M 1999 Durability of reactive powder composites: influence of silica fume on the leaching properties of very low water/binder pastes *J. Cement and Concrete Composites* **21(1)** 1–9.

[4] Allan CLW, Paul AC and Richard B 2007 Simultaneous measurement of shrinkage and temperature of reactive powder concrete at early-age using fiber Bragg grating sensors *J. Cement and Concrete Composites* **29(60)** 490–7.

[5] Cheyrezy M, Maret V and Frouin L 1995 Microstructural analysis of reactive powder concrete *J.Cement and Concrete Research* **25(7)** 1491–500.

[6] Zanni H, Cheyrezy M, Maret V, Philippot S and Nieto P 1996 Investigation of hydration and pozzolanic reaction in reactive powder concrete (RPC) using 29Si NMR *J. Cement and Concrete Research* **26(1)** 93–100.

[7] Feylessoufi A, Villieras F, Michot LJ, De Donato P and Cases JM 1996 Water environment and nanostructural network in a reactive powder concrete *J. Cement and Concrete Composites* **18(1)** 23–9.

[8] Reda MM, Shrive NG and Gillott JE 1999 Microstructural investigation of innovative UHPC *J.Cement and Concrete Research* **29(3)** 323–9.

[9] Bonneau O, Vernetb C, Moranville M and Aitcin P C 2000 Characterization of the granular packing and percolation threshold of reactive powder concrete *J. Cement and Concrete Research* **30(12)** 1861–7.

[10] Matte V, Moranville M, Adenot F, Richet C and Torrenti JM 2000 Simulated microstructure and transport properties of ultra-high performance cement based materials *J. Cement and Concrete Research* **30(12)** 1947–54

[11] Morin V, Cohen-Tenoudji F, Feylessoufi A and Richard P 2002 Evolution of the capillary network in a reactive powder concrete during hydration process *J. Cement and Concrete Research* **32(2)** 1907–14

[12] Pierre-Claude Aitcin 2000 Cements of yesterday and today: concrete of tomorrow *J. Cement and Concrete Research* **30(9)** 1349–59

[13] Cyr MF, Shah SP 2000 Advances in concrete technology. In: Proceedings of the international conference on advances in building technology, 4–6 December, Hong Kong, China; 2002. 17–27

[14] Ming GL, Yung CW and Chui TC 2007 A preliminary study of reactive powder concrete as a new repair material *J. Construction and Building Materials* **21(1)** 182–9

[15] Bonneau Oliver, Poulin Claude and Dugat Jerome 1996 Reactive powder concretes: from theory to practice *J. Concrete International* **18(4)** 47–9

[16] Cheyrezy Marcel 1999 Structural applications of RPC. *Concrete* **33(1)** 20–3

[17] Richard Pierre and Cheyrezy Marcel 1995 Composition of reactive powder concrete *J. Cement and Concrete Research* **25(7)** 1501–11

[18] Feylessoufi A, Crespin M, Dion P, Bergaya F, Van Damme H and Richard P 1997 Controlled rate thermal treatment of reactive powder concretes *J. Advanced Cement Based Materials* **6(1)** 21–7.

[19] Halit Yazıcı 2007 The effect of curing conditions on compressive strength of ultra-high strength concrete with high volume mineral admixtures *J. Building and Environment* **42(5)** 2083–9.

[20] Geetha S and Selvakumar M 2017 High performance concrete with copper slag for marine Environment *J. Materials Today: Proceedings* **4** 3525–3533

Application of lime and GGBS to improve the strength of clayey sand

Muthu Lakshmi S[1,2] Geetha S[1], Selvakumar M[1], Sivanesan N[1] and Sreedharan KR[1]

[1]Department of Civil Engineering, Rajalakshmi Engineering College, Thandalam, Chennai, India.

[3]Email: muthulakshmi.s@rajalakshmi.edu.in

Abstract. Lot of industrial waste by-products such as fly ash, GGBS, silica fume, paper pulp etc. are generated all over the world leading to environmental degradation and pollution. Utilization of the industrial waste by-products in construction industry will at least reduce the quantity of these wastes that are being dumped in landfills leading to wastage of useful land area. But utilization of these industrial wastes as a construction material or as partial replacement in concrete or soil should not have a negative effect on their properties. Thus a thorough experimental investigation should be carried out before these industrial wastes are utilized in the construction industry. In the present experimental work, an attempt has been made to utilize GGBS along with quick lime to study their effect on the strength characteristics of locally available soil. GGBS was mixed with locally available Clayey Sand (SC soil) along with quick lime to investigate their effect on the UCC strength and soaked CBR strength of the soil. Quick lime was varied in percentages of 2%, 4%, 6% and 8% whereas GGBS was varied in percentages of 20%, 25%, 30%, 35% and 40% to determine the ideal percentage of lime and GGBS that should be added to SC soil for which maximum UCC strength and soaked CBR strength was attained. From the experimental results, it was found that for soil mixed with lime and GGBS increase in UCC strength and soaked CBR strength of 2.11 times and 7.44 times that of untreated soil was observed respectively.

1. Introduction

Advancement in technology and industrialization has led to generation of lot of waste by-products all over the world. The different varieties of waste may be classified as industrial waste, agricultural waste, construction and demolition waste, municipal solid waste etc. The different industrial waste by-products that are generated may be listed as fly ash, cenosphere, silica fume, rice husk ash, ground granulated blast furnace slag (GGBS), paper pulp etc. Utilization of industrial wastes in construction industry will reduce the volume of these wastes that are being dumped in landfills thus reducing environmental pollution. Also utilization of such wastes as construction materials will reduce the requirement of conventional materials like coarse aggregates, stones, sand, cement, lime, backfill material etc. thus preserving them for future generations. As per [27], soil stabilisation with by-product materials has proved to be a useful cost effective option. In the current project, an attempt has been made to utilize GGBS along with quick lime to improve the strength characteristics of locally available soil. GGBS is a waste by-product of iron and steel manufacturing industries. GGBS is highly cementitious and high in calcium silicate hydrates (CSH) which is a strength enhancing compound. Generally, GGBS is used in concrete but in recent times GGBS is also being used for stabilizing weak and problematic soil and for improving the strength characteristics of different soil. Utilizing GGBS in the construction industry will thus lead to economical and sustainable construction. The soil used in the experimental work was classified as Clayey Sand (SC soil) as per Indian Standard Soil Classification System (ISCS). SC soil is not a weak soil but still if we can further improve the strength of SC soil by utilizing GGBS and lime we can reduce the quantity of GGBS that is being dumped as waste and can also reduce the requirement of conventional construction materials thus leading to cost-effective construction. SC soil was first mixed with varying lime content to determine the ideal percentage of lime that needs to be added to SC soil for which maximum UCC strength and soaked CBR strength was attained after 3 days curing. SC soil with optimum lime content was then mixed with varying percentages of GGBS to determine the ideal percentage of GGBS that needs to be added to SC soil for which maximum UCC strength and soaked CBR strength was attained after 3 days curing. The experimental results were then compared with that of untreated soil or control specimen (CS) to evaluate the maximum UCC strength and soaked CBR strength gain attained.

Many studies have been conducted to study about the effect of lime on the geotechnical properties of different types of soil. As per [5], compared to hydrated lime, quick lime stabilized soil showed improved performance and was considered to be superior for soil treatment. As per [1], amount of lime required to effectively treat a soil depends on the type of clay mineral present, like for Kaolinite 4 to 6 % lime was required and for illite and montmorillonite about 8 % lime was required for maximum strength gain. As per [19], treatment of Ikpayongo laterite with 4 % lime efficiently modified it for economic stabilization. Reference [8, 13, 17], applied lime treatment to improve and stabilize expansive clays, black cotton soil and observed improvement in their plasticity and swelling characteristics. Reference [2, 4, 6, 8, 11], observed that adding lime and water to soil quickly increased the pH of soil thus causing pozzolanic reaction between free Calcium cations and the dissolved silica and alumina found in soil. This lead to the formation of calcium-silicate-hydrates (CSH) and calcium-aluminate-hydrates (CAH) which are cementitious products similar to that formed in Portland cement, thus contributing to the strength of lime-stabilized soil. As per [10, 12, 14, 16, 18], lime-clay-water interaction happens due to four main processes which causes modification in soil properties. They are hydration of lime, cation exchange between pore fluids and clay minerals, flocculation of clay to form larger clusters and aggregation of the soil matrix by cementitious precipitates.

Few researchers have worked on the application of GGBS along with some additives like lime, cement, carbide slag, reactive magnesia (MgO) etc. for improving the characteristics of different soil types. As per [23], GGBS was widely used to stabilize soils due to its environmental and economic merits. Reference [21] used a binder developed by mixing fly ash and GGBS to stabilise expansive soils and the binder was found to be effective and economic to stabilise expansive soils with lesser amount of lime. As per [22], carbide slag-GGBS was a better binder than cement for the treatment of gypseous soil. Reference [31, 32], utilized reactive magnesia-activated GGBS for soil stabilisation. As per [30], optimum reactive magnesia-GGBS and carbide slag-GGBS were superior to Portland cement. As per [33], soil stabilization using a mixture of soda residue and GGBS increased shear strength as GGBS content and curing time increased. As per [34], GGBS activated with olivine (Mg_2SiO_4) and sodium hydroxide (NaOH) was used to stabilise a clayey soil and the significant strength increase was attributed to the higher reaction degree provided by the NaOH. As per [29], lower temperature curing environment (<5°C) significantly retards the hydration and pozzolanic reactions, thereby inhibiting the strength development. Reference [25, 26], studied the effectiveness of GGBS for providing better stabilization of sulfate bearing soils. As per [15, 35], stabilisation/solidification has emerged as an efficient and cost-effective technology for the treatment of contaminated soils with a blend of hydrated lime and GGBS as the binder.

2. Experimental work and result discussion

In the present work, disturbed soil collected from Vembakkam in Thiruvannamalai district of Tamil Nadu was used. The disturbed soil was collected from a depth of 0.5 m below the ground surface after clearing the surface of vegetation. Basic index property tests such as specific gravity test, liquid limit and plastic limit test, wet sieve analysis and Standard Proctor Compaction (SPC) test were conducted on the disturbed soil as per the Indian Standard [36, 37, 38, 40], results of which are tabulated below in Table 1. The soil was then classified as per ISCS [39] and based on the maximum dry density (MDD) and optimum moisture content (OMC), the soil specimen for Unconfined Compression (UCC) test and soaked CBR test were prepared at 97% relative compaction.

Table 1. Basic Index Properties of Soil.

Soil Test	Result
Specific Gravity	2.6
Liquid Limit	25 %
Plastic Limit	13.8 %
Plasticity Index	11.2 %
Wet Sieve Analysis	Gravel = 0 %
	Sand = 62.6 %
	Silt and Clay = 37.4 %
SPC Test	MDD = 2.01 gm/cc
	OMC = 11.04 %
Soil Classification as per ISCS	SC (Clayey Sand)

It was observed that soil specimen used in the experimental work had a specific gravity of 2.6 and based on the results of wet sieve analysis and consistency limit tests, the soil has been classified as Clayey Sand (SC).

2.1. SC Soil mixed with varying percentage of lime
SC soil was mixed with quick lime in the percentages of 2%, 4%, 6% and 8% and allowed to cure for a period of 3 days. After 3 days curing, UCC test and soaked CBR test were conducted on the soil with varying lime content to determine the ideal percentage of lime to be added to SC soil based on the shear strength and soaked CBR strength gain observed from the test results. Generally, for soaked CBR test the prepared soil specimen is soaked in water for a period of 96 hours or 4 days but in the current experimental work the prepared soil specimen was soaked in water for a period of 3 days for the purpose of comparison of results.

2.1.1. UCC test conducted on SC soil with varying lime content
UCC test was conducted as per IS:2720 (Part 10) – 1991 [41] on SC soil or Control specimen (CS) by preparing the soil specimen at 97% relative compaction based on the MDD and OMC values obtained from SPC test. SC soil was then mixed with varying percentage of quick lime (2%, 4%, 6% and 8%) and the soil lime mixture was allowed to cure for 3 days. Soil and quick lime was first mixed in dry condition and then water equal to OMC was added slowly and mixed with soil and quick lime to form a homogeneous mixture. This soil lime mixture was then kept covered with polythene sheet to prevent air drying of the mixture and allowed to cure for a period of 3 days. After 3 days curing, soil specimen for UCC test was prepared at 97% relative compaction by compacting in a UCC mould. UCC test on CS was carried out immediately without curing as the specimen prepared for UCC test after 3 days curing became very brittle due to loss of moisture. Table 2 shows the UCC strength of SC soil mixed with varying percentage of lime after 3 days curing which is represented graphically in Figure 1.

Table 2. UCC strength of SC soil with varying lime content.

Soil Sample	CS (Soil + 0% lime)	Soil + 2% lime	Soil + 4% lime	Soil + 6% lime	Soil + 8% lime
UCC Strength (kPa)	39.43	54.98	82.38	60.04	55.09

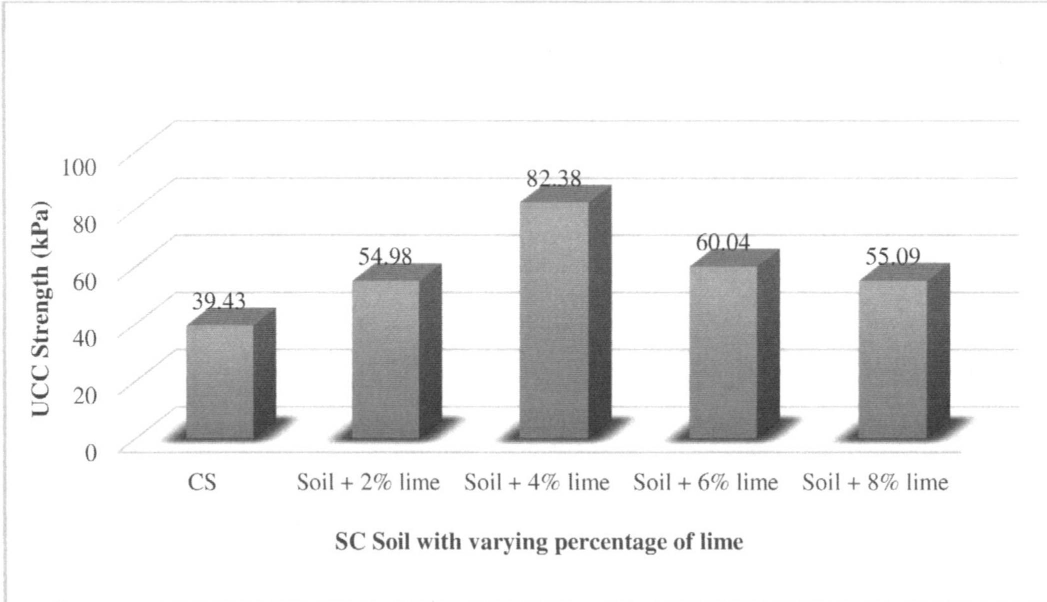

Figure 1. UCC strength of SC soil with varying percentage of lime.

From the results, it can be observed that addition of quick lime to SC soil increased the UCC strength of the soil after 3 days curing. As the percentage of lime increased, UCC strength of SC soil also increased up to 4% lime beyond which the UCC strength decreased. For soil mixed with 4% lime, maximum UCC strength was found to be 82.38 kPa and the increase in UCC strength of 2.09 times the UCC strength of CS was observed. Thus optimum percentage of lime to be added to SC soil is 4% based on the UCC strength attained. Decrease in UCC strength for the soil mixed with lime more than the optimum percentage may be attributed to the fact that, there were no more silica or/and alumina minerals available within the soil to react with the calcium found in excess lime thus resulting in decreased UCC strength of the soil beyond the optimum percentage of 4% lime. According to [7], peak soil strength was attained at an optimum lime content which for many soils lies within the range of 3% to 9%.

2.1.2. Soaked CBR test conducted on SC soil with varying lime content

Soaked CBR test was conducted as per IS:2720 (Part 16) – 1987 [42] on CS and on SC soil with varying lime content (2%, 4%, 6% and 8%) after 3 days soaking in water. To prepare the soil specimen for soaked CBR test, soil and quick lime was first mixed in dry condition and then water equal to OMC was added slowly and mixed with soil and lime to form a homogeneous mixture. This soil lime mixture was compacted in CBR mould at 97% relative compaction by means of dynamic compaction and then soaked in water for a period of 3 days. After 3 days curing period, soaked CBR test was conducted on CS and on soil with varying lime content to determine the ideal percentage of lime to be added to SC soil based on the improvement observed in soaked CBR strength of soil. Table 3 shows the soaked CBR strength of SC soil mixed with varying percentage of lime after 3 days curing which is represented graphically in Figure 2.

Table 3. Soaked CBR value of SC soil with varying percentage of lime.

Soil Sample	CS (Soil + 0% lime)	Soil + 2% lime	Soil + 4% lime	Soil + 6% lime	Soil + 8% lime
Soaked CBR (%)	2.98	2.13	2.56	2.13	2.13

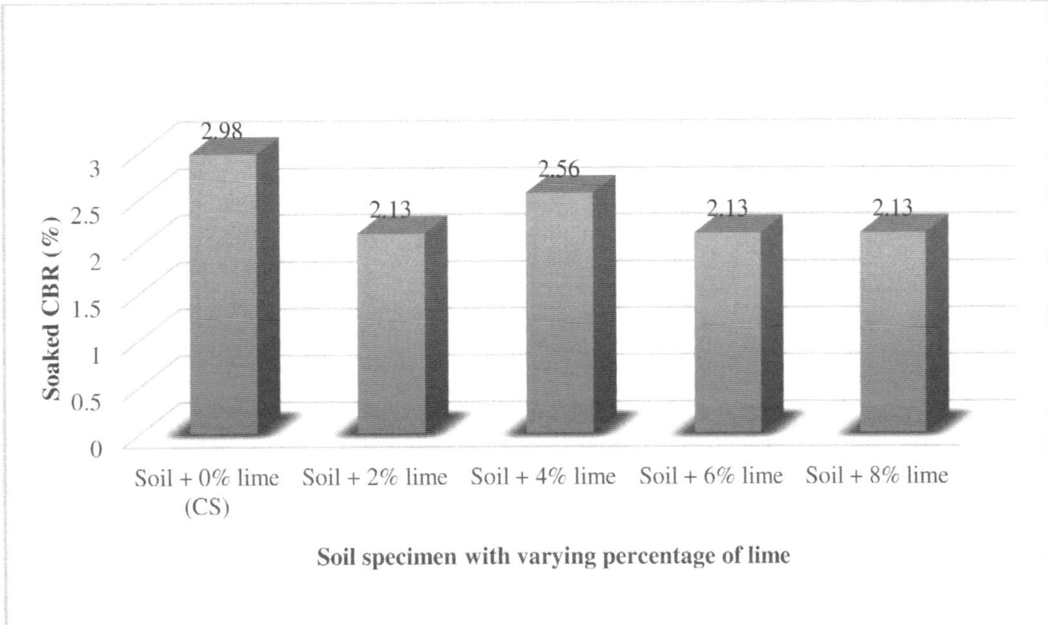

Figure 2. Soaked CBR value of SC soil with varying percentage of lime.

Based on the results of soaked CBR test, it can be observed that CBR value of SC reduced on addition of varying lime content after 3 days curing. This reduction in CBR strength after 3 days curing may be due to the fact that more time is required for the pozzolanic reaction to complete between the silica and/or alumina minerals found in soil and calcium found in lime, leading to the formation of cementitious products like CSH and CAH which are responsible for the strength gain within soil. As per [7, 9], CBR strength of soil increased drastically with increasing curing period. For SC soil mixed with 4% lime, CBR strength of 2.56% was attained and the decrease in CBR value of 0.859 times that of CS was observed.

2.2. SC Soil mixed with 4% lime and varying percentage of GGBS

SC soil with optimum percentage of quick lime (4%) was mixed with varying percentage of GGBS (20%, 25%, 30%, 35% and 40%) and allowed to cure for a period of 3 days. After 3 days curing, UCC test and soaked CBR test were conducted on SC soil with 4% lime and varying GGBS content to determine the ideal percentage of GGBS to be added to SC soil containing 4% lime based on the UCC strength gain and soaked CBR strength gain observed in the experimental work.

2.2.1. UCC test conducted on SC soil with 4% lime and varying GGBS content

Soil specimen for UCC test was prepared by first mixing SC soil, 4% quick lime and varying percentages of GGBS in dry state and then water equal to OMC was added slowly to this dry mixture of soil-lime-GGBS to form a homogenous mixture. Soil-lime-GGBS mixture was then allowed to cure for a period of 3 days by covering it with polythene sheets to avoid air drying of the specimen. After 3 days curing, the soil specimen for UCC test was prepared at 97% relative compaction and the UCC test was conducted as per IS:2720 (Part 10) – 1991 [41]. The results of the UCC test was then compared with that of CS and with soil containing 4% lime. Figure 3 shows the stress strain deformation curve obtained from UCC test for SC soil mixed with 4% lime and varying percentage of GGBS after 3 days curing. Table 4 shows the UCC strength of SC soil mixed with 4% lime and varying percentage of GGBS after 3 days curing which has been represented graphically in Figure 4.

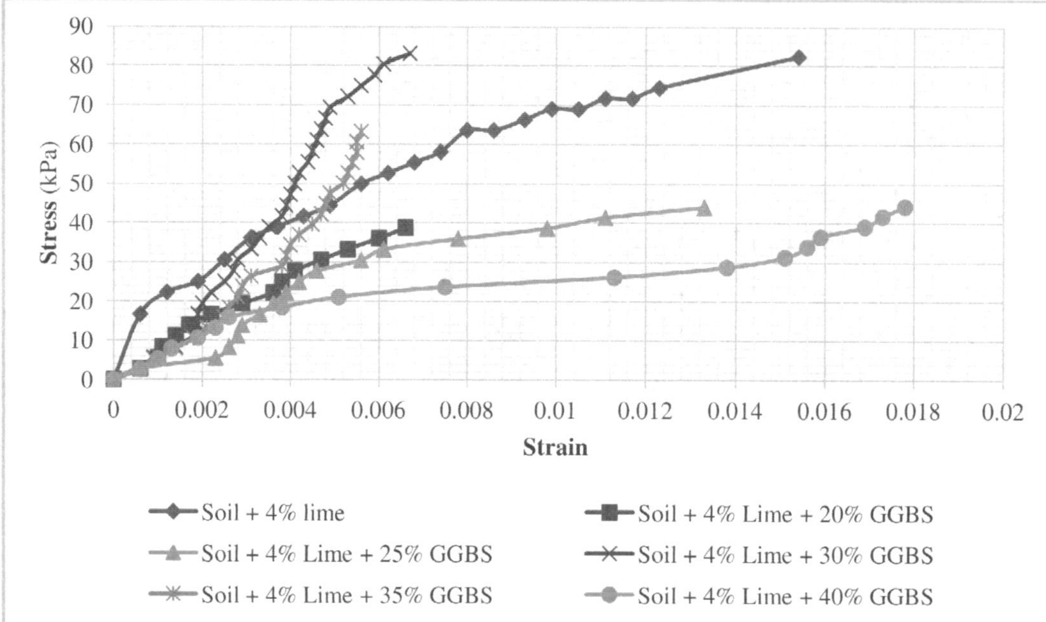

Figure 3. Stress - Strain curve of UCC test conducted on SC soil with 4% lime and varying percentage of GGBS.

Table 4. UCC Strength of SC soil with 4% lime and varying percentage of GGBS.

Soil Specimen	UCC Strength (kPa)	Increase/Decrease in UCC strength w.r.t. CS	Increase/Decrease in UCC strength w.r.t. soil with 4% lime
CS	39.43	-	-
Soil + 4% lime	82.378	2.09 times UCC strength of CS	-
Soil + 4% lime + 20% GGBS	38.79	0.98 times UCC strength of CS	0.47 times UCC strength of soil with 4% lime
Soil + 4% lime + 25% GGBS	44.029	1.12 times UCC strength of CS	0.53 times UCC strength of soil with 4% lime
Soil + 4% lime + 30% GGBS	83.108	2.11 times UCC strength of CS	1.01 times UCC strength of soil with 4% lime
Soil + 4% lime + 35% GGBS	63.271	1.6 times UCC strength of CS	0.77 times UCC strength of soil with 4% lime
Soil + 4% lime + 40% GGBS	44.272	1.12 times UCC strength of CS	0.54 times UCC strength of soil with 4% lime

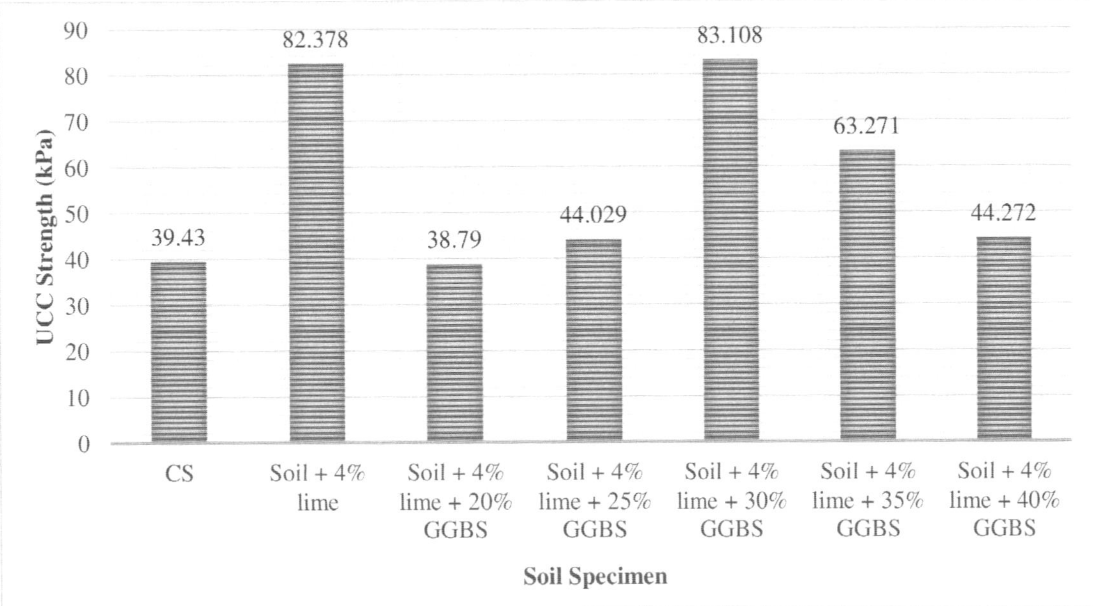

Figure 4. UCC Strength of SC soil with 4% lime and varying percentage of GGBS.

From the results, it can be observed that adding GGBS to SC soil mixed with 4% quick lime was not very effective as far as enhancement in shear strength was concerned. Maximum UCC strength of 83.108 kPa was observed for SC soil mixed with 4% lime and 30% GGBS as compared to the UCC strength of 82.378 kPa observed for SC soil with 4% lime after 3 days curing. Increase in UCC strength of 1.01 times that of soil with 4% lime and 2.11 times that of CS was observed. Decrease in UCC strength on addition of GGBS to SC soil with 4% lime may be attributed to the fact that 3 days curing period may not be sufficient for the pozzolanic reaction to complete between silica, alumina and calcium to form cementitious products like CSH and CAH which are responsible for the strength gain within soil. According to [20], maximum improvement in strength for lime treated soil was achieved for 14 days of curing period and no substantial change in soil strength was observed beyond 14 days curing period. In forthcoming experimental studies on soil treatment with lime and GGBS, the effect of curing period on the strength of soil can be investigated in detail and included in the future scope of the work.

2.2.2. Soaked CBR test conducted on SC soil with 4% lime and varying GGBS content

Soaked CBR test was carried out as per IS:2720 (Part 16) – 1987 [42] for SC soil mixed with 4% quick lime and varying percentage of GGBS (20%, 25%, 30%, 35% and 40%). Soil specimen for soaked CBR test was prepared by mixing the soil, 4% quick lime and varying percentages of GGBS in dry state and then adding water equal to OMC slowly to this dry mixture of soil-lime-GGBS to form a homogenous mixture. Soil specimen for soaked CBR test was then prepared at 97% relative compaction based on the MDD value obtained from SPC test by means of dynamic compaction. Prepared soil specimen was soaked in water for a period of 3 days and after 3 days curing the soaked CBR test was conducted and the results were compared with that of CS and with soil containing 4% lime. Figure 5 shows the load penetration curves obtained from soaked CBR test conducted on SC soil mixed with 4% lime and varying percentage of GGBS after 3 days curing. Table 5 shows the soaked CBR value of SC soil mixed with 4% lime and varying percentage of GGBS after 3 days curing which has been represented graphically in Figure 6.

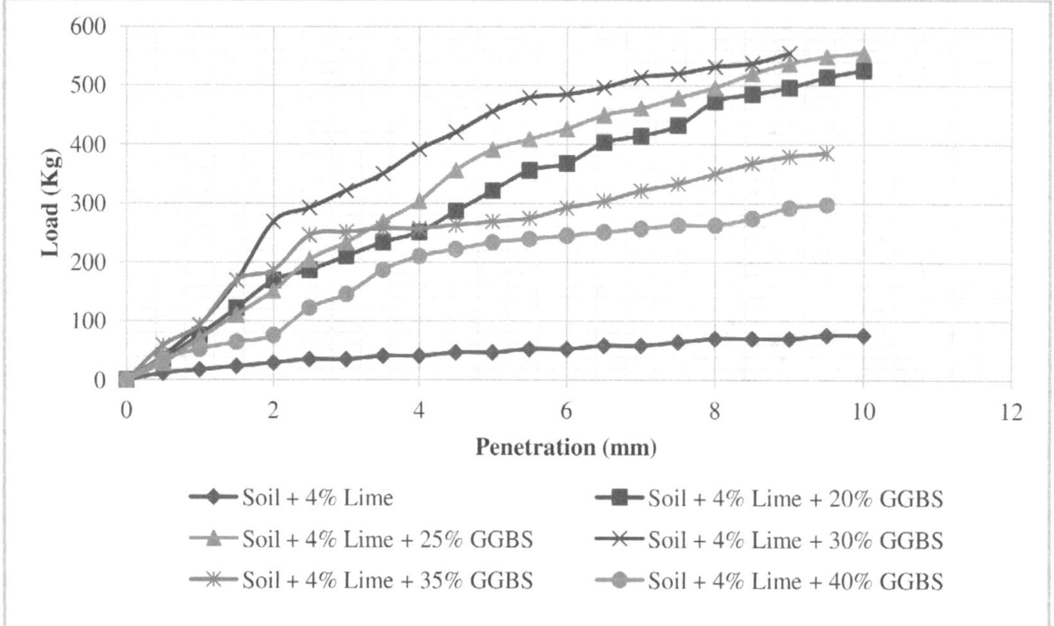

Figure 5. Load Penetration curves of soaked CBR test for SC soil with 4% lime and varying percentage of GGBS.

Table 5. Soaked CBR value of SC soil with 4% lime and varying percentage of GGBS.

Soil Specimen	Soaked CBR values (%)	Increase in soaked CBR w.r.t CS	Increase in soaked CBR w.r.t soil with 4% lime
CS	2.98	-	-
Soil + 4% lime	2.56	0.86 times CBR of CS	-
Soil + 4% lime + 20% GGBS	15.63	5.24 times CBR of CS	6.11 times CBR of soil with 4% lime
Soil + 4% lime + 25% GGBS	19.04	6.39 times CBR of CS	7.44 times CBR of soil with 4% lime
Soil + 4% lime + 30% GGBS	22.17	7.44 times CBR of CS	8.66 times CBR of soil with 4% lime
Soil + 4% lime + 35% GGBS	17.91	6.01 times CBR of CS	7 times CBR of soil with 4% lime
Soil + 4% lime + 40% GGBS	11.37	3.82 times CBR of CS	4.44 times CBR of soil with 4% lime

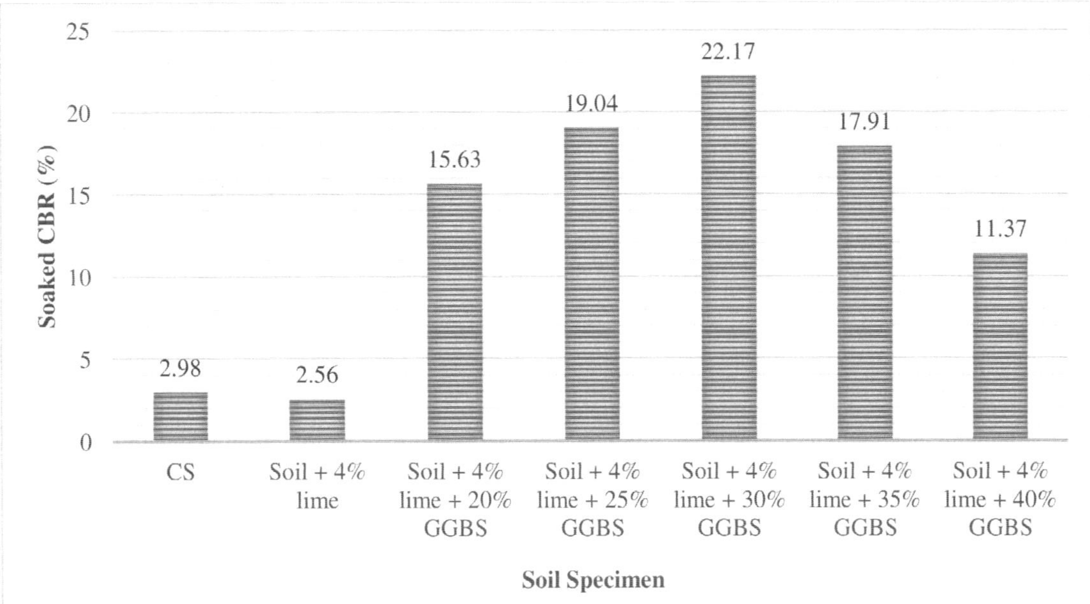

Figure 6. Soaked CBR value of SC soil with 4% lime and varying percentage of GGBS.

From the results, it can be observed that addition of GGBS to SC soil with 4% lime had a positive effect on the CBR strength of the soil. As per [24], the common hydration products in all GGBS-stabilized soils were CSH like compounds. According to [27, 28], lime-activated GGBS was a better binder for clay than cement-activated slag and GGBS required only small quantities of lime for activation to stabilize clay. As the percentage of GGBS increased, CBR value of the SC soil with 4% lime also increased up to 30% GGBS beyond which the CBR value decreased. Maximum CBR strength of 22.17% was attained for SC soil with 4% lime and 30% GGBS and the increase in CBR strength of 7.44 times CBR of CS and 8.66 times CBR of soil with 4% lime was observed. Addition of 4% lime to SC soil decreased the CBR strength to 2.56% from that of 2.98% observed for CS thus concluding that addition of lime to SC soil had a negative effect on the CBR strength of SC soil after 3 days of curing but addition of GGBS along with 4% lime improved the CBR strength drastically within 3 days of curing. Also addition of GGBS to SC soil with 4% lime showed prominent improvement in CBR strength but not much increase was observed in the UCC strength after 3 days curing. This may be due to the presence of excess water in soaked CBR test to complete the pozzolanic reaction between silica, alumina and calcium to form CSH and CAH whereas in case of UCC test, quantity of water equal to OMC added to prepare the soil specimen may not be sufficient to complete the pozzolanic reaction to form CSH and CAH. Thus it can be concluded that addition of GGBS to SC soil with 4% lime is more suitable for enhancing the CBR strength of subgrade soil consisting of SC soil for the design of flexible pavements. Increased CBR strength of the SC subgrade would decrease the thickness of the flexible pavement, thus decreasing the construction cost and also reducing the quantity of materials required in the base and sub-base course of flexible pavement leading to the preservation of natural resources for future generations. As per [3], lime stabilized soil showed significant improvement in CBR strength, which, when incorporated in flexible pavement design, reduced the pavement thickness resulting in substantial savings.

3. Conclusion

Addition of quick lime to SC soil increased the UCC strength but reduced the soaked CBR strength of SC soil after 3 days curing. Maximum UCC strength of 82.38 kPa was attained for SC soil mixed with 4% lime and the increase in UCC strength of 2.09 times that of CS was observed. For SC soil mixed with 4% lime, soaked CBR strength of 2.56% was attained and the decrease in CBR value of 0.859

times that of CS was observed thus concluding that addition of lime to SC soil had a negative effect on the CBR strength of SC soil after 3 days of curing.

For SC soil mixed with 4% lime and 30% GGBS, maximum UCC strength of 83.108 kPa was observed and the increase in UCC strength of 1.01 times that of soil with 4% lime and 2.11 times that of CS was observed. Thus addition of GGBS to SC soil mixed with 4% lime was not very effective as far as enhancement in shear strength was concerned but it had a positive effect on the CBR strength of the soil. Maximum CBR strength of 22.17% was attained for SC soil with 4% lime and 30% GGBS and the increase in CBR strength of 7.44 times CBR of CS and 8.66 times CBR of soil with 4% lime was observed. Thus it can be concluded that addition of GGBS to SC soil with 4% lime is more appropriate for enhancing the CBR strength of SC subgrade soil for the purpose of designing flexible pavements.

4. References

[1] Eades J L and Grim R E 1960 The reaction of hydrated lime with pure clay minerals in Soil Stabilization *U.S. Highway Research Board Buletin* **262** 51 - 63

[2] Jawad I T, Taha M R, Majeed Z H and Khan T A 2014 Soil stabilization using lime: Advantages, disadvantages and proposing a potential alternative *Research Journal of Applied Sciences, Engineering and Technology* **8(4)** 510-520

[3] Qubain B S, Seksinsky E J and Li J 2000 Incorporating subgrade lime stabilization into pavement design *Transportation Research Record* **1721(1)** 3-8

[4] Negi A S, Faizan M, Siddharth D P and Singh R 2013 Soil stabilization using lime *International journal of innovative research in science, engineering and technology* **2(3)** 448-453

[5] Amadi A A, Okeiyi A 2017 Use of quick and hydrated lime in stabilization of lateritic soil: comparative analysis of laboratory data *Geo-Engineering* **8(3)** https://doi.org/10.1186/s40703-017-0041-3

[6] Ghobadi M H, Abdilor Y and Babazadeh R 2014 Stabilization of clay soils using lime and effect of pH variations on shear strength parameters *Bulletin of Engineering Geology and the Environment* **73 (2)** 611-619

[7] Liu M D, Pemberton S and Indraratna B 2010 A study of the strength of lime treated soft clays *International Symposium and Exhibition on Geotechnical and Geosynthetics Engineering: Challenges and Opportunities in Climate Change* Bangkok Thailand

[8] Stoltz G, Cuisinier O and Masrouri F 2014 Weathering of a lime-treated clayey soil by drying and wetting cycles *Engineering Geology* **181** 281-289

[9] Kavak A 1996 The behavior of lime stabilized clays under cyclic loading Ph.D. dissertation Bogazici University

[10] Croft J B 1968 The structures of soils stabilized with cementitious agents *Engineering Geology* **2(2)** 63-80

[11] Haas S and Ritter H J 2019 Soil improvement with quicklime–long-time behaviour and carbonation *Road Materials and Pavement Design* **20(8)** 1941-1951

[12] Bell F G 1996 Lime stabilization of clay minerals and soils *Engineering Geology* **42** 223-237

[13] Harish G R 2017 Studies on stabilization of black cotton soil using lime *International Research Journal of Engineering and Technology* **4(6)** 1725-1727

[14] Porbaha A, Shibuya S and Kishida T 2000 State of the art in deep mixing technology, Part III: Geomaterial Characterization, *Ground Improvement* **3** 91-110

[15] Kogbara R B and Al-Tabbaa A 2011 Mechanical and leaching behaviour of slag-cement and lime-activated slag stabilised/solidified contaminated soil *Science of the Total Environment* **409(11)** 2325-2335

[16] Broderick G P and Daniel D E 1990 Stabilizing compacted clay against chemical attack *ASCE J Geotech Eng Div.* **116(10)** 1549–1567

[17] Schanz T and Elsawy M B 2015 Swelling characteristics and shear strength of highly expansive clay–lime mixtures: A comparative study *Arabian Journal of Geosciences* **8(10)** 7919-7927

[18] Clare K E and Crunchley A E 1957 Laboratory experiments in the stabilization of clays with hydrated lime *Geotechnique* **7** 97–111

[19] Joel M and Edeh J E 2015 Comparative analysis of cement and lime modification of Ikpayongo laterite for effective and economic stabilization *Journal of emerging trends in engineering and applied sciences* **6(1)** 49-56
[20] Sahoo J P and Pradhan P K 2010 Effect of lime stabilized soil cushion on strength behaviour of expansive soil *Geotechnical and Geological Engineering* **28(6)** 889-897
[21] Sharma A K and Sivapullaiah P V 2017 Swelling behaviour of expansive soil treated with fly ash–GGBS based binder *Geomechanics and Geoengineering* **12(3)** 191-200
[22] Li W, Yi Y and Puppala A J 2019 Utilization of carbide slag-activated ground granulated blastfurnace slag to treat gypseous soil *Soils and Foundations* **59(5)** 1496-1507
[23] Wu H L, Jin F, Bo Y L, Du Y J and Zheng J X 2018 Leaching and microstructural properties of lead contaminated kaolin stabilized by GGBS-MgO in semi-dynamic leaching tests *Construction and Building Materials* **172** 626-634
[24] Yi Y, Liska M, Jin F and Al-Tabbaa A 2016 Mechanism of reactive magnesia–ground granulated blastfurnace slag (GGBS) soil stabilization *Canadian Geotechnical Journal* **53** (5) 773-782
[25] Celik E and Nalbantoglu Z 2013 Effects of ground granulated blastfurnace slag (GGBS) on the swelling properties of lime-stabilized sulfate-bearing soils *Engineering Geology* **163** 20-25
[26] Tasong W A, Wild S and Tilley R J 1999 Mechanisms by which ground granulated blastfurnace slag prevents sulphate attack of lime-stabilised kaolinite *Cement and concrete research* **29(7)** 975-982
[27] Obuzor G N, Kinuthia J M and Robinson R B 2012 Soil stabilisation with lime-activated-GGBS—A mitigation to flooding effects on road structural layers/embankments constructed on floodplains *Engineering Geology* **151** 112-119
[28] Obuzor G N, Kinuthia J M and Robinson R B 2011 Enhancing the durability of flooded low-capacity soils by utilizing lime-activated ground granulated blastfurnace slag (GGBS) *Engineering Geology* **123(3)** 179-186
[29] Islam S, Haque A and Wilson S A 2014 Effects of curing environment on the strength and mineralogy of lime-GGBS–treated acid sulphate soils *Journal of materials in civil engineering* **26(5)** 1003-1008
[30] Yi Y, Zheng X, Liu S and Al-Tabbaa A 2015 Comparison of reactive magnesia-and carbide slag-activated ground granulated blastfurnace slag and Portland cement for stabilisation of a natural soil *Applied Clay Science* **111** 21-26
[31] Yi Y, Liska M and Al-Tabbaa A 2014 Properties of two model soils stabilized with different blends and contents of GGBS, MgO, lime, and PC *Journal of Materials in Civil Engineering* **26(2)** 267-274
[32] Wang F, Cai Z and Xu J 2020 Effects of curing temperature and dosage on the performance of GGBS-MgO-CaO in stabilizing/solidifying heavy metal–contaminated site soil *Journal of Testing and Evaluation* **48(5)** 3515-3524
[33] He J, Wang X Q, Su Y, Li Z X and Shi X K 2019 Shear strength of stabilized clay treated with soda residue and ground granulated blast furnace slag *Journal of Materials in Civil Engineering* **31(3)** 06018029 https://doi.org/10.1061/(ASCE)MT.1943-5533.0002629
[34] Fasihnikoutalab M H, Pourakbar S, Ball R J, Unluer C and Cristelo N 2020 Sustainable soil stabilisation with ground granulated blast-furnace slag activated by olivine and sodium hydroxide *Acta Geotechnica*, **15(7)**1981-1991
[35] Kogbara R B, Yi Y and Al-Tabbaa A 2011 Process envelopes for stabilisation/solidification of contaminated soil using lime–slag blend *Environmental Science and Pollution Research* **18(8)** 1286-1296
[36] IS: 2720 (Part III/Sec 2) – 1980, Indian Standard Methods of Test for Soils: Determination of Specific Gravity - Fine, Medium and Coarse Grained Soils.
[37] IS:2720 (Part 5) - 1985, Indian Standard Method of Test for Soils: Determination of Liquid and Plastic Limit.

IOP Conf. Series: Materials Science and Engineering **989** (2020) 012028 doi:10.1088/1757-899X/989/1/012028

[38] IS:2720 (Part 4) – 1985, Indian Standard Methods of Test for Soils: Grain Size Analysis.
[39] IS:1498 - 1970, Indian Standard Classification and Identification of Soils for General Engineering Purposes.
[40] IS:2720 (Part VII) – 1980, Indian Standard Methods of Test for Soils: Determination of Water Content - Dry Density Relation using Light Compaction.
[41] IS:2720 (Part 10) – 1991, Indian Standard Methods of Test for Soils: Determination of Unconfined Compressive Strength.
[42] IS:2720 (Part 16) – 1987, Indian Standard Methods of Test for Soils: Laboratory Determination of CBR.

IOP Conf. Series: Materials Science and Engineering **989** (2020) 012029 doi:10.1088/1757-899X/989/1/012029

Earthquake data analysis and data visualization of Maharashtra state, India from 1912 to 2009 using R programming

Vaishnavi Battul[1,3] Helen Santhi M[1] and Malathi G[2]

[1]School of Civil Engineering, Vellore Institute of Technology, Chennai, TamilNadu, India
[2] School of Computer Science and Engineering, Vellore Institute of Technology ,Chennai, Tamil Nadu, India

[3]Email: vaishnavi.2019@vitstudent.ac.in

Abstract. This paper presents the seismic data analysis and visualization using R programming for the state of Maharashtra, India. Maharashtra has witnessed a significant number of earthquakes ranging from low to high in magnitude. The past earthquake data of a particular location are useful to the civil and structural engineers for the safe and efficient planning, designing and construction of seismic resistant buildings in that locality. In this paper the past earthquake data of Maharashtra from the year 1912 to 2009 are collected and analyzed. There are 53 earthquake data with magnitudes ranging from 3.6 to 7.2. This analysis shows the relationship between magnitude and depth, intensity and PGA, visualization of classification of earthquakes depends on magnitude and intensity.

1. Introduction

Data analysis and data visualization are being used in almost all fields of science, engineering and business. In order to understand and make conclusions on any project, interpretation of data is the key function which majorly depends on how efficiently the data analysis and visualization are carried out. Many researchers have started analyzing earthquake data worldwide [1, 4, 5] and contributing towards the design of seismic resistant infrastructure. Earthquake is one of the major disasters threatening the world and the assessment of risk involved is to be evaluated to protect the life of people and their possessions. In India, most of the parts are vulnerable to earthquake and the loss of life and damages to the infrastructure are huge in some of the past earthquakes. The ill effects of earthquakes can be minimized by understanding the characteristics of earthquakes using the past seismic history of any particular region in the country. The seismic data available can be analysed [7, 9, 10] visualized [2, 3, 6, 8] and interpreted to protect the life and to reduce the damages to the infrastructure from future earthquakes using sophisticated computer tools. In this paper, data analysis and data visualization of earthquake data between 1912 and 2009 in Maharashtra State, India is carried out using R software. The graphical representation of the earthquake data in the study area gives better understanding of the site specific characteristics of the past earthquakes.

2. Data analysis and data visualization with ggplot using R

Following are the packages and libraries that are loaded to attain useful outputs related to seismic activity in Maharastra.

```
install.packages("dplyr")
install.packages("ggmap")
install.packages("ggplot2")
library(dplyr) library(ggmap)
library(ggplot2)
```

IOP Conf. Series: Materials Science and Engineering **989** (2020) 012029 doi:10.1088/1757-899X/989/1/012029

2.1 Maharashtra earthquake dataset

The past earthquake data of Maharashtra State are collected from the year 1912 to 2009 and the following steps are followed to read and view the data (Figure 1).

Input:

maha <- read.csv("maha.csv", stringsAsFactors = FALSE)

summary() function

Input: summary(maha)

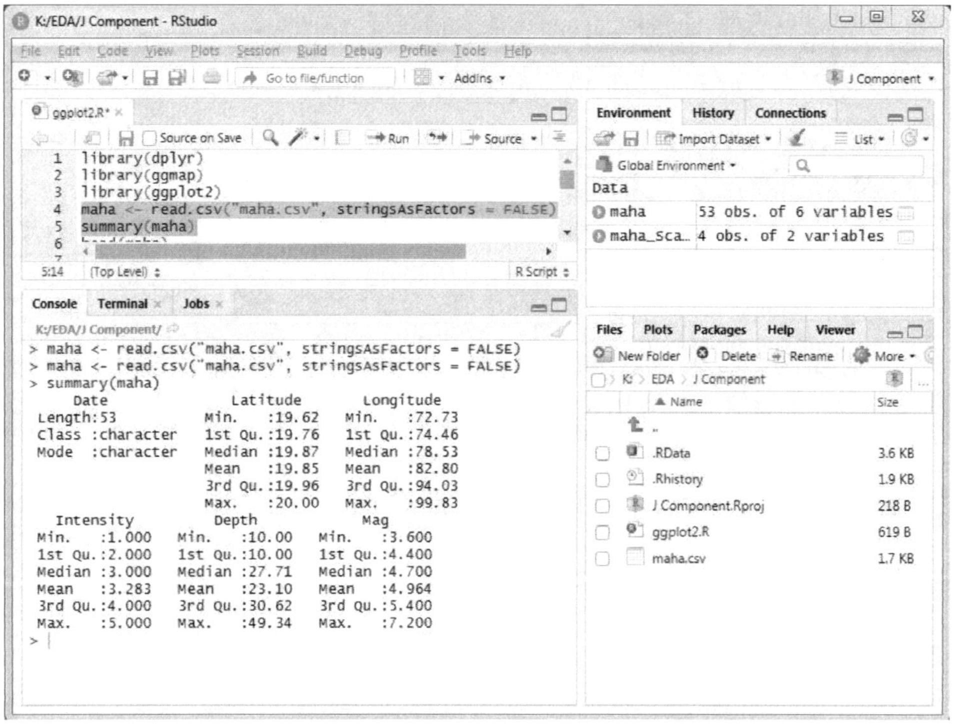

Figure 1. Data downloaded ready to read and view.

head() function

- Input:
  head(maha)

It displays the data in Tabular form with headings to the columns

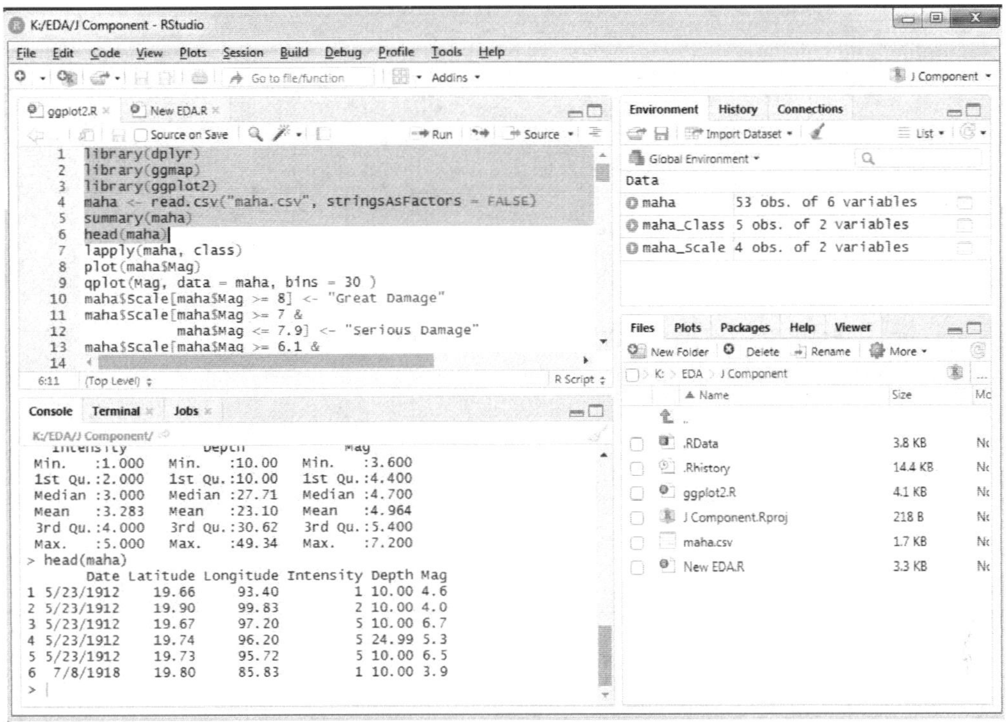

Figure 2. Data arranged in columns.

lapply() function

- To display the earthquake data the above outputs from the summary () and head() functions are helpful.
- The lapply() command is useful to see classes of data.
- Input
  lapply(maha, class)

2.2 Maharashtra earthquake plot

Figure 3 depicts the plot of the magnitude of each earthquake. It is seen that the plot is scattered because of the frequency of occurrence of earthquakes. From the plot it is recognized that most of the earthquakes are lower scaled ones.

Input

#Plot of Magnitude of each
Earthquake plot(maha$Mag)

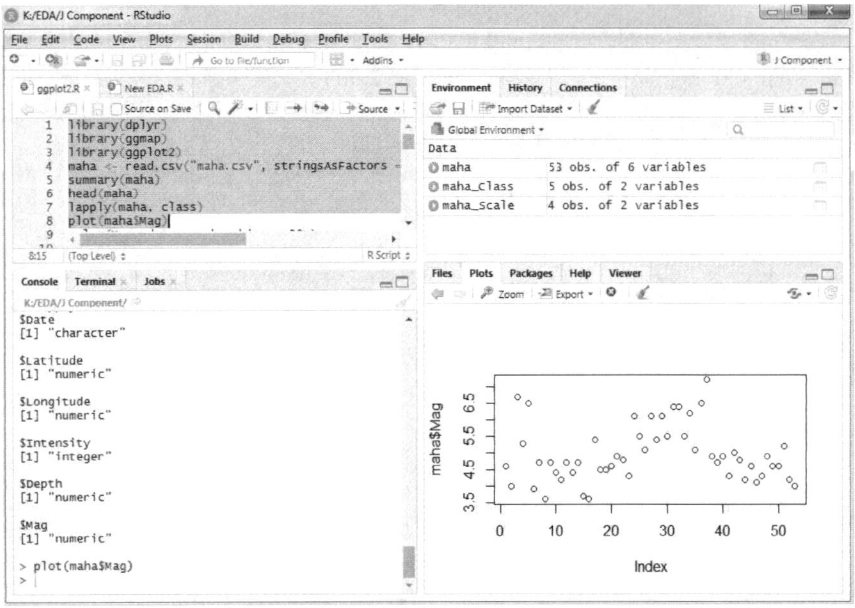

Figure 3. Earthquake magnitude plot (Scattered).

The above details can also be presented in another way of plotting as shown Figure 4 that would be easily understandable, **qplot()**.

```
#qplot of Magnitude and Total Count
qplot(Mag, data = maha, bins = 50)
```

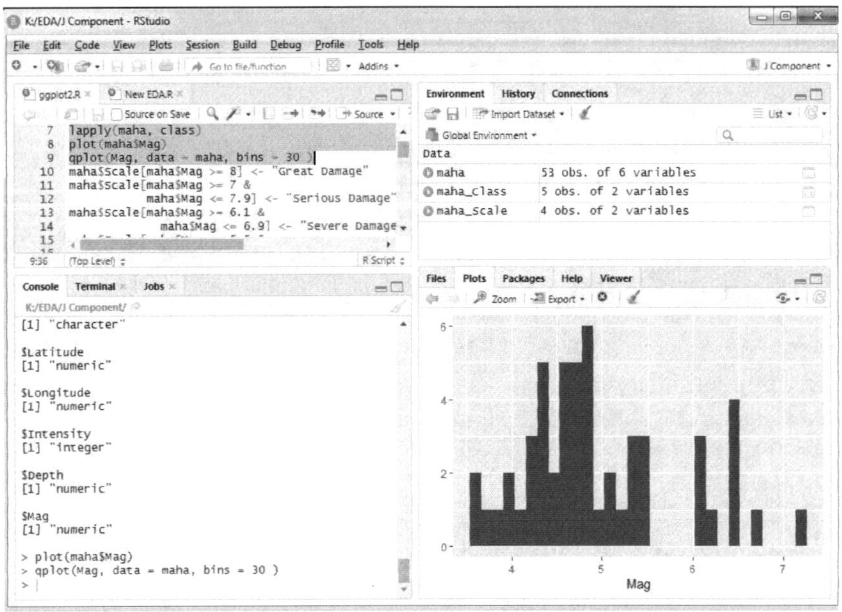

Figure 4. Earthquake magnitude plot (Bar chart).

2.3 Scale of earthquake magnitude

Generally the earthquake magnitude is scaled based on the severity of the earthquakes. Table 1 shows the various levels of earthquake scale adopted in the study and it is found that the damage level falls under minor damage.

Table 1. Damage level for different earthquake magnitude.

Magnitude	Damage level
< 2.5	Limited damage represents very small damage
2.5 to 5.4	Minor damage represents small damage
5.5 to 6.0	Slight damage represents noticeable damage
6.1 to 6.9	Severe damage represents highly noticeable damage
7.0 to 7.9	Serious damage represents very severe damage
>8.0	Great damage represents very high damage

The damage levels of earthquake in Maharastra state can be visualized using the following steps and the plot is shown in figure 5.

```
ggplot(data = maha_Scale, aes(x = Scale, y = n, fill = Scale)) +
geom_bar(stat = "identity") +

xlab("Earthquake Magnitude by Scale")
+ ylab("Total Earthquakes") +

ggtitle("Maharashtra Earthquake Magnitude Scale") +
theme(axis.text.x = element_text(angle = 45, hjust = 1)) +
theme(plot.title = element_text(hjust = 0.5))
```

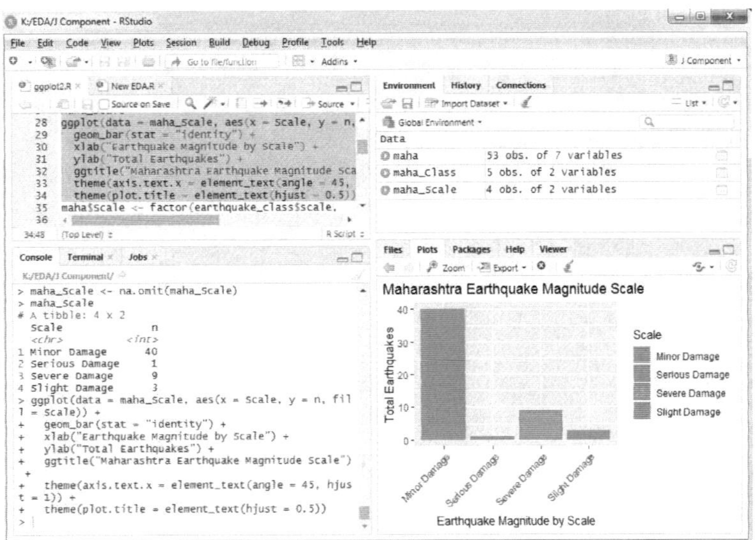

Figure 5. Visualization of earthquake damage levels.

2.4 Earthquake Scatter plot of magnitude and depth

Earthquakes represented in terms of magnitude and depth would be useful to check the influence od depth on magnitude of any earthquake. Here the earthquakes are classified based on the depth of origin of earthquake as follows. If the depth is from 0 to 70 km, it is shallow earthquake; from 70 to 300 km, it is intermediate earthquake; from 300 to 700 km, it is deep earthquake.

In order to find the influence of depth on the earthquake magnitude, a scatter plot is produced. Here the magnitude is scaled between zero and eight. From the plot, it can be seen that the correlation between the depth and the magnitude is positive.

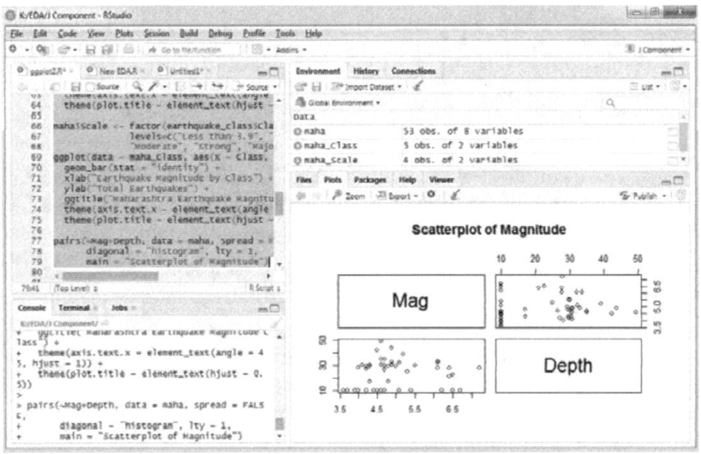

Figure 6. Depth versus magnitude plot.

2.5 get_map() and ggmap() for mapping earthquakes

Now the data is ready and for mapping, get_map() and ggmap() functions are used.

```
(map <- get_map(c(left = 72.25345, bottom = 15.321458, right =
100.099334, top = 21.559652)))
ggmap(map)

ggmap(map, extent = "device")
+ geom_point(data = maha,
aes(x = Longitude, y = Latitude, color = Mag, size = Depth)) +
scale_colour_gradient(low = "yellow", high = "red")
```

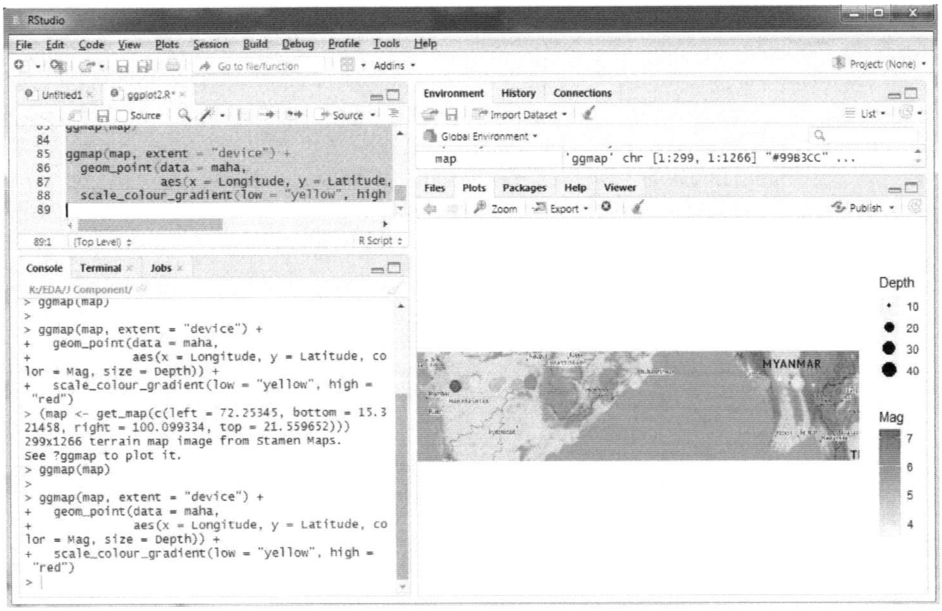

Figure 7. Mapping of earthquakes.

3. Regression model

Multiple regression analysis is carried out with three independent variables such as epicentral distance, depth and magnitude of earthquake and peak ground acceleration (PGA) as dependant variable. From the intercept and coefficient values, the mathematical equation is obtained as below.

$$Y = 1.963736194 \times 10^{-3} X_1 - 2.874770724 \times 10^{-4} X_2 + 7.174167726 \times 10^{-1} X_3 - 1.639089517 \times 10^{-1}$$

$$R^2 = 0.8054246502$$

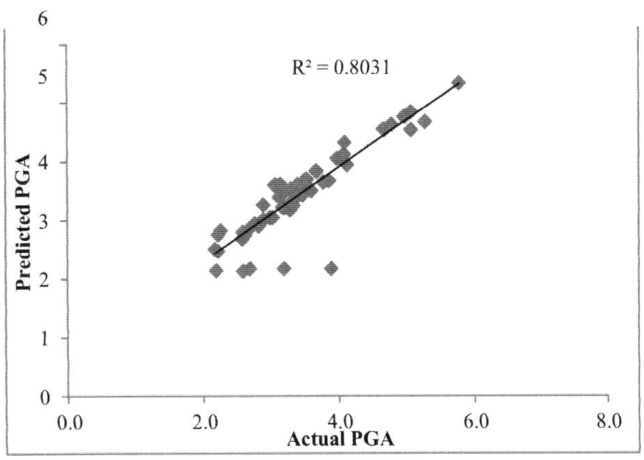

Figure 8. Correlation between actual and predicted PGA

4. Conclusion

The past earthquake data of Maharashtra state are collected, analyzed and visualized using R programming to understand the earthquake activities easily and meaningfully. Also the equation for PGA is arrived at from multiple regression analysis to predict the earthquake size. Similar kind of investigation can be conducted for all the vulnerable regions in the country to predict the characteristics of earthquakes and to design safer infrastructure.

IOP Conf. Series: Materials Science and Engineering **989** (2020) 012029 doi:10.1088/1757-899X/989/1/012029

5. References

[1] Admasu F and Toennies K 2010 Anisotropic 3D seismic features for robust horizons correlation across faults *Proceed.IEEE Inte. Con. Image Pro.* (Carolina USA) 221

[2] Chourasia A, Cutchin S and Aagaard B 2008 Visualizing the ground motions of the 1906 San Francisco earthquake comp *Geosciences* **34(12)** 1798

[3] Li W and Wang D 2012 Visualization system of massive 2D seismic data *Inform.Engg.Appl* **15(4)** 748

[4] Liu R C, Guo H Q and Yuan X R 2014 Seismic structure extraction based on multi-scale sensitivity Analysis *J. Visual.***17 (3)**157

[5] Ma R A, Zhang A H and Yang J Y 2005 A fast volume rendering method for 3D seismic volume data *J.Com.-Aided Des.& Com. Gra.***17(5)** 1008

[6] Naka T, Yamada M, Endo M, Miyazaki S and Hasegawa J 2006 Visualization of seismic-center distribution data for earthquake prediction Proc. of Nicograph Intl

[7] Patel D, Giertsen C, Thurmond J, Gjelberg J and Groller M E 2008 The seismic analyzer-interpreting and illustrating 2D seismic data *IEEE Trans. Visual.Com.Gra.***14 (6)** 1571

[8] Tung-Ju Hsieh, Cheng-Kai Chen and Kwan-Liu Ma 2010 Visualizing field-measured seismic data Proc. of IEEE Pacific Visualization Symposium 65

[9] Y in L, Li XL, Zheng W, Yin Z, Song L, Ge L and Zeng Q 2019 Fractal dimension analysis for seismicity spatial and temporal distribution in the circum-pacific seismic belt *J.Earth Sys. Sci.***128 (1)** 22

[10] Yu Y, Jia Z, Cao L, Zhao J D, Liu Z W and Liu J L 2018 Fast density-based clustering algorithm for location big data *J. Soft.* **29 (8)** 2470

IOP Conf. Series: Materials Science and Engineering **989** (2020) 012030 doi:10.1088/1757-899X/989/1/012030

Seismic performance assessment of a typified school building damaged during the 2019 Albanian earthquakes

Marjo HYSENLLIU[1], Hüseyin BILGIN[2,3]

[1]Polytechnic University of Tirana, Department of Civil Engineering, 1001, Tirana, Albania

[2]EPOKA University, Faculty of Engineering and Architecture, 1032, Tirana, Albania

[3] Email:hbilgin@epoka.edu.al

Abstract. This study presents the seismic performance evaluation of a typified masonry school building, Shkolla 9-Vjeçare "26 Nëntori", which was damaged by the recent earthquakes which hit Albanian territory on November 26, 2019. This building was constructed in 1985 by following the template school projects in Albanian construction practice. The current situation of the building was evaluated by considering the provisions of modern codes i.e. Eurocode 6, Eurocode 8 and Turkish building Code. Analytical model of the school building was developed using the experimental test results conducted on the brick units and mortar. Nonlinear static pushover analyses were executed to estimate the seismic capacity, the performance point, and the damage limit state levels. The performance based assessment rules were used for this objective. Deficiencies and weak points were identified by the detailed examination of the pushover curves. Results showed that selected school designed per pre-modern code is far from satisfying the desired criteria, suggesting that urgent response and necessary measures should be put into action.

1. Introduction

The school buildings should be seismic resistant in earthquake-prone countries whose closure may cause disruption of community life and hampering education immediately after the earthquakes. Recent devastating earthquakes in Albania and the world have highlighted deficient seismic performance of school buildings [1, 2].

Throughout the history, earthquakes in Albanian territory, like in various earthquake-prone regions of our world, have damaged built environment causing several fatalities and economic loses [3-8]. In this context, structures' vulnerability is a crucial notion to concentrate on to decrease the consequences of devastating events. Usually, unreinforced masonry [URM] structures represent a poorer earthquake response compared to reinforced concrete (RC) structures because of low strength and stiffness of their elements. Therefore, new seismic guidelines comprise suggestions and commentaries aimed at reducing their earthquake vulnerability. However, a significant portion of the current masonry construction has been designed and constructed per older code requirements.

The main building typology in the Albanian building stock consist of URM buildings, RC framed systems and prefabricated concrete wall systems. The majority of current masonry buildings are built following the guidance of the Albanian technical codes and regulations. These codes were initially published and put into force as a legal document in 1963 and latest amendment was done in 1989 and still in law. Most of the existing URM buildings in the country, like in several European countries were constructed considering previous seismic guidelines when earthquake forces were not mandatory, or the design was to reduce the degree of seismic loads. Nonetheless, its conformity conditions were not as clear as those suggested by recent modern seismic codes like Eurocode 6 and Eurocode 8 from European practice. This caused the lack of seismic issues in building design procedures. Several construction companies have recently introduced European norms in their projects. However, design of structures must follow the Albanian technical codes and guidelines according to Albanian legislation in the field of construction [9-10]. Today, Eurocode norms in construction sector are national standards that are used voluntarily [11].

The URM buildings suffered the most by the November 26, 2019 Durrës Earthquake due to several reasons including the poor construction quality, old construction age, interventions, poor workmanship, the design code of the time, lack of maintenance and insufficient repair after former damaging earthquakes. The object of this study, Shkolla 9-Vjeçare Publike "26 Nentori" school in Albanian capital city Tirana also suffered considerable damage during November 26, 2019 Durrës Earthquake sequences. In the on-site studies, the general condition of the building, the presence of observable settlements on the ground and the crack situation in the structure were determined and damage relieves were prepared. The necessary information has been taken from the blueprints and on-site visits to the building.

This aim of this paper is to assess the earthquake performance of a typified low-rise URM school building, constructed following the guidance of pre-modern seismic codes of Albanian Construction practice. To reveal the properties of low-rise current masonry buildings, a damaged URM school building during November 26, 2019 Earthquake was chosen and modelled by using the TREMURI [12]. Structural characteristics such as component sizes, material kinds and loading conditions of the building were obtained from their architectural and structural projects and site inspections on studied building in several Albanian cities. Mechanical features were revealed from experimental tests and used for inelastic analysis. Pushover analyses have been performed to obtain the seismic capacities, the performance points and the damage level states according to Eurocode 8 by using 3muri software [12].

2. Seismicity of the region

Albanian neighborhood is in a quite complicated seismo-tectonic area and susceptible to earthquakes. Faulting regions in East Albania are typically characterized by the impact of normal faulting. Into the Albanian orogeny are evidenced two transverse and three longitudinal active fault zones (Figure 1).

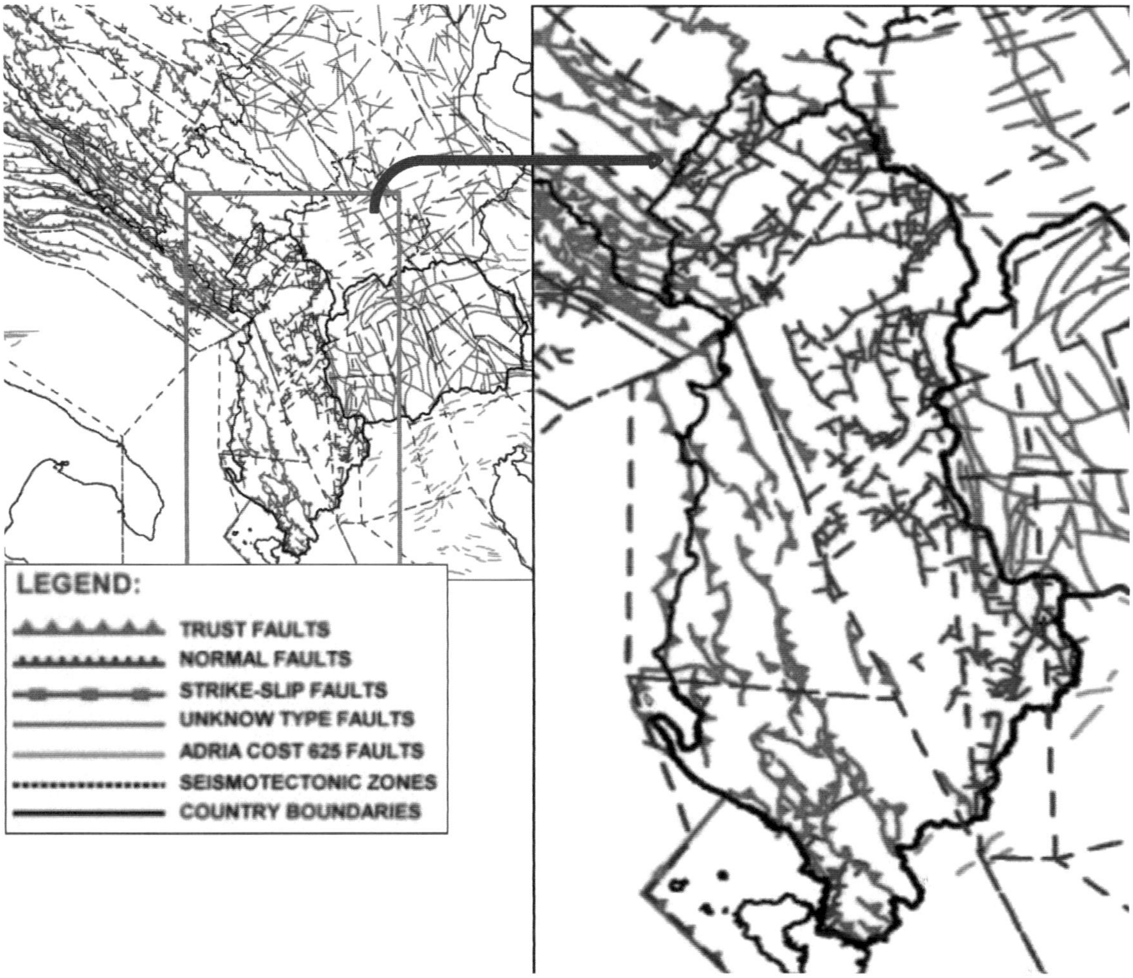

Figure 1. Active faults in Albania (From NATO SfP Project No. 983054 (BSHAP)).

The 2019 Durres earthquake affected zones are induced by Northwest-Southeast striking reverse faults. Blue lines represent the faults triggered for the period of middle Pleistocene-Holocene, the green lines triggered during Pliocene-Lower Pleistocene and the red lines to faults initiated during Pre-Pliocene age. Based on the tectonic structure of the earthquake-affected area and the fault plane solutions of the November 26, 2019 Durrës earthquake, it can be revealed that there is a correlation between fault plane explanations and the geological data on the field.

Albanian seismicity can be described by severe seismic micro activity ($1.0 < M \leq 3.0$), several minor earthquakes ($3.0 < M \leq 5.0$), few mid-sized earthquakes ($5.0 < M \leq 7.0$) and very rarely by powerful earthquakes ($M > 7.0$).

2.1. Seismic hazard maps of Albania

Based on the Albanian seismic map's zonation, it can be stated that the induced intensities from this earthquake under consideration, are in the boundaries stipulated by the Seismic Zonation Map (Figure 2).

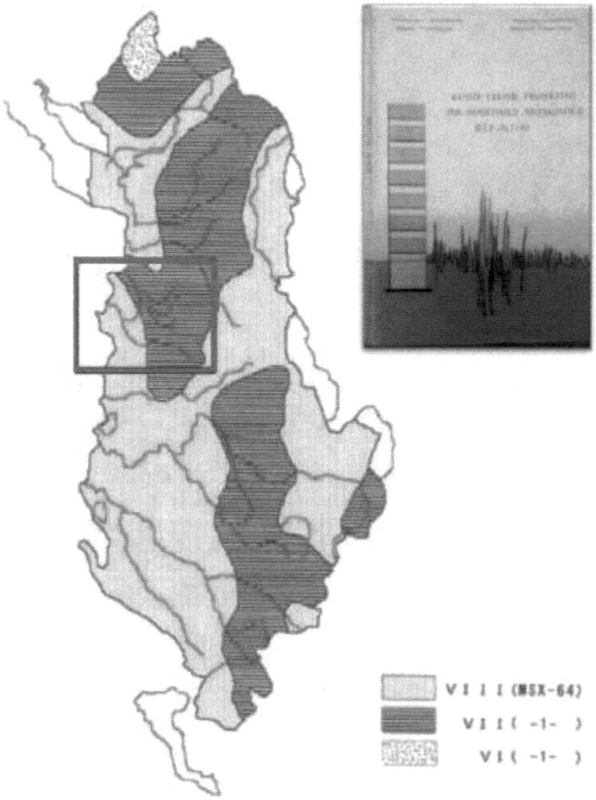

Figure 2. Seismic zonation map of Albania.

2.2. The November 26, 2019 Earthquake and its consequences on Albanian Buildings
A strong earthquake hit the central western part of Albania on November 26, 2019. It was measured as Mw 6.4. Its epicenter was situated off-shore NW Durres, around 7-km North of the city and 30 km from the capital Tirana. Its focal depth was about 10 km [USGS, 2019]. According to the focal plane explanations given by numerous seismological organizations and studies, the main shock was produced by the launch of a NW-SE striking reverse fault. The aftershock sequence until 30 November included 100s of aftershocks, 22 with magnitude larger than M=4.0 and 4 with magnitude larger than M=5.0 (aftershocks by USGS until December 1, 2019).

Regarding the impact on the building stock, the main shock and the following aftershocks induced moderate to severe destruction to buildings in Durrës, Tirana and numerous dwellings of the wider neighborhood. The most earthquake-affected regions are found in the city of Durrës and the town of Thumanë at the central-western Albania. More specifically, Durrës city is located along the coastal part of the central-western Albania in a distance of 30 km West from the Tirana, while Thumanë in a distance of about 25 km NW from the capital city of Tirana. Damage was also observed in Laç town, Fushë-Krujë town, Kamëz as well as capital city, Tirana. Building damage was concentrated along NW-SE (Figure 3.). This direction overlaps along with the strike of the seismogenic faults as it is obtained from the fault plane solutions provided by several seismological institutes and observatories [INGV, 2019 and USGS, 2019].

IOP Conf. Series: Materials Science and Engineering **989** (2020) 012030 doi:10.1088/1757-899X/989/1/012030

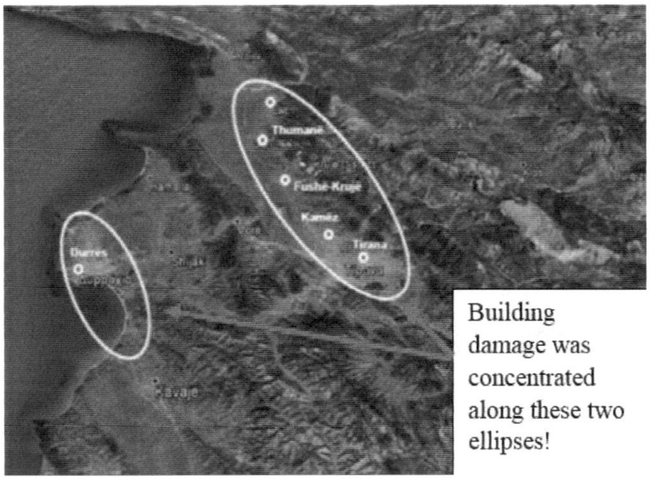

Figure 3. Earthquake-affected area during the November 26, 2019 Durres Earthquake.

3. Current status of the school building and the load bearing system and its adequacy according to seismic guidelines

The Shkolla 9-Vjeçare "26 Nëntori" school and sport hall were built and located in the city of Tirana. General views of the hospital buildings are given in Figure 4.

Figure 4. Front view of the main building under investigation.

The primary dimensions of the load bearing system of the school building were determined with in-situ site investigations and compared with the blueprints of the school designs obtained from national archive. Analyses were carried out according to the prepared load bearing system dimensions. Since the buildings under investigation are old ones, limited number of architectural drawings or details of the initial conditions of the building were reached. Therefore, a detailed inspection of the existing structures was extracted. In these plans; the location and dimensions of the walls, windows and doors were determined. Based on the measurements obtained, structural floor plans of the existing structures were prepared, and structural models were developed accordingly for seismic analysis in computer environment. This building was designed as three stories and did not experience a serious structural intervention since its design. From the site survey and inspections, the floor plans of the blocks were generated and used for detailed seismic analysis (Figure 5).

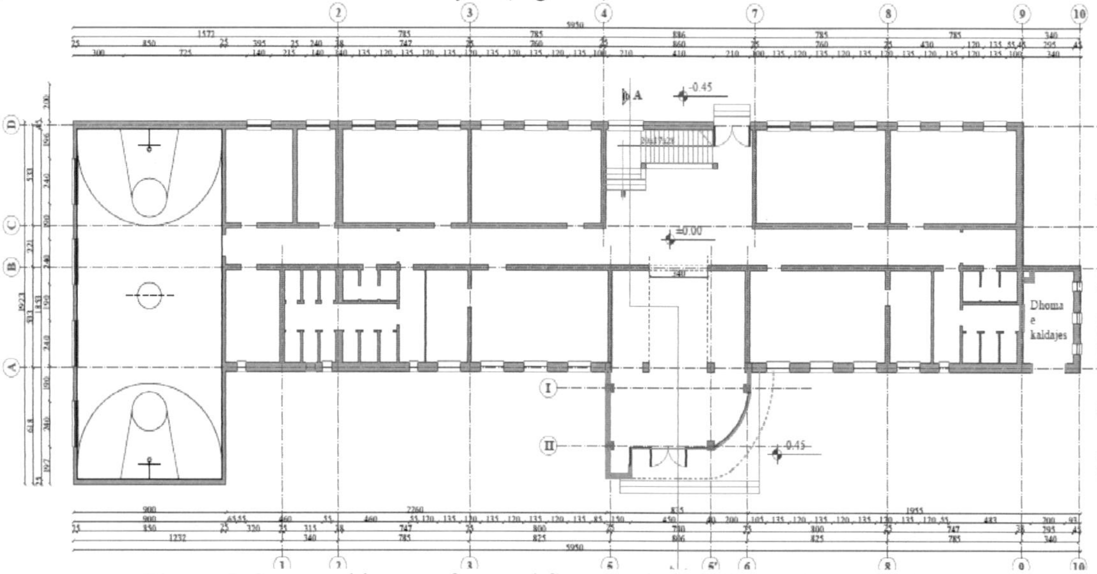

Figure 5. Structural layout of ground floor (main building and sport hall).

To determine the type of wall material used in the masonry structures and the way the walls are built in the corner area, the covering material was removed and examined. Mortar and solid silicate bricks are used for wall construction.

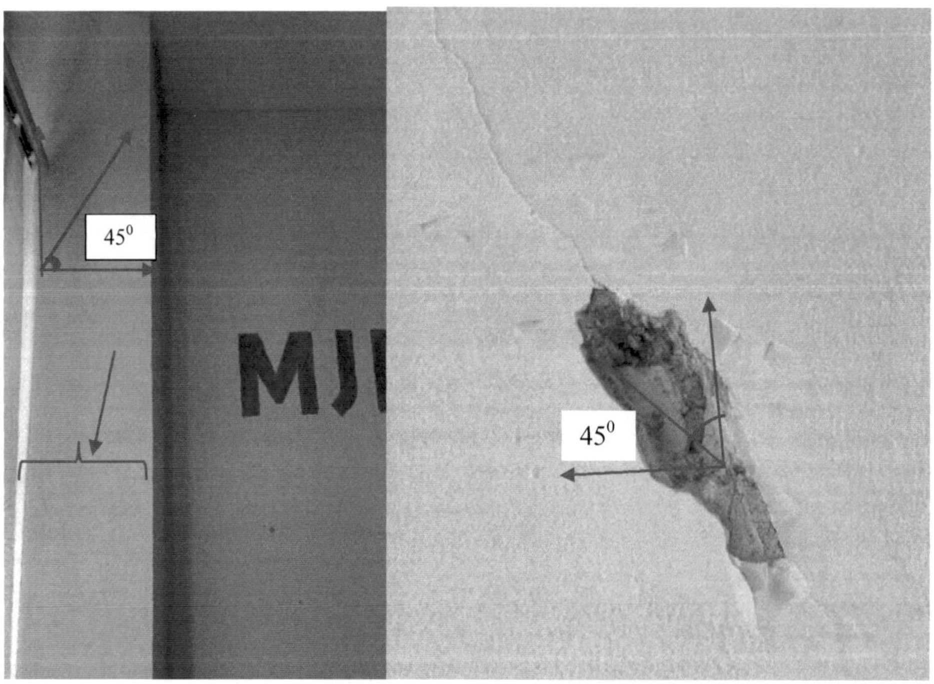

Figure 6. Damage on pier element on the corner and typical shear crack on the load-bearing wall.

During the inspection inside the buildings, a considerable damage was observed on the main building walls. We observed a 450 inclined shear crack on a number of walls on the ground floor (Figure 6.). These structural cracks were formed on the 45 cm thick wall, which was subsequently removed in the upper floors to make additional space, occurred during the Durres earthquake of 26 November 2019.

4. Pushover analyses and damage assessment

The nonlinear static pushover analysis consists of the use of gravity loads and a pre-defined horizontal load pattern. Gravity loading is in place throughout lateral push. Lateral loads are monotonically applied in a step-by-step fashion. The lateral loads are proportional to the product of mass and the first mode-shape amplitude at every story levels. In pushover analysis, the behavior of buildings is measured using a pushover (capacity) curve that symbolizes the relationship between the base shear force and roof displacement. This approach in not only very practical but also gives easy visual graphic of the condition of the building.

4.1. Determination of the material characteristics of the school building

For the analytical modelling of the school building, there are two major issues to be thought: accurate description of the computer model and nonlinear characteristics of materials. URM is a compound building material which comprises of masonry units and mortar as bonding element. Brick and stone are the traditional elements of masonry components, in this case are brick parts. Subjected to the vertical and horizontal loads, load bearing of masonry is considered as the assemblage of the masonry components and mortar is affected by several factors including water absorption, durability, compressive, shear and flexural strengths, and thermal expansion of both materials.

Building block is comprised of two main parts, namely load-bearing blocks, and rigid diaphragms. The blocks are rigid with cavities and the diagrams are typically constructed by continuous RC slabs. For the construction of the school building, calcium silicate solid bricks and cement mortar was used. For the mathematical modelling and the seismic analyses of the school building, material properties obtained from site investigations and experimental test were used. To accurately characterize the strength and structural integrity of the building, inherent characteristics of the material were obtained from the laboratory tests with destructive methods. Samples were taken from the building for material testing. Then, tests were executed on samples to define the compressive strength of the solid bricks and the obtained results are used for seismic analyses.

Building block is comprised of two main parts, namely load-bearing blocks, and rigid diaphragms. The representation of masonry wall blocks in current buildings is a rather complicated procedure. In this paper, mechanical properties were taken by guidelines of the building regulations, blueprints of the original design and the experimental tests on samples from the building. In the case when experimental data is not reached, numerous equations are recommended by various codes and guidelines to evaluate the compressive strength of the masonry walls. In this paper, Eurocode 6 [15] criteria are followed:

Table 1. *Masonry wall data used as Input for analytical modelling.*

Building Type	URM
Compressive strength f_k (MPa)	4.0
Mortar strength f_m (MPa)	5.00
Fracture energy (Tensile) G_t (N/mm)	0.1
Shear strength f_t (N/mm^2) as per EC 6	0.29
E (MPa) as per EC 6	4000
G (MPa) *as per EC 6*	1400
Poisson ratio v	0.2

IOP Conf. Series: Materials Science and Engineering **989** (2020) 012030 doi:10.1088/1757-899X/989/1/012030

Following the Eurocode 6 guidelines [CEN, 2005] together with the appropriate material properties gained from laboratory tests are calculated the necessary input data for mathematical modeling of the building [Table.1]

4.2. Mathematical modeling

Modelling masonry buildings is not an easy job due to the inelastic response of masonry and lack of experimental data concerning the inherent characteristics of masonry structural components. In-situ investigations were performed to obtain the necessary information about the geometry and structural aspects of the model. From the national archives of Albania, the original designs were obtained and checked comparing to the actual status of the current buildings. To define the physical and mechanical characteristics of the materials, laboratory tests were performed on samples taken from the building. Masonry behavior under different loadings may be very complex. To model the behavior of masonry, numerous theories and analytical models are suggested in literature [Lourenço, 2010]. Each of these techniques necessitates the implementation of unique constitutive simulations. Due to the complexity of the case study building, some assumptions on the material characteristics and the need of having high performance computers to process the analyses, macro-modelling approach was considered in this study. Nonlinear static analyses (pushover) have been performed by 3Muri software package that is a multi-purpose FE program committed for the linear and nonlinear analyses of masonry buildings. "26 Nentori" School building block is analytically modelled, and seismic load bearing capacities are estimated by using pushover analysis in TREMURI software (Figure. 7).

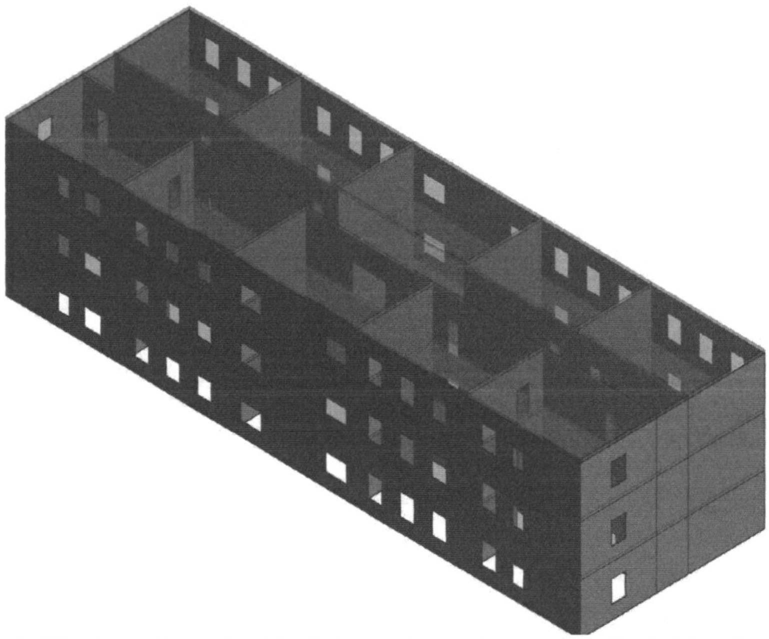

Figure 7. 3D view of the school building mathematical models (3MURI software).

During the modelling, the load bearing components are considered the masonry walls, while the floors are considered as rigid elements, which distribute the lateral loading effects between the connected walls.

Damage limit states are mathematical characterization of performance levels by a suitable damage value characterizing the seismic performance with proper damage levels. Generally, displacements and deformation quantities are used to characterize the structural response while defining the damage limit states. In this study, they were defined according to Eurocode 8 [17]. Capacity assessment of the school building is done according to Eurocode 8. Limited Damage (LD), Significant Damage (SD) and Near Collapse (NC) limit states are considered for the performance assessment.

4.3. Seismic analyses of the school building

The building is pushed by 24 different loading cases and the pushover curves are plotted for the most critical loading pattern (Figure 8 and 10).

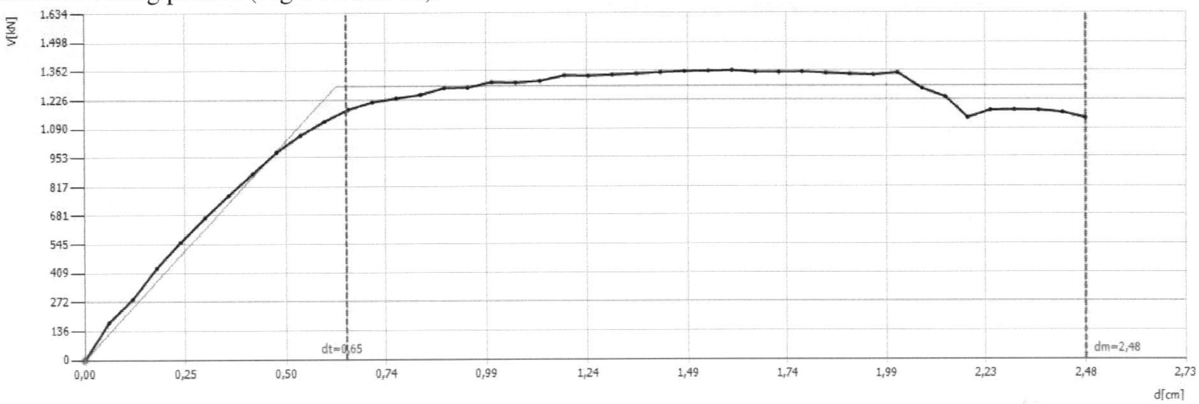

Figure 8. Pushover curve for x- direction.

Damage levels are identified for each damage states specified in Eurocode under pushover loading in x- direction (Figure 9.) and pushover loading in y- direction (Figure 11).

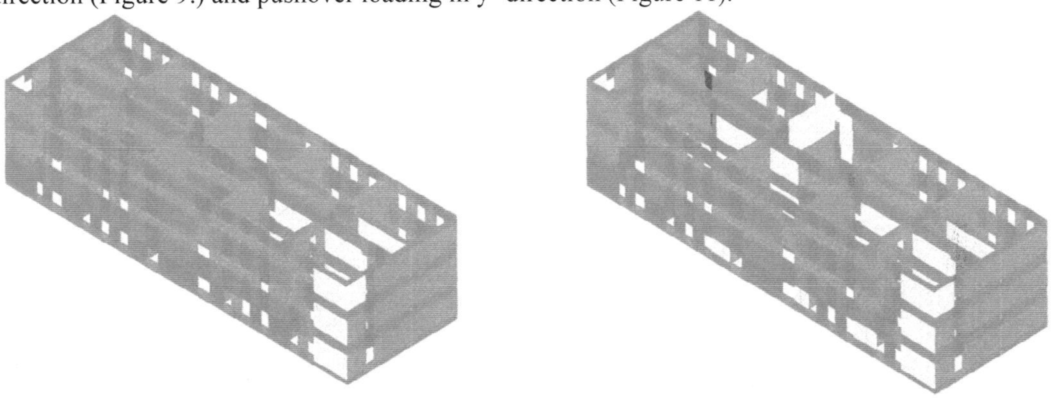

a) Limited Damage (LD) state. b) Significiant Damage (SD) state.

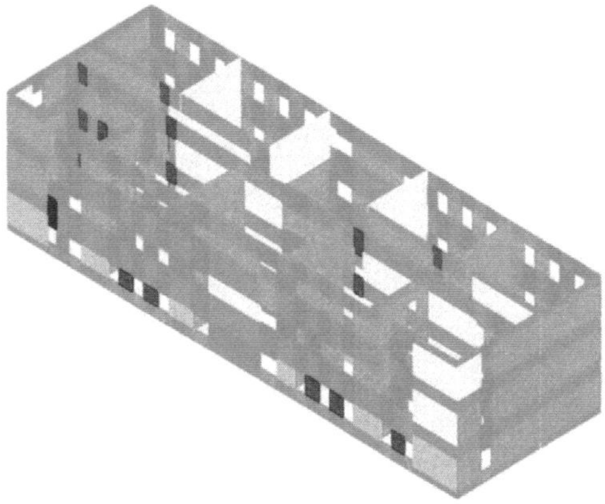

c) No collapse (NC) damage state.

d) Legend (description of the damage type).

Figure 9. Damage limit state levels according to EC 8 under pushover x- direction.

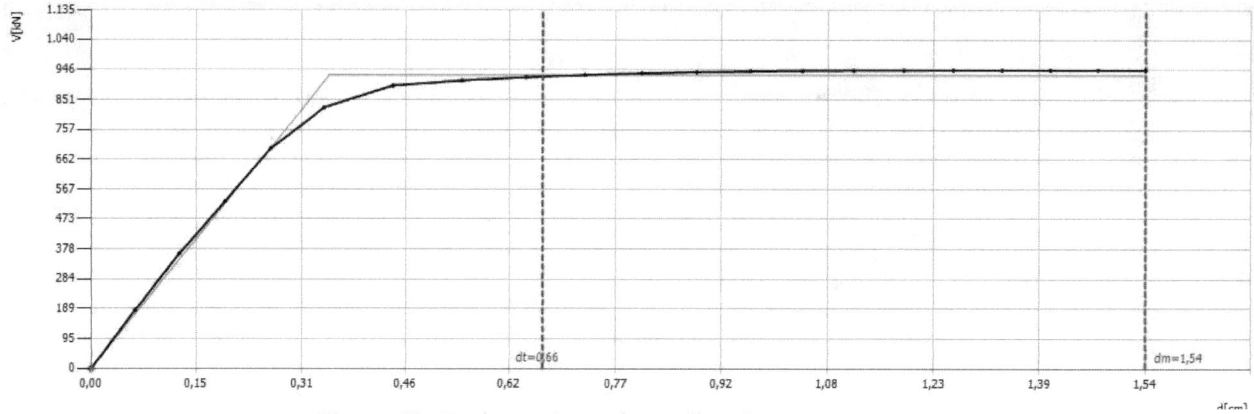

Figure 10. Pushover curve for y- direction.

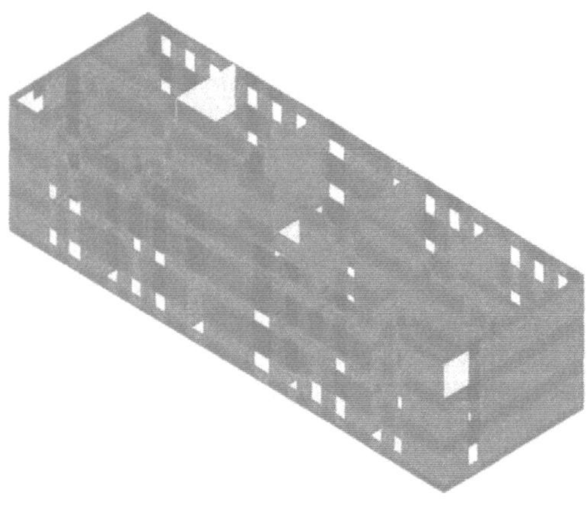

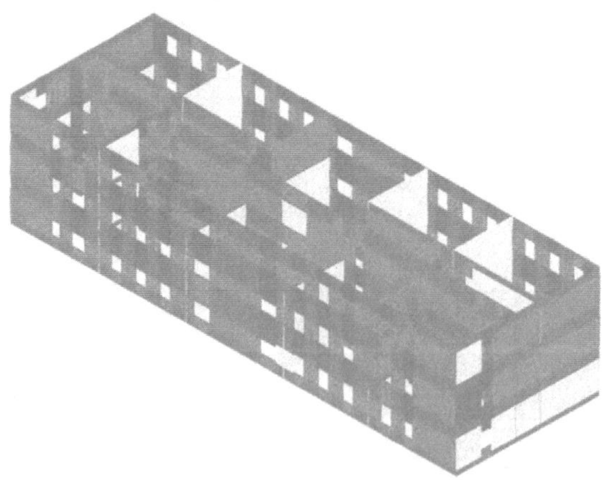

a) Limited Damage (LD) state.

b) Significiant Damage (SD) state.

IOP Conf. Series: Materials Science and Engineering **989** (2020) 012030 doi:10.1088/1757-899X/989/1/012030

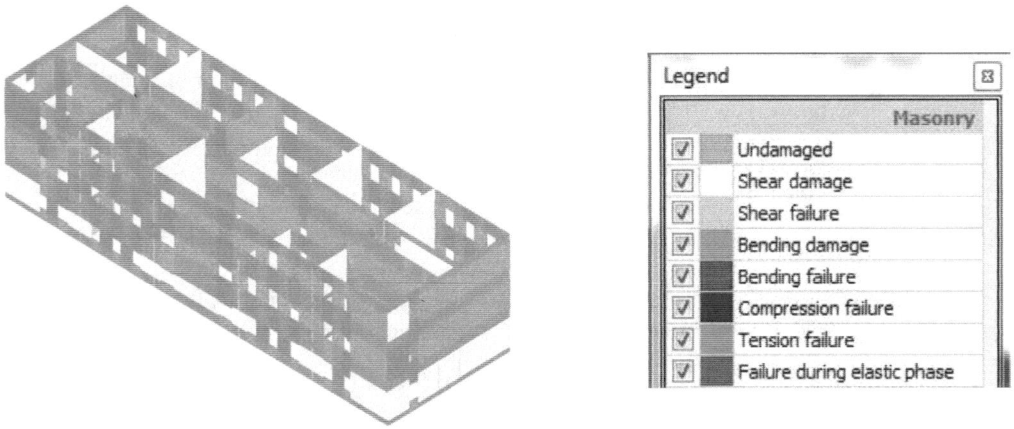

c) No collapse (NC) damage state. d) Legend (description of the damage type)

Figure 11. Damage limit state levels according to EC 8 under pushover y- direction.

Table 2-3 defines the peak ground acceleration (PGA) limits that can be sustained by building for the damage limit states according to EC 8.

Table 2. PGA limit states in x- direction (1 m/s^2 = 0.101972 g).

Limit state	PGA [m/s2]	α
NC	1,759	2,513
SD	1,401	2,803
DL	0,686	2,286

Table 3. PGA limit states in y- direction (1 m/s^2 = 0.101972 g).

Limit state	PGA [m/s2]	α
NC	1,272	1,817
SD	1,020	2,041
DL	0,494	1,646

Due to the low material quality and consequently the insufficient amount of stiffness, school building's seismic capacities in both directions are relatively small compared to the similar buildings for this type. Pushover analyses results and the damage patterns observed during the damage inspection confirms the induced damage during the November 26, 2019 earthquake for the school building.

5. Conclusions

The school structure used as 26 Nentori school block was examined on site after November 26, 2019 Earthquakes. The masonry structures, which were built as pre-2000 schools, were originally designed as 3- floors. The current situation of the buildings was evaluated by considering the provisions of earthquake resistant design rules and Eurocode regulation for masonry structures under seismic loads. The seismic capacity of the school building was assessed with TREMURI software, which uses a structural model based on macro-modelling method for the load bearing walls.

The type of soil that forms the foundation of the area is classified as "Dense gravel or medium dense sand and gravel" according to the laboratory test results that has an allowable load bearing capacity of 180 kPa. According to the EC 8, the soil class is considered in the seismic analysis as Type

C. The study area is in one of the earthquake-prone zones of the country according to the Albanian probabilistic seismic hazard map.

The provisions of earthquake-resistant building design regulations on masonry structures and the suitability of the building are summarized in Table 2. Pushover analyses were conducted, and the damaged regions with the corresponding reasons are marked on the 3-D analytical model. Seismic capacity of the school on both transverse directions are determined according to EC-8.

During the Durres earthquake of 6.4 magnitude on November 2019, slight-moderate damage was observed. However, this earthquake is smaller than design earthquake and it is classified as a medium intensity earthquake. School buildings are buildings that need to be preserved in terms of their location, history and construction. School buildings to be used immediately after the earthquake should not be damaged. Considering the actual status of these school buildings, urgent need needs to be put into action.

Conflict of interest

No conflict of interest was declared by the authors.

6. References

[1] Organization for Economic Co-operation and Development (OECD) Report 2004 Keeping schools safe in earthquakes *Proceedings of the ad hoc experts' group meeting on earthquake safety in schools*, Paris, 9-11 February.

[2] Bilgin H and Hysenlliu M 2020 Comparison of near and far-fault ground motion effects on low and mid-rise masonry buildings *Journal of Building Engineering* **30** https://doi.org/10.1016/j.jobe.2020.101248

[3] Magenes G and Penna A 2011 Seismic design and assessment of masonry buildings in Europe: Recent Research and Code Development Issues *9th Australia's Masonry Conference* Queenstown 583–603

[4] Yılmaz S, Tama YS and Bilgin H 2013 Seismic performance evaluation of unreinforced masonry school buildings in Turkey *Journal Vibration Control* **19 (16)** 2421-2433

[5] Komur M, Karabork T and Deneme I 2011 Nonlinear dynamic analysis of isolated and fixed-base reinforced concrete structures *Gazi University Journal of Science* **24 (3)** 463-475

[6] Mwafy, Aman and Elkholy S 2016 Performance assessment and prioritization of mitigation approaches for pre-seismic code structures *Advances in Structural Engineering* **20(6)** 917-939

[7] Bilgin H and Huta E 2018 Earthquake performance assessment of low and mid-rise buildings: Emphasis on URM buildings in Albania *Earthquakes and Structures* **14(6)** 599-614

[8] Estêvão J, Ferreira M, Morales-Esteban A, Martinez-A´lvarez, F, Fazendeiro, Sa´ L and Requena-Garcı´a-Cruz MV 2018 Earthquake resilient schools in Algarve (Portugal) and Huelva (Spain) *16th European Conference Earthquake Engineering; Thessaloniki*, 1–11.

[9] KTP-9-78 Albanian masonry design code (1978).

[10] KTP-N2-89 Albanian seismic design code (1989).

[11] Bilgin H and Frangu I 2017 Predicting the seismic performance of typical R/C healthcare facilities: emphasis on hospitals *International Journal of Advanced Structural Engineering* **9(3)** 277-292

[12] Lagomarsino S, Penna A, Galasco A and Cattari S 2013 TREMURI program: an equivalent frame model for the nonlinear seismic analysis of masonry buildings *Engineering Structures* **56** 1787-99

[13] Nato Sfp – 983054 Harmonization of seismic hazard maps for the Western Balkan countries (Bshap), (2007-2010).

[14] Aliaj S, Kociu S, Muco B and Sulstarova E 2010 Seismicity, seismo-tectonics and seismic hazard assessment in Albania *The Academy of Sciences of Albania*, ISBN: 978-99956-10-26-5.

[15] Eurocode 6 Design of Masonry Structures - Part 1-1: General rules for buildings-rules for reinforced and unreinforced masonry *European Committee for Standardization*, Brussels, (2005).

[16] Lourenco P, Mendes N, and Marques R 2009 Earthquake design and assessment of masonry structures: review and applications. *In book: Trends in Civil and Structural Engineering Computing, Chapter: 4* Publisher: Saxe-Coburg Publications Editors: B.H.V. Topping, L.F. Costa Neves, R.C. Barros, 10.4203/csets.22.4.

[17] Eurocode 8, Design of structures for earthquake resistance. Part 3: Assessment and retrofitting of buildings, European Committee for Standardization, Brussels, (2005).

Comparative study of conventional bricks with Padobe

Yugasini S [1,2] Uma Maguesvari M[1], Muthaiyan P [1] and Ammaiappan M [1]

[1]Department of Civil Engineering, Rajalakshmi Engineering College, Chennai and 602 105, India.

[2] Email:yugasini.s@rajalakshmi.edu.in

Abstract. In this study the properties of conventional brick compared with the brick made with Padobe. Padobe is a mixture of clay, paper pulp and water .Here we have added GGBS into this mixture. A composite material of the brick that having moderate physical and mechanical characteristics, which mainly helps the business industry like as construction. The addition of paper pulp and GGBS in clay helps in the reduction of environmental issues and cost of construction. Paper pulp is kept as constant and the other materials like clay and GGBS is done with various proportions such as 70:20:10, 60:3:10, 50:40:10 and 40:50:10 .An experimental study has been conducted for optimize the mix of Padobe depending on the compressive strength.

1. Introduction

Due to the huge requirement has been fixed on construction elements business specifically in the recent decade due to the raising community it causes a persistent scarcity on building elements, designer have been confronted to turn the manufacturing wastes to appropriate construction materials. It is huge time we wish to find the best substitute which must be cost effective as well as eco friendly. The maximum waste paper is assembled from various region a causes certain environmental issues. This article presents an experimental investigation which analyse the suitable utilize of waste paper and GGBS for providing an energy retaining, cost - effective and less weight blended brick as a construction element. [1] Due to the lesser weight of Padobe, the self weight of the structure is reduced. It do not expand or contract due to the presence of fiber content in the Papercrete [2] Increment of paper pulp essence in the mix compression strength is gradually reducing. [3] The sun dried Papercrete bricks gives that increase in compression strength whereas water cured bricks infers the reduction in the number of curing days. [4] It is light in weight when compared to conventional brick. Padobe bricks are better sound absorbent. The density of Padobe brick is lesser than the normal brick. By sing this paper pulp waste material in brick decreases the landfill. [5] The water absorption is greater when the essence of GGBS is lesser and the compression strength is low when the GGBS essence is lesser. [6]Papercrete Bricks neither consumes energy nor emits pollutants and hence gives an economical choice to design GREEN Buildings. The water retention of Papercrete brick was found to be greater than 20%, this makes it not worthy for external walls. [7] Paper pulp and clay results in a porous blended material, it infers high crushing and shear strength when compared to other low density material.

1.1. Objectives

- To determine the characteristics of material used.

- To compare the compression strength, , hardness, soundness , efflorescence, dry density, water absorption value of Padobe brick with normal brick.
- To replace normal bricks by an innovative and sustainable material (Papercrete and GGBS).

2. Materials used

2.1. Clay

Clay consists of huge proportion of fine fragments and colloidal material. Clay is used as the binding material for Padobe. The physical properties of clay such as plastic limit, liquid limit and plasticity index are determined and tabulated in table1.The image of the clay is shown in figure 1.Atterberg limit test was conducted and observed the limit of plastic and liquid.

Figure 1. Clay.

2.2. Paper pulp

The waste paper was collected and cannot be utilized directly .The papers were weighed and soaked in a tank for two to three days or till the papers becomes into a paste. Waste Paper is collected from the industry, which was shown in figure 2.

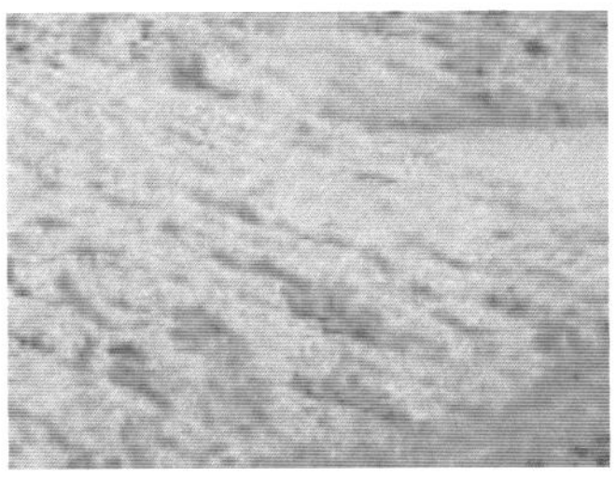

Figure 2. Paper pulp.

2.3. GGBS

Ground granulated blast furnace slag is a by-product procured in the making treatment of iron in blast kiln. This process generates a lustrous and granular material which has pozzolanic characteristics. GGBS in off white colour in appearance which was given in figure 3.

Figure 3. GGBS.

2.4. *Water*

Water is an essential constituent of Padobe as it vigorously takes part in the synthetic response with clay. It shall be free from impurities. pH value of water lies between six and seven.

3. Mix proportion of Padobe

The proportioning of mix used in Padobe is calculated by trail mix. Padobe mix contains clay, paper pulp, GGBS and water. Weigh batching was followed in this study. Therefore the material was measured in kg. GGBS was mixed with clay in a dry condition thoroughly until homogeneous color is obtained. Next the paper pulp is added in drenching condition then water is added to obtain adequate consistency

Table 1. Mix design proportion.

Mix Id	Clay (%)	Paper pulp (%)	GGBS (%)	Weight of clay (kg)	Weight of paper pulp (kg)	Weight of GGBS (kg)
P_1	70	10	20	2.81	0.19	0.36
P_2	60	10	30	2.41	0.19	0.54
P_3	50	10	40	2.01	0.19	0.72
P_4	40	10	60	1.61	0.19	0.90

3.1. *Mixing and casting*

Each and every ingredients of Padobe is mixed together before adding water to the mix as given in figure 4. The clay starts to experience a reaction as soon as paper pulp and water is added to the mixture, which acts as a bonding agent. The operation of specimen casting and drying was mentioned in figure 5. Once the specimen was casted then it is dried and burnt as shown in figure 6.

Figure 4. Mixing of Clay and GGBS.

Figure 5. Casting and Drying of Padobe.

Figure 6. After burning of Padobe.

4. Results and discussion

This stretch generally examines the performance of Padobe brick with normal brick.

4.1. Weight of padobe

The weight of the normal brick is compared with the Padobe brick and tabulated in the table 2.The weight of the Padobe is less when compared to the normal brick.

Table 2. Weight comparison of Padobe brick with normal brick.

Mix Id	Conventional brick (kg)	Padobe brick(kg)
P_1		2.13
P_2		1.97
P_3	3.05	1.77
P_4		1.63

4.2. Water absorption test

Water Absorption experiment is carried out to determine the water absorbent percentage. The readings were tabulated in table 3.

Table 3. Comparison of water absorption of Padobe brick with normal brick.

Mix Id	Water absorption (%)
P_1	25
P_2	23
P_3	20
P_4	18

The Figure 7 shows that the water absorption slight decreases with the increase in the addition of clay, GGBS and constant proportion of paper pulp.

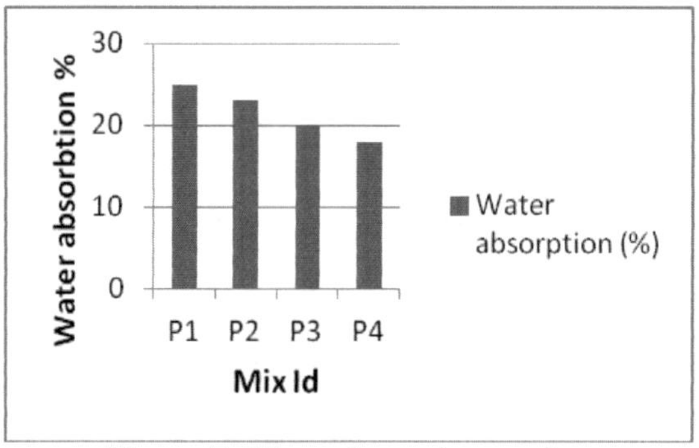

Figure 7. Water absorption results.

4.3. Compression test

Compression test conducted on the Padobe bricks in the compression testing machine and the readings are tabulated in the table 4.The crushing strength of normal brick is 7.06 N/mm^2. It is coming under the category of second class. While comparing the Padobe crushing vale with normal brick, a steady increment in different mixes P1, P2 , P3 and P4such as 2.124%, 4.81%, 7.5% and 3.25%. Among these vales obtained P3 achieves maximum crushing strength.

Table 4. Compression strength comparison of Padobe brick with normal brick.

Mix Id	**Compression strength (N/mm^2))**
P_1	7.21
P_2	7.40
P_3	7.59
P_4	7.29

Figure 8 indicates that the compression strength of the Padobe brick varies with a addition of paper pulp and GGBS. The compression strength increases up to 50 % clay, 40 % GGBS and 10 % paper pulp and then reduces.

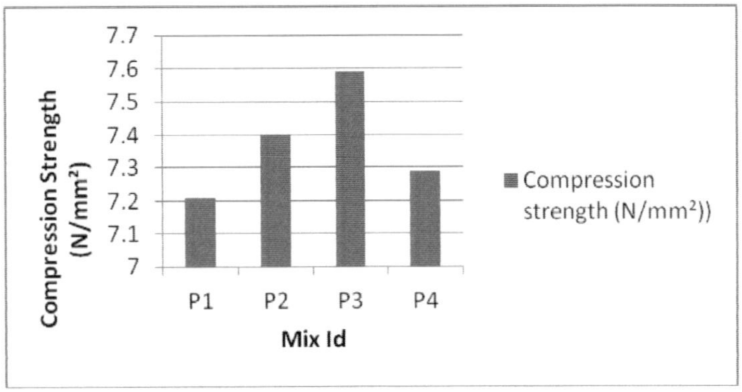

Figure 8. Compression test results.

4.4. Testing of Efflorescence

This experiment is used to determine the presence of soluble salts (Figure 9). It is conducted in a tub filled with water at a depth of 2.5 cm with a temperature range of 68° F to 86 ° F. Next the normal and Padobe bricks are immersed in the water until the water is absorbed completely and then it is dried in shade. The crushing strength is maximum at P_3. So considering P_3 for the efflorescence experiment .It is concluded that the normal and Padobe bricks have less than ten percentage of the revealed area of brick is wrapped with a slim deposit.

Figure 9. Presence of soluble salts.

4.5. Testing of Soundness

In this experiment two P_3 specimens were taken with same fraction and they were smacked with each other. A clear ringing sound was generated which indicates the bricks obtained quality results .The figure 10 shows the soundness test carried out.

Figure 10. Soundness test.

5. Conclusion

From the above experimental studies, following conclusions were made

- Padobe brick may be a substitute source in the future building and construction industry.
- The utilization of Padobe bricks uses the paper pulp due to this it decreases the landfills and pollution. The crushing strength of Padobe is maximum at p3 is 7.5% when compared to other mixes.
- The compression strength increases up to certain percentage of the mixture of clay, paper pulp and GGBS and then it reduces
- The weight of the normal brick is more when compared to the Padobe.
- Since the water absorption is high, these bricks cannot be used for water logging and external walls. This range lies between second class brick.

6. References

[1] Liladia Y, Malviya S and Baig A 2020 Utilzation of paper to produce bricks *International Research Journal of Engineering and Technology* **7(2)** 1259-1263

[2] Ramesh Kumar G B 2018 Papercrete – A sustainable building material *International Journal of Pure and Applied Mathematics* **119** 87-93

[3] Myriam M D, Rahul V, Abhilash C and Pavan M K 2017 Papercrete bricks an alternative sustainable building materials *International Journal of Engineering Research and Applications* **7(3)** 09-14

[4] Siva Prasad G V S, Padmanabha R, Swathi M, Kiran Kumar P D V , Praveen Raja T and Naveen M 2015 Study and behavior of some properties of papercrete brick with modular brick *International Journal of Engineering Research-Online* **3(3)** 267-275

[5] Anandaraju K , Jose Ravindra Raj B and Vijaya Sarathy R 2015 Experimental investigation of Papercrete bricks *International Journal of Machine and Construction Engineering* **2 (2)** ISSN (online): 2394-3025

[6] Dasgupta S and Das S K 2002 Paper pulp waste – A new source of raw material for the synthesis of a porous ceramic composite *Bulletin of Material Science* **25(5)**381-385

[7] IS 3495 PART -1(1992) – Guidelines for compression strength of brick

[8] IS 3495 PART -2(1992) – Guidelines for Efflorescence

[9] IS 3495 PART -3(1992) – Guidelines for Water Absorption Test

IOP Conf. Series: Materials Science and Engineering **989** (2020) 012032 doi:10.1088/1757-899X/989/1/012032

Experimental studies on pervious geopolymer concrete

Uma Maguesvari M [1,2], Muthaiyan P [1], Yugasini S [1] and Ammaiappan M [1]

[1]*Department of Civil Engineering, Rajalakshmi Engineering College, Chennai and 602 105, India.*

[2]*Email:umamaguesvari.m@rajalakshmi.edu.in*

Abstract. The aim of the investigation is to assess the features of the utilization of a geo polymer binder for producing permeable geopolymer concrete. Materials used for producing pervious geopolymer concrete are fly ash, ground granulated blast furnace slag with the proportion of 60:40, sodium silicate solution, sodium hydroxide and coarse aggregate. The strength, functional characteristics and its relationship of the geo polymer pervious concrete were investigated. Compressive strength of geo polymer permeable concrete has enhanced along with the rise in density of Pervious geo polymer concrete. Permeability and voids exhibit a decreasing pattern with the increase in molarity and alkali binder ratio. Based on the strength attainment in the previous geo polymer concrete, it has used as a sub - base and base course of a flexible pavement layers.

1. Introduction

Geopolymer concrete has the potential to reduce carbon emission globally and lead to a sustainable development to form an important contributor towards environmentally sustainable construction and building products industry [1]. "Pervious concrete" as PC describes different category of concrete that has high porousness and used for uniform run application that permits water from precipitation and different sources to enter through it, by minimising the overflow from the location and permitting water to recharge. Pervious concrete has attracted greater interest in concrete enterprise in pursuit of environmental safety and sustainability. Due to a porosity of 15–30% and a permeability of 0.14 –1. 24 cm/s, PC can infiltrate stormwater and decrease the quantity of stormwater runoff. It has structurally weaker than the traditional concrete. The Compressive strength of normal concrete ranges between 17 to 40 MPa, but for pervious concrete the attainment is from 2.8 to 28 MPa [2]. Geopolymer concrete can diminish carbon emission all around and bring about a maintainable improvement to shape, a basic supporter toward naturally feasible creation and developing product endeavours [3]. Past exploration has been directed to diminish the reliance on conventional Portland cement (OPC) as a significant material development because of its effect on the earth [4,5]. Because of the high vitality prerequisites in the creation of OPC, it is assessed that each ton of OPC brings about a huge amount of CO_2 discharges [6]. In the midst of greenhouse gases, CO_2 give rise to huge amount of impact on Earth-wide temperature. Furthermore, concrete creation and coming about outflows are relied upon to increment by 100% from the current level by 2020[7]. Geopolymer concrete is an increasingly supportable option to OPC concrete because of utilization of safe waste materials and decreased creation of CO_2. Furthermore, geopolymer concrete exhibits better execution relative than OPC [8]. Geopolymer is synthesized from substances of geological beginning. Geopolymer concrete proposed more eco- friendly replacement to Portland cement [9].

The characteristics of "pervious geopolymer concrete - PGC" have assessed, use of binder as high calcium fly ash gives satisfactory mechanical properties for fabricating geopolymer pervious concrete [10]. It has developed pervious geo polymer concrete using recycled aggregate. They concluded that recycled aggregates have given acceptable results for producing pervious geopolymer concrete using same binder [11]. It has revealed geo polymer concrete exhibit high compressive strength, low shrinkage, acid resistance, fire resistance and low thermal conductivity characteristics [6]. PC will be employed in the development of park floors, areas with light-weight traffic, walkways, tennis courts, floors of green buildings etc. Its porous nature of structure has the potentials utilization as thermal insulating, acoustic absorption, and water purification materials. The geo polymeric materials reveal better mechanical properties than Ordinary Portland cement products [12]. Attainment of strength for geopolymer concrete utilizing low calcium fly ash have not shown large variations by rising the binder content from 450 to 550 kg/m^3 [13]. Various curing conditions too influence on the strength of geopolymer concrete [14]. Attempts have also been made use of GGBS as binder and fly ash as aggregate on mechanical characteristics of geopolymer concrete [15-16]. Cement industry produces about a ton of carbon di-oxide while manufacturing one ton of OPC, which contributes major effect on climate change. This can be reduced by incorporating potential binder waste materials like Fly ash, GGBS, silica fume, Rice hush ash etc. In this aspect geopolymer concrete has the potential to reduce the carbon di-oxide emission [17]. Consequently, In the frame work of the present investigation is to use geopolymer binder for producing PGC. The mechanical characteristics viz density, compressive strength and functional properties namely permeability, porosity and abrasion resistance of pervious geo polymer concrete was studied and formulate the interrelationship between the strength, permeability and voids of PGC and its potential applications.

2. Experimental program

2.1. Materials and characteristics

Materials used for making PBG were fly ash and granulated blast furnace slag as geo polymeric binder materials, Gravel as coarse aggregate, maximum size of the aggregate as 9 mm, sodium hydroxide flakes and sodium silicate solution. Sodium hydroxide is used to activate sodium silicate. The characteristics of fly ash and GGBFS are given in Table 1, coarse aggregate is given in Table 2, Figure 1 represents the materials.

Table 1. Characteristics of Fly ash and GGBFS.

Characteristics	Fly ash	GGBFS
Consistency	40 %	36 %
Initial Setting Time	38 min	55 min
Specific Gravity	2.45	2.78
Soundness	0	0

Table 2. Characteristics of Aggregate.

Characteristics	Value
Crushing Value	15%
Impact Value	17%
Specific Gravity	2.70
Water Absorption	0.1%

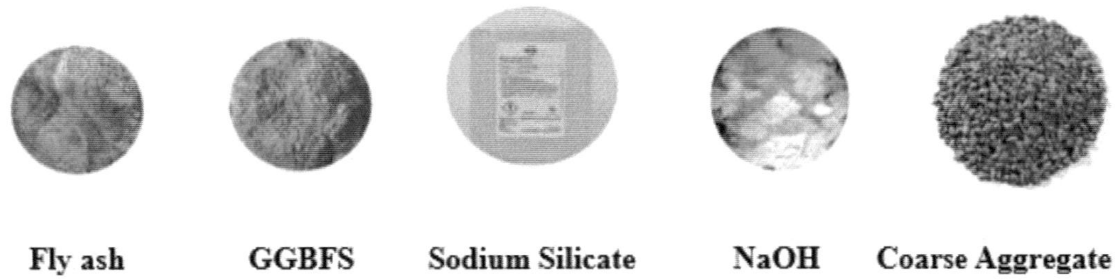

Figure 1. Materials used.

2.2. Preparation of alkaline liquid

Alkaline liquid is produced by blending of the sodium hydroxide solution and sodium silicate at room temperature. At the point when the solution combined the both begin to respond polymerisation happens, it releases huge part of warmth because it is prescribed not to disturb for around 24 hours. It is fit to be used as binding agent.

2.3. Mix design

Polymeric binder of FA and GGBFS with the composition of 60:40, Granite as coarse aggregate with the maximum size of 9mm. Polymeric binder content of 250 Kg/m^3 , aggregate to binder ratio as 6.9, a constant NS/NH proportion as 1 and the alkaline liquid/binder (AL/B) ratio of 0.4,0.45 and 0.5 with the molarity of 10, 15 and 20 were used for preparing the mixes of making pervious geopolymer concrete. Totally nine mixes were used. The FA and the GGBS were initially mixed then coarse aggregate were added and blended for 2 to 3 min then alkaline liquid is poured and blended thoroughly around 5 min, geopolymer concrete mixtures were transferred to mould of cylinder of 100 diameter ,200 mm height and 100 mm cube for casting the specimen and left it for air curing, after 24 hours samples were removed from mould and kept it in the room temperature until testing.

3. Results and discussion

3.1. Density

Densities of PGC for various alkaline liquid/binder (40, 45 and 50) for different molar concentration are shown in Fig. 2. As the AL/B content increases, the density also increases continuously, but very gently for 10 molar concentrations. There is a continuous improvement in density with increase in molar concentration from 10 to 15M, but the same trend is followed for 20M with decreasing percentage. Density of PGC varies from 1832 to 1918 Kg/m^3.

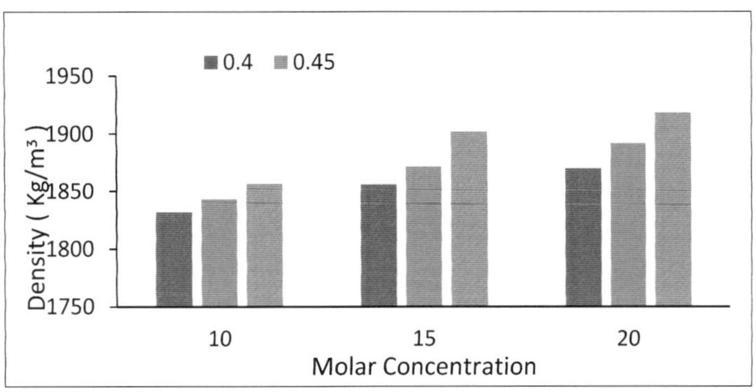

Figure 2. Density of PGC for different alkaline liquid/binder and molarity.

3.2. Compressive strength

Crushing strength of PGC for several alkaline liquid/binder (40, 45 and 50) for different molarity were presented in Figure. 3. The crushing strength of PGC is increases upon rise in addition of alkaline liquid / binder content. It ranges from 11.65 to 13.43 MPa. Average raise in compressive strength is marginal to slightly marginal with the increase in AL/B content for molarity 10. Similar, trend is followed for molar concentrations 15 and 20. Based on the strength attainments in the previous geopolymer concrete, it satisfied the requirements of flexible pavement layers such as sub - base and base course according to Indian codes.

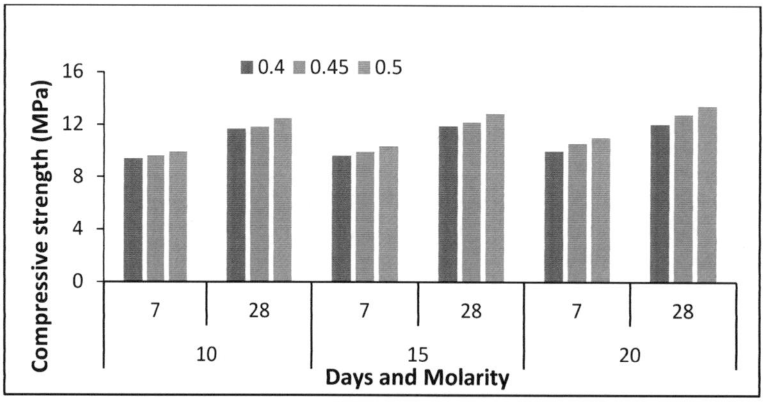

Figure 3. Compressive strength of PGC for different alkaline liquid/binder and molarity.

3.3. Permeability

Permeability is determined by falling head method according to the procedure given in ACI522R-10 [2]. Permeability of PGC for various alkaline liquid/binder (35, 40 and 45) for different molar concentrations is shown in Fig. 4. Permeability of geo polymer pervious concrete decreases with increase in AL/B content, it follows similar trend for various molar concentrations 10 to 20 with the increment of 5M. The co-efficient of permeability of PGC varies between 0.654 and 1.742 cm/s.

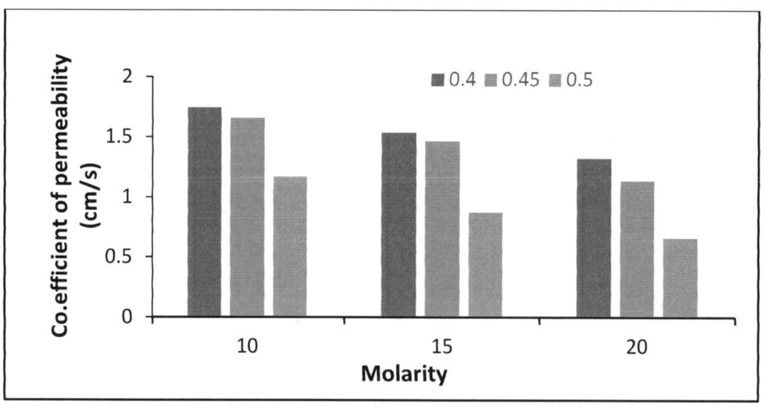

Figure 4. Permeability of pervious geopolymer concrete for various alkaline liquid/binder and molarity.

Abrasion Resistance of PGC for various alkaline liquid/binder (35, 40 and 45) for different molar concentrations is illustrated in Fig. 5. The abrasive value of PGC has enhanced with elevated amount of addition of alkaline liquid / binder content. Average increase in the percentage is marginal to slightly marginal with the increase in AL/B content for molarity 10. Similar, trend is followed for molar concentrations 15 and 20. It ranges from 27 % to 40%.

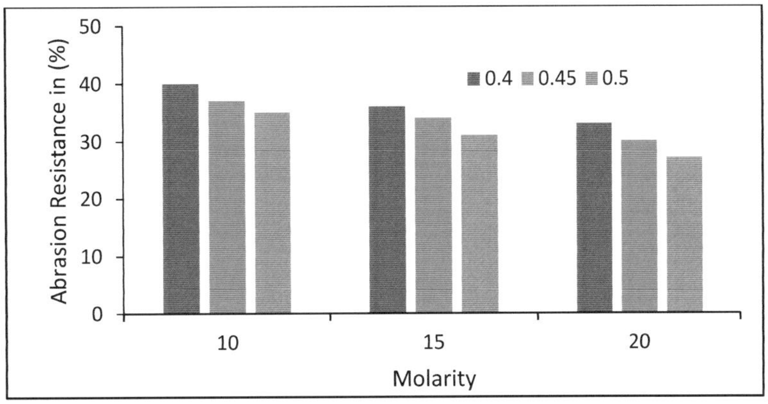

Figure 5. Abrasion resistance of pervious geopolymer concrete for various alkaline liquid/binder and molarity.

3.4. Compressive strength and voids relationship

The regression correlation for the permeability and compressive strength is represented in Figure. 6. In view of compressive strength falling pattern occurs with respect to increment in voids present in the mixture. The represented relationship is linear and given as $y = -0.4483x + 21.657$; $R^2 = 0.9032$. Where x- voids (%) and y- Compressive strength (MPa).

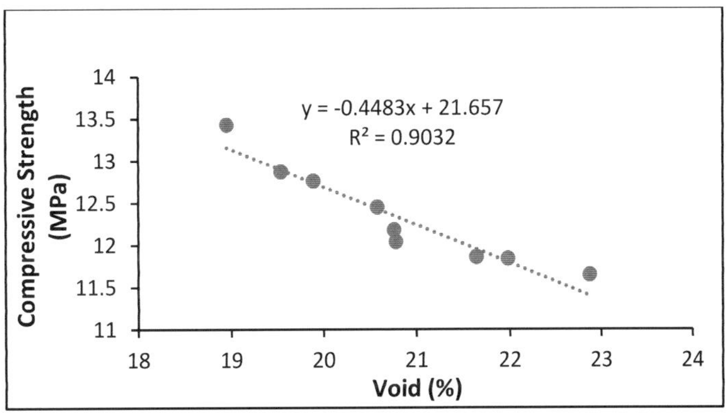

Figure 6. Relation between the voids and the compressive strength of geopolymer pervious concrete.

3.5. Voids and permeability

Permeability and voids of PBC decreases with increase in AL/B content, it follows similar trend for various molar concentrations 10 to 20 with the increment of 5M. The permeability of geo polymer pervious concrete varies from 0.654 to 1.742 cm/s. The void content follows the similar trend of permeability of geo polymer pervious concrete. The voids present in pervious geo polymer concrete differ between 18.95% - 22.87%. The correlation between void and permeability is linear as illustrated in Figure 7 and given as $y = 0.2793x - 4.5207$; $R^2 = 0.918$. Where x – void in % and y- Permeability (cm/s).

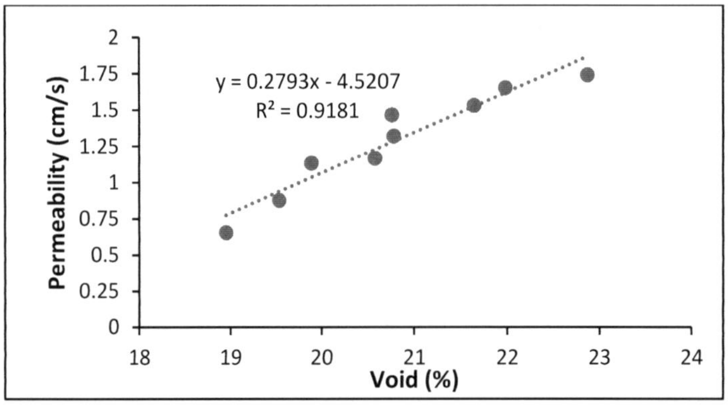

Figure 7. Relation between the voids and permeability of geopolymer pervious concrete.

4. Conclusions

This investigation assesses the characteristics of pervious geo polymer concrete produced from utilisation of waste materials as geo polymeric binder. Compressive strength of geo polymer pervious concrete varies between 11.65 to 13.43 MPa. Compressive strength enhanced by the use of increment in AL/B for the mixes of same molar concentration. Permeability exhibit a converse pattern while contrast with compressive strength of pervious geo polymer concrete. Permeability of pervious geo polymer concrete varies between 0.654 to 1.742 cm/s. The abrasion shows a similar trend to compressive strength and it ranges from 27% to 40 %. The voids of pervious geo polymer concrete decreases with increase in AL/B content. Based on the strength attainments in the previous geopolymer concrete, it satisfied the requirements of flexible pavement layers such as sub - base and base course according to Indian codes.

5. References

[1] Miller S A, Horvath A, Monteiro P J M 2016 Readily implementable techniques can cut annual CO_2 emissions from the production of concrete by over 20%. *Environ. Res. Lett.* **11** 074029

[2] ACI, American Concrete Institute Committee 522, Report on Pervious Concrete, 2010. ACI 522R–10.

[3] Andrew R M 2017 Global CO_2 emissions from cement production *Earth Syst. Sci. Data* 1-52

[4] Stajanca M and Estokova A 2012 Environmental impacts of cement production Tech. Univ. Kosice, *Civ. Eng. Fac. Inst. Archit. Eng.* 296–302

[5] Hasanbeigi A, Menke C and Price L 2010 The CO_2 abatement cost curve for the Thailand cement industry *Journal of Cleaner Production* **18** 1509–1518

[6] Duxson P, Fernández-Jiménez A, Provis J L, Lukey G C, Palomo A and Deventer J S J 2007 Geopolymer technology: The current state of the art *J. Mater. Sci.* **42** 2917–2933

[7] Naik T 2008 Sustainability of concrete construction. *ASCE Pract Period Struct Des Constr* **13(2)** 98–103

[8] Mehta P K and Walters M 2008 Roadmap to a sustainable concrete construction industry *Constr Specifier* **61** 48–57

[9] Mehta A and Siddique R 2016 An overview of geopolymers derived from industrial byproducts *Constr. Build. Mater.* **127** 183–198

[10] Tawatchi Tho–in and Vanchai Sata 2012 Pervious high calcium fly ash geopolymer concrete *Construction and Building Materials* **30** 366 – 371

[11] Vanchai Sata, Ampol Wongsa and Prinya Chindaprasrit 2013 Properties of pervious geopolymer concrete using recycled aggregates *Construction and Building Materials* **42** 33 – 39

[12] Kamal Neupane, Des Chalmers and Paul Kidd 2018 High strength geopolymer concrete - properties, advantages and challenges *Advances in Materials* **7** 15- 25
[13] Ramesh Babu C and Ganesan 2017 A study on the strength development of geopolymer concrete using fly ash *International Journal of Engineering and Technology* **6** 163 – 167
[14] Sai Krishna Teja T, Sai Pradeep P and Vasugi V 2015 Influence of curing conditions on strength development of geopolymer concrete *International Journal of Applied Engineering Research* **10(57)** 589-605
[15] Venkata Suresh G, Pavan Kumar Reddy P, and Karthikeyan J 2017 Effect of GGBS and fly ash aggregates on properties of geopolymer concrete *Journal of Structural Engineering* **43**, 436 – 444
[16] Shyam Verma and Vasugi V 2015 Experimental study on geopolymer pervious concrete using copper slag *International Journal of Applied Engineering Research* **10(33)** 24273-24276
[17] Davidovits J 1989 Geopolymers and geopolymeric new materials *Journal of Thermal Analysis* **35(2)** 429-441

AUTHOR INDEX

Institute of Physics
Dirac House, Temple Back
Bristol BS1 6BE UK

ISSN: 1757-8981
ISBN 978-1-7138-3301-7

9 781713 833017

CURRAN ASSOCIATES INC.
proceedings.com